Ford Transit Owners Workshop Manual

I M Coomber

Models covered
All Ford Transit ('Mk 3') models with in-line petrol engines
80, 100, 115, 120, 130, 160 & 190 models
1593 cc & 1993 cc

Does not cover Diesel or V6 petrol engine models, 4-speed (A4LD) automatic transmission or Transit 4 x 4. Specialist bodywork or conversions not covered

(1468-11S1) ABCD

Haynes Publishing Group
Sparkford Nr Yeovil
Somerset BA22 7JJ England

Haynes Publications, Inc
861 Lawrence Drive
Newbury Park
California 91320 USA

Acknowledgements

Thanks are due to the Champion Sparking Plug Company Limited who supplied the illustrations showing spark plug conditions, to Holt Lloyd Limited who supplied the illustrations showing bodywork repair, and to Duckhams Oils who provided lubrication data. Certain other illustrations are the copyright of the Ford Motor Company Limited, and are used with their permission. Thanks are also due to Sykes-Pickavant Limited who supplied some of the workshop tools, and all those people at Sparkford who assisted in the production of this manual.

© **Haynes Publishing Group 1991**

A book in the **Haynes Owners Workshop Manual Series**

Printed in the USA

All rights reserved. No part of this book may be reproduced or transmitted in any form or by any means, electronic or mechanical, including photocopying, recording or by any information storage or retrieval system, without permission in writing from the copyright holder.

ISBN 1 85010 468 9

British Library Cataloguing in Publication Data
Coomber, Ian, *1943-*
   Ford Transit '86 to '89
   1. Motor vans. Maintenance & repair.
   I. Title II. Series
   629.28'73
   ISBN 1-85010-468-9

Whilst every care is taken to ensure that the information in this manual is correct, no liability can be accepted by the authors or publishers for loss, damage or injury caused by any errors in, or omissions from, the information given.

Restoring and Preserving our Motoring Heritage

Few people can have had the luck to realise their dreams to quite the same extent and in such a remarkable fashion as John Haynes, Founder and Chairman of the Haynes Publishing Group.

Since 1965 his unique approach to workshop manual publishing has proved so successful that millions of Haynes Manuals are now sold every year throughout the world, covering literally thousands of different makes and models of cars, vans and motorcycles.

A continuing passion for cars and motoring led to the founding in 1985 of a Charitable Trust dedicated to the restoration and preservation of our motoring heritage. To inaugurate the new Museum, John Haynes donated virtually his entire private collection of 52 cars.

Now with an unrivalled international collection of over 210 veteran, vintage and classic cars and motorcycles, the Haynes Motor Museum in Somerset is well on the way to becoming one of the most interesting Motor Museums in the world.

A 70 seat video cinema, a cafe and an extensive motoring bookshop, together with a specially constructed one kilometre motor circuit, make a visit to the Haynes Motor Museum a truly unforgettable experience.

Every vehicle in the museum is preserved in as near as possible mint condition and each car is run every six months on the motor circuit.

Enjoy the picnic area set amongst the rolling Somerset hills. Peer through the William Morris workshop windows at cars being restored, and browse through the extensive displays of fascinating motoring memorabilia.

From the 1903 Oldsmobile through such classics as an MG Midget to the mighty 'E' type Jaguar, Lamborghini, Ferrari Berlinetta Boxer, and Graham Hill's Lola Cosworth, there is something for everyone, young and old alike, at this Somerset Museum.

Haynes Motor Museum

Situated mid-way between London and Penzance, the Haynes Motor Museum is located just off the A303 at Sparkford, Somerset (home of the Haynes Manual) and is open to the public 7 days a week all year round, except Christmas Day and Boxing Day.

Telephone 01963 440804.

Contents

	Page
Acknowledgements	2
About this manual	5
Introduction to the Ford Transit	5
General dimensions, weights and capacities	6
Jacking and towing	7
Buying spare parts and vehicle identification numbers	9
General repair procedures	10
Tools and working facilities	11
Conversion factors	13
Safety first!	14
Routine maintenance	15
Recommended lubricants and fluids	21
Fault diagnosis	22
Chapter 1 Engine	26
Chapter 2 Cooling, heating and ventilation systems	52
Chapter 3 Fuel and exhaust systems	63
Chapter 4 Ignition system	76
Chapter 5 Clutch	83
Chapter 6 Manual gearbox and automatic transmission	88
Chapter 7 Propeller shaft	144
Chapter 8 Rear axle	149
Chapter 9 Braking system	159
Chapter 10 Suspension and steering	179
Chapter 11 Bodywork	212
Chapter 12 Electrical system	232
Index	280

Ford Transit short wheelbase Van

Ford Transit GL 9-seater Bus

About this manual

Its aim

The aim of this manual is to help you get the best value from your vehicle. It can do so in several ways. It can help you decide what work must be done (even should you choose to get it done by a garage), provide information on routine maintenance and servicing, and give a logical course of action and diagnosis when random faults occur. However, it is hoped that you will use the manual by tackling the work yourself. On simpler jobs it may even be quicker than booking the car into a garage and going there twice, to leave and collect it. Perhaps most important, a lot of money can be saved by avoiding the costs a garage must charge to cover its labour and overheads.

The manual has drawings and descriptions to show the function of the various components so that their layout can be understood. Then the tasks are described and photographed in a step-by-step sequence so that even a novice can do the work.

Its arrangement

The manual is divided into twelve Chapters, each covering a logical sub-division of the vehicle. The Chapters are each divided into Sections, numbered with single figures, eg 5; and the Sections into paragraphs (or sub-sections), with decimal numbers following on from the Section they are in, eg 5.1, 5.2, 5.3 etc.

It is freely illustrated, especially in those parts where there is a detailed sequence of operations to be carried out. There are two forms of illustration: figures and photographs. The figures are numbered in sequence with decimal numbers, according to their position in the Chapter – eg Fig. 6.4 is the fourth drawing /illustration in Chapter 6. Photographs carry the same number (either individually or in related groups) as the Section or sub-section to which they relate.

There is an alphabetical index at the back of the manual as well as a contents list at the front. Each Chapter is also preceded by its own individual contents list.

References to the 'left' or 'right' of the vehicle are in the sense of a person in the driver's seat facing forwards.

Unless otherwise stated, nuts and bolts are removed by turning anti-clockwise, and tightened by turning clockwise.

Vehicle manufacturers continually make changes to specifications and recommendations, and these, when notified, are incorporated into our manuals at the earliest opportunity.

Whilst every care is taken to ensure that the information in this manual is correct, no liability can be accepted by the authors or publishers for loss, damage or injury caused by any errors in, or omissions from, the information given.

Project vehicle

The vehicle used in the preparation of this manual, and appearing in many of the photographic sequences was a 1989 short wheelbase 2.0 litre panel van.

Introduction to the Ford Transit

The latest Ford Transit range of models was introduced in February 1986, and is the third generation of this popular range of vehicles. This manual covers the 1.6 and 2.0 litre petrol engine models.

The most obvious and distinctive features of the new range is the wedge-front styling to improve the aerodynamics.

Both the 1.6 and 2.0 litre engines fitted to the models covered are single overhead camshaft, water-cooled design with a crossflow cylinder head and cast iron cylinder block. The engine is mounted in-line and, depending on model, is coupled to a four-speed or five-speed manual gearbox or an automatic transmission. The MT75 five-speed gearbox fitted to some 2.0 litre models is a new type designed by Ford and has synchro mesh on all gears (including reverse).

The steering and suspension on the short wheelbase (LCX) models is new, being rack-and-pinion type with MacPherson strut independent front suspension. The long wheelbase (LCY) models continue with the worm and nut steering gear and beam axle with leaf springs as used on earlier models.

Telescopic double-acting shock absorbers are fitted to the front and rear suspension on all models.

Drive to the rear wheels is by a two-piece propeller shaft to the rear axle, and this is suspended by leaf springs on each side.

All models are fitted with dual circuit servo-assisted brakes with discs at the front and self-adjusting drum brakes at the rear.

Seemingly innumerable variations of body types and styles of Transit are available, from the Pick-up to the Caravanette. All body styles and variants are mounted on long or short wheelbase chassis as necessary.

The Transit, given regular maintenance at the specified intervals, will give reliable service over a long period, and is therefore worth looking after.

General dimensions, weights and capacities

Dimensions

	Short wheelbase (LCX)	Long wheelbase (LCY)
Overall length	4606 mm (181.3 in)	5358 mm (210.9 in)
Overall width	1938 mm (76.3 in)	1972 mm (77.6 in)
Wheelbase	2815 mm (110.8 in)	3020 mm (118.9 in)

Weights

The kerb weights given are for a vehicle with a full fuel tank and basic equipment. The vehicle types are standard panel vans with side-loading door. For other models and non-standard types, consult a Ford dealer for details.

1.6 litre:
- 80 model .... 1266 kg (2791 lb)
- 100 model .... 1268 kg (2795 lb)

2.0 litre:
- 80 model .... 1286 kg (2835 lb)
- 100 model .... 1288 kg (2840 lb)
- 100L model .... 1310 kg (2888 lb)
- 100L (long wheelbase) model .... 1478 kg (3259 lb)
- 120 model .... 1458 kg (3214 lb)
- 130 model .... 1498 kg (3303 lb)
- 160 model .... 1501 kg (3309 lb)
- 190 model .... 1511 kg (3331 lb)

Capacities

Engine oil:
- With filter .... 3.75 litres (6.6 pints)
- Without filter .... 3.25 litres (5.7 pints)

Fuel system:
- Fuel tank .... 68 litres (15 gallons)

Cooling system:
- With heater .... 8.4 litres (14.8 pints)
- Without heater .... 7.2 litres (12.6 pints)

Transmission:
- 4-speed (type F) .... 1.45 litres (2.55 pints)
- 4-speed (type G) .... 1.98 litres (3.48 pints)
- 4-speed (type G with overdrive) .... 2.5 litres (4.4 pints)
- 5-speed (type N) .... 1.5 litres (2.64 pints)
- 5-speed (MT75) .... 1.25 litres (2.2 pints)
- Automatic transmission (including converter oil cooler) .... 6.3 litres (11.08 pints)

Rear axle:
- Type F .... 1.4 litres (2.4 pints)
- Type H .... 2.7 litres (4.7 pints)
- Type G .... 1.7 litres (3.0 pints)

Jacking and towing

Jacking points

To change a wheel in an emergency, use the jack supplied with the vehicle. Ensure that the roadwheel nuts are released before jacking up the vehicle.

The type of jack and its location is dependent on model. On short wheelbase models a scissor jack is supplied, and this is located at the points shown in the accompanying figures. On long wheelbase models, a pillar jack is supplied and this is located as indicated in the appropriate figures. When jacking up under a roadspring, position the jack as close to the axle as possible. Ensure that the jack is fully engaged at the lift point before raising the vehicle. Wherever possible, jack up on a firm level surface. Fully apply the handbrake, and engage reverse gear (or set lever at P on automatic transmission models). Chock the roadwheels on the side opposite that being raised. Whenever the rear of the vehicle is being raised (for whatever purpose), do not jack up under the centre of the axle when the vehicle is fully loaded. Do not raise the vehicle higher than is necessary (photos).

The jack supplied with the vehicle is not suitable for use when raising the vehicle for maintenance or repair operations. For this work, use a trolley or heavy duty bottle jack of suitable capacity. Also support the vehicle with axle stands.

When changing a wheel on vehicles fitted with six-stud wheels, note that the wheel nuts on the left-hand side of the vehicle have a *left-hand thread,* ie they undo *clockwise.* These nuts and studs are marked with the letter L.

When fitting twin rear wheels, make sure that the wheel mating surfaces are clean.

Retighten the nuts to the specified torque wrench setting (Chapter 10).

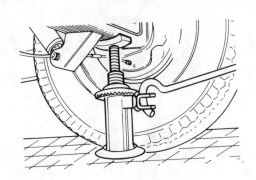

Jack location at front of vehicle on short wheelbase models

Alternative jack locations at front of vehicle on long wheelbase models. If jacking under spring, keep close to the axle

Jack location at rear of vehicle on short wheelbase models

Jack location at rear of vehicle on long wheelbase models

Jacking and towing

Spare wheel holder bolt

Spare wheel and holder

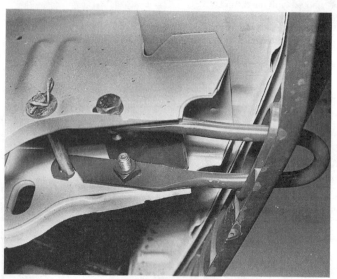
Towing eye – front

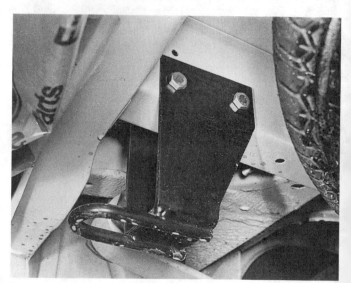
Towing eye – rear

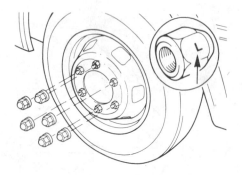
Left-hand thread wheelnut identification mark (arrowed) on six-stud wheels

Towing

Towing eyes are provided at the front and rear of the vehicle (photos).

When being towed the ignition switch should be in position II so that the steering lock is released and the direction indicators, horn and stop lights are operational. On automatic transmission models the selector lever must be in neutral (N), the towing speed must not exceed 30 mph, and the towing distance must not exceed 30 miles (50 km). For longer distances the propeller shaft should be removed or the rear of the vehicle lifted clear of the ground.

Push or tow starting is not possible on vehicles fitted with automatic transmission.

Remember that if the vehicle is to be towed and the engine is not running, there will be no servo assistance to the brakes and therefore additional pressure will be required to operate them.

If towing another vehicle, attach the tow-rope to the towing eyes.

Buying spare parts and vehicle identification numbers

Buying spare parts

Spare parts are available from many sources, for example: Ford garages, other garages and accessory shops, and motor factors. Our advice regarding spare part sources is as follows:

Officially appointed Ford garages – This is the best source of parts which are peculiar to your vehicle and are otherwise not generally available (eg complete cylinder heads, internal gearbox components, badges, interior trim etc). It is also the only place at which you should buy parts if your vehicle is still under warranty – non-Ford components may invalidate the warranty. To be sure of obtaining the correct parts it will always be necessary to give the storeman your vehicle's identification number, and if possible, to take the 'old' part along for positive identification. Remember that many parts are available on a factory exchange scheme – any parts returned should always be clean! It obviously makes good sense to go straight to the specialists on your vehicle for this type of part for they are best equipped to supply you.

Other garages and accessory shops – These are often very good places to buy materials and components needed for the maintenance of your vehicle (eg oil filters, spark plugs, bulbs, drivebelts, oil and greases, touch-up paint, filler paste etc). They also sell general accessories, usually have convenient opening hours, charge lower prices and can often be found not far from home.

Motor factors – Good factors will stock all of the more important components which wear out relatively quickly (eg clutch components, pistons, valves, exhaust systems, brake cylinders/pipes/hoses/seals /shoes and pads etc). Motor factors will often provide new or reconditioned components on a part exchange basis – this can save a considerable amount of money.

Vehicle identification numbers

Although many individual parts, and in some cases sub-assemblies, fit a number of different models, it is dangerous to assume that just because they look the same, they are the same. Differences are not always easy to detect except by serial numbers. Make sure therefore, that the appropriate identity number for the model or sub-assembly is known and quoted when a spare part is ordered.

The vehicle identification plate is located in the cab footwell and lists the vehicle type and axle weight (photo).

The engine number is located on the right-hand side of the crankcase towards the front.

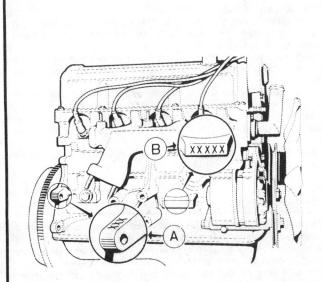

Engine serial number (B) and engine code (A) locations

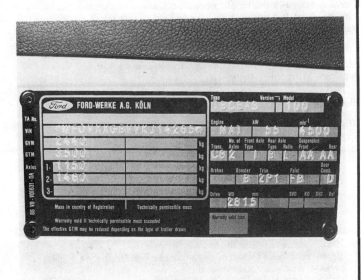

Vehicle identification plate

General repair procedures

Whenever servicing, repair or overhaul work is carried out on the vehicle or its components, it is necessary to observe the following procedures and instructions. This will assist in carrying out the operation efficiently and to a professional standard of workmanship.

Joint mating faces and gaskets

Where a gasket is used between the mating faces of two components, ensure that it is renewed on reassembly, and fit it dry unless otherwise stated in the repair procedure. Make sure that the mating faces are clean and dry with all traces of old gasket removed. When cleaning a joint face, use a tool which is not likely to score or damage the face, and remove any burrs or nicks with an oilstone or fine file.

Make sure that tapped holes are cleaned with a pipe cleaner, and keep them free of jointing compound if this is being used unless specifically instructed otherwise.

Ensure that all orifices, channels or pipes are clear and blow through them, preferably using compressed air.

Oil seals

Whenever an oil seal is removed from its working location, either individually or as part of an assembly, it should be renewed.

The very fine sealing lip of the seal is easily damaged and will not seal if the surface it contacts is not completely clean and free from scratches, nicks or grooves. If the original sealing surface of the component cannot be restored, the component should be renewed.

Protect the lips of the seal from any surface which may damage them in the course of fitting. Use tape or a conical sleeve where possible. Lubricate the seal lips with oil before fitting and, on dual lipped seals, fill the space between the lips with grease.

Unless otherwise stated, oil seals must be fitted with their sealing lips toward the lubricant to be sealed.

Use a tubular drift or block of wood of the appropriate size to install the seal and, if the seal housing is shouldered, drive the seal down to the shoulder. If the seal housing is unshouldered, the seal should be fitted with its face flush with the housing top face.

Screw threads and fastenings

Always ensure that a blind tapped hole is completely free from oil, grease, water or other fluid before installing the bolt or stud. Failure to do this could cause the housing to crack due to the hydraulic action of the bolt or stud as it is screwed in.

When tightening a castellated nut to accept a split pin, tighten the nut to the specified torque, where applicable, and then tighten further to the next split pin hole. Never slacken the nut to align a split pin hole unless stated in the repair procedure.

When checking or retightening a nut or bolt to a specified torque setting, slacken the nut or bolt by a quarter of a turn, and then retighten to the specified setting.

Locknuts, locktabs and washers

Any fastening which will rotate against a component or housing in the course of tightening should always have a washer between it and the relevant component or housing.

Spring or split washers should always be renewed when they are used to lock a critical component such as a big-end bearing retaining nut or bolt.

Locktabs which are folded over to retain a nut or bolt should always be renewed.

Self-locking nuts can be reused in non-critical areas, providing resistance can be felt when the locking portion passes over the bolt or stud thread.

Split pins must always be replaced with new ones of the correct size for the hole.

Special tools

Some repair procedures in this manual entail the use of special tools such as a press, two or three-legged pullers, spring compressors etc. Wherever possible, suitable readily available alternatives to the manufacturer's special tools are described, and are shown in use. In some instances, where no alternative is possible, it has been necessary to resort to the use of a manufacturer's tool and this has been done for reasons of safety as well as the efficient completion of the repair operation. Unless you are highly skilled and have a thorough understanding of the procedure described, never attempt to bypass the use of any special tool when the procedure described specifies its use. Not only is there a very great risk of personal injury, but expensive damage could be caused to the components involved.

Tools and working facilities

Introduction

A selection of good tools is a fundamental requirement for anyone contemplating the maintenance and repair of a motor vehicle. For the owner who does not possess any, their purchase will prove a considerable expense, offsetting some of the savings made by doing-it-yourself. However, provided that the tools purchased meet the relevant national safety standards and are of good quality, they will last for many years and prove an extremely worthwhile investment.

To help the average owner to decide which tools are needed to carry out the various tasks detailed in this manual, we have compiled three lists of tools under the following headings: *Maintenance and minor repair, Repair and overhaul,* and *Special.* The newcomer to practical mechanics should start off with the *Maintenance and minor repair* tool kit and confine himself to the simpler jobs around the vehicle. Then, as his confidence and experience grow, he can undertake more difficult tasks, buying extra tools as, and when, they are needed. In this way, a *Maintenance and minor repair* tool kit can be built-up into a *Repair and overhaul* tool kit over a considerable period of time without any major cash outlays. The experienced do-it-yourselfer will have a tool kit good enough for most repair and overhaul procedures and will add tools from the *Special* category when he feels the expense is justified by the amount of use to which these tools will be put.

It is obviously not possible to cover the subject of tools fully here. For those who wish to learn more about tools and their use there is a book entitled *How to Choose and Use Car Tools* available from the publishers of this manual.

Maintenance and minor repair tool kit

The tools given in this list should be considered as a minimum requirement if routine maintenance, servicing and minor repair operations are to be undertaken. We recommend the purchase of combination spanners (ring one end, open-ended the other); although more expensive than open-ended ones, they do give the advantages of both types of spanner.

Combination spanners - 10, 11, 12, 13, 14 & 17 mm
Adjustable spanner - 9 inch
Engine sump/gearbox/rear axle drain plug key
Spark plug spanner (with rubber insert)
Spark plug gap adjustment tool
Set of feeler gauges
Brake adjuster spanner
Brake bleed nipple spanner
Screwdriver - 4 in long x $\frac{1}{4}$ in dia (flat blade)
Screwdriver - 4 in long x $\frac{1}{4}$ in dia (cross blade)
Combination pliers - 6 inch
Hacksaw (junior)
Tyre pump
Tyre pressure gauge
Grease gun
Oil can
Fine emery cloth (1 sheet)
Wire brush (small)
Funnel (medium size)

Repair and overhaul tool kit

These tools are virtually essential for anyone undertaking any major repairs to a motor vehicle, and are additional to those given in the *Maintenance and minor repair* list. Included in this list is a comprehensive set of sockets. Although these are expensive they will be found invaluable as they are so versatile - particularly if various drives are included in the set. We recommend the $\frac{1}{2}$ in square-drive type, as this can be used with most proprietary torque wrenches. If you cannot afford a socket set, even bought piecemeal, then inexpensive tubular box spanners are a useful alternative.

The tools in this list will occasionally need to be supplemented by tools from the *Special* list.

Sockets (or box spanners) to cover range in previous list
Torx sockets (male and female)
Reversible ratchet drive (for use with sockets)
Extension piece, 10 inch (for use with sockets)
Universal joint (for use with sockets)
Torque wrench (for use with sockets)
'Mole' wrench - 8 inch
Ball pein hammer
Soft-faced hammer, plastic or rubber
Screwdriver - 6 in long x $\frac{5}{16}$ in dia (flat blade)
Screwdriver - 2 in long x $\frac{5}{16}$ in square (flat blade)
Screwdriver - 1$\frac{1}{2}$ in long x $\frac{1}{4}$ in dia (cross blade)
Screwdriver - 3 in long x $\frac{1}{8}$ in dia (electrician's)
Pliers - electrician's side cutters
Pliers - needle nosed
Pliers - circlip (internal and external)
Cold chisel - $\frac{1}{2}$ inch
Scriber
Scraper
Centre punch
Pin punch
Hacksaw
Valve grinding tool
Steel rule/straight-edge
Allen keys (inc. splined/Torx type if necessary)
Selection of files
Wire brush (large)
Axle-stands
Jack (strong trolley or hydraulic type)

Special tools

The tools in this list are those which are not used regularly, are expensive to buy, or which need to be used in accordance with their manufacturers' instructions. Unless relatively difficult mechanical jobs are undertaken frequently, it will not be economic to buy many of these tools. Where this is the case, you could consider clubbing together with friends (or joining a motorists' club) to make a joint purchase, or borrowing the tools against a deposit from a local garage or tool hire specialist.

The following list contains only those tools and instruments freely available to the public, and not those special tools produced by the vehicle manufacturer specifically for its dealer network. You will find

Tools and working facilities

occasional references to these manufacturers' special tools in the text of this manual. Generally, an alternative method of doing the job without the vehicle manufacturers' special tool is given. However, sometimes, there is no alternative to using them. Where this is the case and the relevant tool cannot be bought or borrowed, you will have to entrust the work to a franchised garage.

> *Valve spring compressor (where applicable)*
> *Piston ring compressor*
> *Balljoint separator*
> *Universal hub/bearing puller*
> *Impact screwdriver*
> *Micrometer and/or vernier gauge*
> *Dial gauge*
> *Stroboscopic timing light*
> *Dwell angle meter/tachometer*
> *Universal electrical multi-meter*
> *Cylinder compression gauge*
> *Lifting tackle*
> *Trolley jack*
> *Light with extension lead*

Buying tools

For practically all tools, a tool factor is the best source since he will have a very comprehensive range compared with the average garage or accessory shop. Having said that, accessory shops often offer excellent quality tools at discount prices, so it pays to shop around.

There are plenty of good tools around at reasonable prices, but always aim to purchase items which meet the relevant national safety standards. If in doubt, ask the proprietor or manager of the shop for advice before making a purchase.

Care and maintenance of tools

Having purchased a reasonable tool kit, it is necessary to keep the tools in a clean serviceable condition. After use, always wipe off any dirt, grease and metal particles using a clean, dry cloth, before putting the tools away. Never leave them lying around after they have been used. A simple tool rack on the garage or workshop wall, for items such as screwdrivers and pliers is a good idea. Store all normal wrenches and sockets in a metal box. Any measuring instruments, gauges, meters, etc, must be carefully stored where they cannot be damaged or become rusty.

Take a little care when tools are used. Hammer heads inevitably become marked and screwdrivers lose the keen edge on their blades from time to time. A little timely attention with emery cloth or a file will soon restore items like this to a good serviceable finish.

Working facilities

Not to be forgotten when discussing tools, is the workshop itself. If anything more than routine maintenance is to be carried out, some form of suitable working area becomes essential.

It is appreciated that many an owner mechanic is forced by circumstances to remove an engine or similar item, without the benefit of a garage or workshop. Having done this, any repairs should always be done under the cover of a roof.

Wherever possible, any dismantling should be done on a clean, flat workbench or table at a suitable working height.

Any workbench needs a vice: one with a jaw opening of 4 in (100 mm) is suitable for most jobs. As mentioned previously, some clean dry storage space is also required for tools, as well as for lubricants, cleaning fluids, touch-up paints and so on, which become necessary.

Another item which may be required, and which has a much more general usage, is an electric drill with a chuck capacity of at least $\frac{5}{16}$ in (8 mm). This, together with a good range of twist drills, is virtually essential for fitting accessories such as mirrors and reversing lights.

Last, but not least, always keep a supply of old newspapers and clean, lint-free rags available, and try to keep any working area as clean as possible.

Spanner jaw gap comparison table

Jaw gap (in)	Spanner size
0.250	$\frac{1}{4}$ in AF
0.276	7 mm
0.313	$\frac{5}{16}$ in AF
0.315	8 mm
0.344	$\frac{11}{32}$ in AF; $\frac{1}{8}$ in Whitworth
0.354	9 mm
0.375	$\frac{3}{8}$ in AF
0.394	10 mm
0.433	11 mm
0.438	$\frac{7}{16}$ in AF
0.445	$\frac{3}{16}$ in Whitworth; $\frac{1}{4}$ in BSF
0.472	12 mm
0.500	$\frac{1}{2}$ in AF
0.512	13 mm
0.525	$\frac{1}{4}$ in Whitworth; $\frac{5}{16}$ in BSF
0.551	14 mm
0.563	$\frac{9}{16}$ in AF
0.591	15 mm
0.600	$\frac{5}{16}$ in Whitworth; $\frac{3}{8}$ in BSF
0.625	$\frac{5}{8}$ in AF
0.630	16 mm
0.669	17 mm
0.686	$\frac{11}{16}$ in AF
0.709	18 mm
0.71	$\frac{3}{8}$ in Whitworth; $\frac{7}{16}$ in BSF
0.748	19 mm
0.750	$\frac{3}{4}$ in AF
0.813	$\frac{13}{16}$ in AF
0.820	$\frac{7}{16}$ in Whitworth; $\frac{1}{2}$ in BSF
0.866	22 mm
0.875	$\frac{7}{8}$ in AF
0.920	$\frac{1}{2}$ in Whitworth; $\frac{9}{16}$ in BSF
0.938	$\frac{15}{16}$ in AF
0.945	24 mm
1.000	1 in AF
1.010	$\frac{9}{16}$ in Whitworth; $\frac{5}{8}$ in BSF
1.024	26 mm
1.063	$1\frac{1}{16}$ in AF; 27 mm
1.100	$\frac{5}{8}$ in Whitworth; $\frac{11}{16}$ in BSF
1.125	$1\frac{1}{8}$ in AF
1.181	30 mm
1.200	$\frac{11}{16}$ in Whitworth; $\frac{3}{4}$ in BSF
1.250	$1\frac{1}{4}$ in AF
1.260	32 mm
1.300	$\frac{3}{4}$ in Whitworth; $\frac{7}{8}$ in BSF
1.313	$1\frac{5}{16}$ in AF
1.390	$\frac{13}{16}$ in Whitworth; $\frac{15}{16}$ in BSF
1.417	36 mm
1.438	$1\frac{7}{16}$ in AF
1.480	$\frac{7}{8}$ in Whitworth; 1 in BSF
1.500	$1\frac{1}{2}$ in AF
1.575	40 mm; $1\frac{5}{8}$ in Whitworth
1.614	41 mm
1.625	$1\frac{5}{8}$ in AF
1.670	1 in Whitworth; $1\frac{1}{8}$ in BSF
1.688	$1\frac{11}{16}$ in AF
1.811	46 mm
1.813	$1\frac{13}{16}$ in AF
1.860	$1\frac{1}{8}$ in Whitworth; $1\frac{1}{4}$ in BSF
1.875	$1\frac{7}{8}$ in AF
1.969	50 mm
2.000	2 in AF
2.050	$1\frac{1}{4}$ in Whitworth; $1\frac{3}{8}$ in BSF
2.165	55 mm
2.362	60 mm

Conversion factors

Length (distance)
Inches (in)	X	25.4	= Millimetres (mm)	X	0.0394	= Inches (in)
Feet (ft)	X	0.305	= Metres (m)	X	3.281	= Feet (ft)
Miles	X	1.609	= Kilometres (km)	X	0.621	= Miles

Volume (capacity)
Cubic inches (cu in; in³)	X	16.387	= Cubic centimetres (cc; cm³)	X	0.061	= Cubic inches (cu in; in³)
Imperial pints (Imp pt)	X	0.568	= Litres (l)	X	1.76	= Imperial pints (Imp pt)
Imperial quarts (Imp qt)	X	1.137	= Litres (l)	X	0.88	= Imperial quarts (Imp qt)
Imperial quarts (Imp qt)	X	1.201	= US quarts (US qt)	X	0.833	= Imperial quarts (Imp qt)
US quarts (US qt)	X	0.946	= Litres (l)	X	1.057	= US quarts (US qt)
Imperial gallons (Imp gal)	X	4.546	= Litres (l)	X	0.22	= Imperial gallons (Imp gal)
Imperial gallons (Imp gal)	X	1.201	= US gallons (US gal)	X	0.833	= Imperial gallons (Imp gal)
US gallons (US gal)	X	3.785	= Litres (l)	X	0.264	= US gallons (US gal)

Mass (weight)
Ounces (oz)	X	28.35	= Grams (g)	X	0.035	= Ounces (oz)
Pounds (lb)	X	0.454	= Kilograms (kg)	X	2.205	= Pounds (lb)

Force
Ounces-force (ozf; oz)	X	0.278	= Newtons (N)	X	3.6	= Ounces-force (ozf; oz)
Pounds-force (lbf; lb)	X	4.448	= Newtons (N)	X	0.225	= Pounds-force (lbf; lb)
Newtons (N)	X	0.1	= Kilograms-force (kgf; kg)	X	9.81	= Newtons (N)

Pressure
Pounds-force per square inch (psi; lbf/in²; lb/in²)	X	0.070	= Kilograms-force per square centimetre (kgf/cm²; kg/cm²)	X	14.223	= Pounds-force per square inch (psi; lbf/in²; lb/in²)
Pounds-force per square inch (psi; lbf/in²; lb/in²)	X	0.068	= Atmospheres (atm)	X	14.696	= Pounds-force per square inch (psi; lbf/in²; lb/in²)
Pounds-force per square inch (psi; lbf/in²; lb/in²)	X	0.069	= Bars	X	14.5	= Pounds-force per square inch (psi; lbf/in²; lb/in²)
Pounds-force per square inch (psi; lbf/in²; lb/in²)	X	6.895	= Kilopascals (kPa)	X	0.145	= Pounds-force per square inch (psi; lbf/in²; lb/in²)
Kilopascals (kPa)	X	0.01	= Kilograms-force per square centimetre (kgf/cm²; kg/cm²)	X	98.1	= Kilopascals (kPa)
Millibar (mbar)	X	100	= Pascals (Pa)	X	0.01	= Millibar (mbar)
Millibar (mbar)	X	0.0145	= Pounds-force per square inch (psi; lbf/in²; lb/in²)	X	68.947	= Millibar (mbar)
Millibar (mbar)	X	0.75	= Millimetres of mercury (mmHg)	X	1.333	= Millibar (mbar)
Millibar (mbar)	X	0.401	= Inches of water (inH₂O)	X	2.491	= Millibar (mbar)
Millimetres of mercury (mmHg)	X	0.535	= Inches of water (inH₂O)	X	1.868	= Millimetres of mercury (mmHg)
Inches of water (inH₂O)	X	0.036	= Pounds-force per square inch (psi; lbf/in²; lb/in²)	X	27.68	= Inches of water (inH₂O)

Torque (moment of force)
Pounds-force inches (lbf in; lb in)	X	1.152	= Kilograms-force centimetre (kgf cm; kg cm)	X	0.868	= Pounds-force inches (lbf in; lb in)
Pounds-force inches (lbf in; lb in)	X	0.113	= Newton metres (Nm)	X	8.85	= Pounds-force inches (lbf in; lb in)
Pounds-force inches (lbf in; lb in)	X	0.083	= Pounds-force feet (lbf ft; lb ft)	X	12	= Pounds-force inches (lbf in; lb in)
Pounds-force feet (lbf ft; lb ft)	X	0.138	= Kilograms-force metres (kgf m; kg m)	X	7.233	= Pounds-force feet (lbf ft; lb ft)
Pounds-force feet (lbf ft; lb ft)	X	1.356	= Newton metres (Nm)	X	0.738	= Pounds-force feet (lbf ft; lb ft)
Newton metres (Nm)	X	0.102	= Kilograms-force metres (kgf m; kg m)	X	9.804	= Newton metres (Nm)

Power
Horsepower (hp)	X	745.7	= Watts (W)	X	0.0013	= Horsepower (hp)

Velocity (speed)
Miles per hour (miles/hr; mph)	X	1.609	= Kilometres per hour (km/hr; kph)	X	0.621	= Miles per hour (miles/hr; mph)

Fuel consumption*
Miles per gallon, Imperial (mpg)	X	0.354	= Kilometres per litre (km/l)	X	2.825	= Miles per gallon, Imperial (mpg)
Miles per gallon, US (mpg)	X	0.425	= Kilometres per litre (km/l)	X	2.352	= Miles per gallon, US (mpg)

Temperature
Degrees Fahrenheit = (°C x 1.8) + 32   Degrees Celsius (Degrees Centigrade; °C) = (°F - 32) x 0.56

*It is common practice to convert from miles per gallon (mpg) to litres/100 kilometres (l/100km),
where mpg (Imperial) x l/100 km = 282 and mpg (US) x l/100 km = 235

Safety first!

Professional motor mechanics are trained in safe working procedures. However enthusiastic you may be about getting on with the job in hand, do take the time to ensure that your safety is not put at risk. A moment's lack of attention can result in an accident, as can failure to observe certain elementary precautions.

There will always be new ways of having accidents, and the following points do not pretend to be a comprehensive list of all dangers; they are intended rather to make you aware of the risks and to encourage a safety-conscious approach to all work you carry out on your vehicle.

Essential DOs and DON'Ts

DON'T rely on a single jack when working underneath the vehicle. Always use reliable additional means of support, such as axle stands, securely placed under a part of the vehicle that you know will not give way.

DON'T attempt to loosen or tighten high-torque nuts (e.g. wheel hub nuts) while the vehicle is on a jack; it may be pulled off.

DON'T start the engine without first ascertaining that the transmission is in neutral (or 'Park' where applicable) and the parking brake applied.

DON'T suddenly remove the filler cap from a hot cooling system – cover it with a cloth and release the pressure gradually first, or you may get scalded by escaping coolant.

DON'T attempt to drain oil until you are sure it has cooled sufficiently to avoid scalding you.

DON'T grasp any part of the engine, exhaust or catalytic converter without first ascertaining that it is sufficiently cool to avoid burning you.

DON'T allow brake fluid or antifreeze to contact vehicle paintwork.

DON'T syphon toxic liquids such as fuel, brake fluid or antifreeze by mouth, or allow them to remain on your skin.

DON'T inhale dust – it may be injurious to health (see *Asbestos* below).

DON'T allow any spilt oil or grease to remain on the floor – wipe it up straight away, before someone slips on it.

DON'T use ill-fitting spanners or other tools which may slip and cause injury.

DON'T attempt to lift a heavy component which may be beyond your capability – get assistance.

DON'T rush to finish a job, or take unverified short cuts.

DON'T allow children or animals in or around an unattended vehicle.

DO wear eye protection when using power tools such as drill, sander, bench grinder etc, and when working under the vehicle.

DO use a barrier cream on your hands prior to undertaking dirty jobs – it will protect your skin from infection as well as making the dirt easier to remove afterwards; but make sure your hands aren't left slippery. Note that long-term contact with used engine oil can be a health hazard.

DO keep loose clothing (cuffs, tie etc) and long hair well out of the way of moving mechanical parts.

DO remove rings, wristwatch etc, before working on the vehicle – especially the electrical system.

DO ensure that any lifting tackle used has a safe working load rating adequate for the job.

DO keep your work area tidy – it is only too easy to fall over articles left lying around.

DO get someone to check periodically that all is well, when working alone on the vehicle.

DO carry out work in a logical sequence and check that everything is correctly assembled and tightened afterwards.

DO remember that your vehicle's safety affects that of yourself and others. If in doubt on any point, get specialist advice.

IF, in spite of following these precautions, you are unfortunate enough to injure yourself, seek medical attention as soon as possible.

Asbestos

Certain friction, insulating, sealing, and other products – such as brake linings, brake bands, clutch linings, torque converters, gaskets, etc – contain asbestos. *Extreme care must be taken to avoid inhalation of dust from such products since it is hazardous to health.* If in doubt, assume that they *do* contain asbestos.

Fire

Remember at all times that petrol (gasoline) is highly flammable. Never smoke, or have any kind of naked flame around, when working on the vehicle. But the risk does not end there – a spark caused by an electrical short-circuit, by two metal surfaces contacting each other, by careless use of tools, or even by static electricity built up in your body under certain conditions, can ignite petrol vapour, which in a confined space is highly explosive.

Always disconnect the battery earth (ground) terminal before working on any part of the fuel or electrical system, and never risk spilling fuel on to a hot engine or exhaust.

It is recommended that a fire extinguisher of a type suitable for fuel and electrical fires is kept handy in the garage or workplace at all times. Never try to extinguish a fuel or electrical fire with water.

Note: *Any reference to a 'torch' appearing in this manual should always be taken to mean a hand-held battery-operated electric lamp or flashlight. It does NOT mean a welding/gas torch or blowlamp.*

Fumes

Certain fumes are highly toxic and can quickly cause unconsciousness and even death if inhaled to any extent. Petrol (gasoline) vapour comes into this category, as do the vapours from certain solvents such as trichloroethylene. Any draining or pouring of such volatile fluids should be done in a well ventilated area.

When using cleaning fluids and solvents, read the instructions carefully. Never use materials from unmarked containers – they may give off poisonous vapours.

Never run the engine of a motor vehicle in an enclosed space such as a garage. Exhaust fumes contain carbon monoxide which is extremely poisonous; if you need to run the engine, always do so in the open air or at least have the rear of the vehicle outside the workplace.

If you are fortunate enough to have the use of an inspection pit, never drain or pour petrol, and never run the engine, while the vehicle is standing over it; the fumes, being heavier than air, will concentrate in the pit with possibly lethal results.

The battery

Never cause a spark, or allow a naked light, near the vehicle's battery. It will normally be giving off a certain amount of hydrogen gas, which is highly explosive.

Always disconnect the battery earth (ground) terminal before working on the fuel or electrical systems.

If possible, loosen the filler plugs or cover when charging the battery from an external source. Do not charge at an excessive rate or the battery may burst.

Take care when topping up and when carrying the battery. The acid electrolyte, even when diluted, is very corrosive and should not be allowed to contact the eyes or skin.

If you ever need to prepare electrolyte yourself, always add the acid slowly to the water, and never the other way round. Protect against splashes by wearing rubber gloves and goggles.

When jump starting a car using a booster battery, for negative earth (ground) vehicles, connect the jump leads in the following sequence: First connect one jump lead between the positive (+) terminals of the two batteries. Then connect the other jump lead first to the negative (–) terminal of the booster battery, and then to a good earthing (ground) point on the vehicle to be started, at least 18 in (45 cm) from the battery if possible. Ensure that hands and jump leads are clear of any moving parts, and that the two vehicles do not touch. Disconnect the leads in the reverse order.

Mains electricity and electrical equipment

When using an electric power tool, inspection light etc, always ensure that the appliance is correctly connected to its plug and that, where necessary, it is properly earthed (grounded). Do not use such appliances in damp conditions and, again, beware of creating a spark or applying excessive heat in the vicinity of fuel or fuel vapour. Also ensure that the appliances meet the relevant national safety standards.

Ignition HT voltage

A severe electric shock can result from touching certain parts of the ignition system, such as the HT leads, when the engine is running or being cranked, particularly if components are damp or the insulation is defective. Where an electronic ignition system is fitted, the HT voltage is much higher and could prove fatal.

Routine maintenance

Maintenance is essential for ensuring safety and desirable for the purpose of getting the best in terms of performance and economy from your vehicle. Over the years the need for periodic lubrication – oiling, greasing, and so on – has been drastically reduced if not totally eliminated. This has unfortunately tended to lead some owners to think that because no such action is required, components either no longer exist, or will last for ever. This is certainly not the case; it is essential to carry out regular visual examination as comprehensively as possible in order to spot any possible defects at an early stage before they develop into major expensive repairs.

When carrying out any oil or fluid level checks, the vehicle must be standing on level ground, and a suitable time interval should have elapsed since the vehicle was run to allow the oils and fluids to have cooled and settled. As the various checks are made, refer to the respective Chapter concerned for further details where necessary.

Every 250 miles (400 km) or weekly – whichever comes first

Engine (Chapter 1)
Check the oil level and top up if necessary.

Cooling system (Chapter 2)
Check the coolant level and top up if necessary.

General
Check the front and rear lights (including indicators), also the horn for satisfactory operation
Check the fluid level in the windscreen washer reservoir (and tailgate reservoir if applicable), and top up with a screen wash such as Turtle Wax High Tech Screen Wash
Check the tyre pressures and adjust if necessary

Every 6000 miles (10 000 km) – additional

Engine (Chapter 1)
Clean the oil filler cap in fuel and allow to dry (or blow dry with an air line)
Check for oil, fuel and water leaks
Check the condition and security of vacuum hoses
Change the engine oil and renew the filter
Check and adjust the valve clearances (if necessary) if using un-leaded petrol

Cooling system (Chapter 2)
Check the condition of drivebelts and adjust if necessary

Fuel system (Chapter 3)
Check the slow running adjustment (adjust slow running mixture only at first 6000 miles/10 000 km)

Brakes (Chapter 9)
Check the hydraulic fluid level in the reservoir and top up if neces-sary. A slight drop due to wear of the disc pads is acceptable, but if regular topping up is required the leak should be located and rectified
Check hydraulic lines for leakage
Check servo vacuum hose for condition and security
Check disc pads and rear brake shoes for wear

Suspension and steering (Chapter 10)
Check tyres for condition and wear
Check wheel nuts for tightness
Check suspension U-bolt nuts for tightness and tighten to the specified torque wrench setting if required
Check the steering free play (LCY models)

Bodywork (Chapter 11)
Check the seatbelts for satisfactory condition and operation

Every 12 000 miles (20 000 km) – additional

Engine (Chapter 1)
Check and if necessary adjust the valve clearances
Check the exhaust system for leaks

Ignition system (Chapter 4)
Renew the spark plugs

Manual gearbox (Chapter 6)
Check and, if necessary, top up the oil level
Lubricate the clutch cable at the lever connection with grease

Automatic transmission (Chapter 6)
Lubricate the shift cable and linkage, and shaft lever linkage
Check and, if necessary, top up the transmission fluid level
Have the brake band adjustment checked by a Ford dealer

Propeller shaft and final drive (Chapters 7 and 8)
Inspect the propeller shaft universal joints and the centre bearing for excessive wear. At the same time, check for any signs of oil leakage from the transmission rear oil seal and the rear axle pinion shaft oil seal. Lubricate the joints (where applicable)
Check the rear axle oil level and top up if required

Braking system (Chapter 9)
Check the operation of the brake fluid level warning light
Lubricate the handbrake linkages

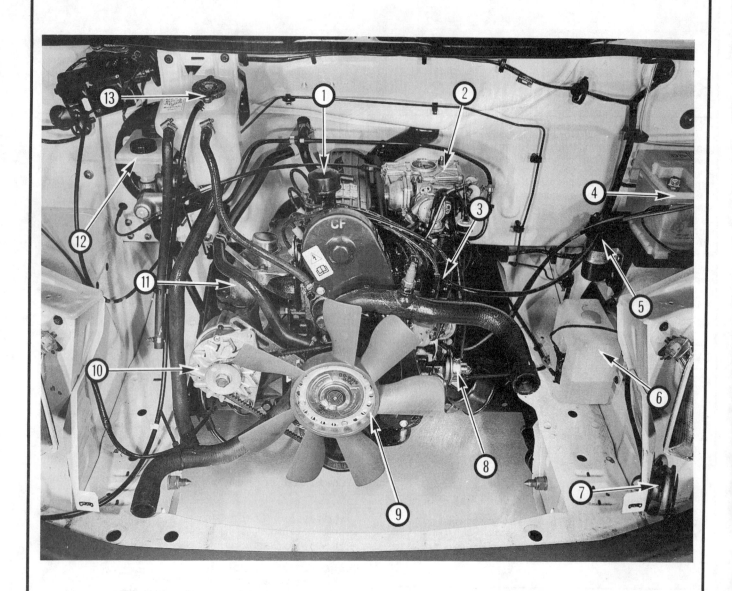

Engine and associate components (radiator and front panel removed for clarity) 2.0 litre engine shown

1 Engine oil filler
2 Carburettor (filter removed)
3 Ignition distributor
4 Battery
5 Ignition coil
6 Washer reservoir
7 Horn
8 Fuel pump
9 Cooling fan
10 Alternator
11 Exhaust manifold
12 Master cylinder reservoir
13 Coolant expansion tank and filler cap

Underside view of engine and vehicle front section

1 Engine sump
2 Starter motor
3 Engine mounting
4 Front brake calliper
5 Gearbox mounting
6 Propeller shaft and front universal joint
7 Exhaust system
8 Gearbox (MT75 type shown)
9 Suspension arm
10 Tie rod
11 Steering gear (rack and pinion)

Underside view of vehicle centre section

1. Handbrake cable compensator unit
2. Rear leaf spring front end mounting point
3. Brake pressure control valve
4. Exhaust system
5. Propeller shaft central bearing
6. Fuel tank filler hose
7. Fuel tank

Underside view of vehicle rear section

1 Shock absorber
2 Rear axle
3 Leaf spring and rear shackle mounting
4 Spare wheel and retainer
5 Rear silencer
6 Rear brake drum
7 Handbrake cable (right-hand)
8 Propeller shaft/pinion flange joint

Suspension and steering (Chapter 10)
Check all linkages and balljoints for wear and damage
Check the condition of the steering rack gaiters (where applicable)
Lubricate the steering kingpins and bushes (beam axle models only)
Check the torque of the roadwheel nuts

Bodywork (Chapter 11)
Lubricate all hinges and catches
Check the underbody for corrosion and damage
Extended wheelbase models: Have chassis extension bolts checked for security by a Ford dealer.

Electrical system (Chapter 12)
Check the operation of all electrical equipment and lights
Clean battery terminals, check them for tightness, and apply petroleum jelly

Every 24 000 miles (40 000 km) – additional

Engine (Chapter 1)
Renew the crankcase emission vent valve

Fuel system (Chapter 3)
Renew the air filter element

Ignition system (Chapter 4)
Lubricate distributor and clean distributor cap and HT leads
Check condition of distributor cap, rotor and HT leads

Suspension and steering (Chapter 10)
Check and adjust if necessary front wheel bearings

Every 36 000 miles (60 000 km) or two years whichever occurs first

Cooling system (Chapter 2)
Flush the cooling system and fill with new antifreeze/corrosion inhibitor
Check cooling system pressure cap and seal – renew if necessary

Every 36 000 miles (60 000 km) or three years whichever occurs first

Engine (Chapter 1)
Renew the camshaft driveshaft. **Note:** *Although not specified by the manufacturer, it is advisable to renew the belt as a precautionary measure against possible high mileage failure*

Braking system (Chapter 9)
Renew the hydraulic brake fluid and check the condition of the visible rubber components of the brake system

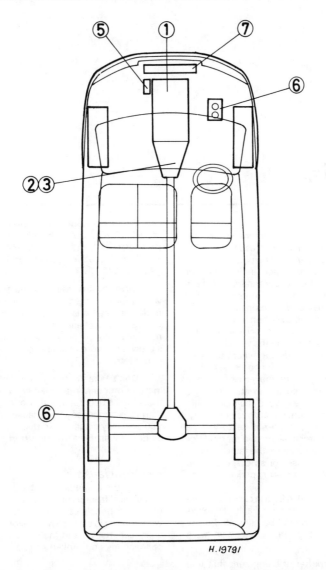

Recommended lubricants and fluids

Component or system	Lubricant type/specification	Duckhams recommendation
1 Engine	Multigrade engine oil, viscosity range SAE 10W/30 to 20W/50, to API SF/CC or better	Duckhams QXR, Hypergrade, or 10W/40 Motor Oil
2 Manual gearbox 4-speed	Gear oil, viscosity SAE 80 EP, to Ford spec SQM-2C 9008-A	Duckhams Hypoid 80
5-speed	Gear oil, viscosity SAE 80 EP, to Ford spec ESD-M2C 175-A and ESD-M2C 186A	Duckhams Hypoid 75W/90S
3 Automatic transmission	ATF to Ford spec SQM-2C 9010-A	Duckhams D-Matic
4 Final drive	Hypoid gear oil, viscosity SAE 90 EP to Ford spec SQM-2C 9002-AA or 9003-AA	Duckhams Hypoid 90S
5 Power-assisted steering	ATF to Ford spec SQM-2C 9010-A	Duckhams D-Matic
6 Brake hydraulic system	Brake fluid to Ford spec Amber SAM-6C 9103-A	Duckhams Universal Brake and Clutch Fluid
7 Cooling system	Soft water and antifreeze to Ford spec SSM-97B 9103-A	Duckhams Universal Antifreeze and Summer Coolant

Fault diagnosis

Introduction

The vehicle owner who does his or her own maintenance according to the recommended schedules should not have to use this section of the manual very often. Modern component reliability is such that, provided those items subject to wear or deterioration are inspected or renewed at the specified intervals, sudden failure is comparatively rare. Faults do not usually just happen as a result of sudden failure, but develop over a period of time. Major mechanical failures in particular are usually preceded by characteristic symptoms over hundreds or even thousands of miles. Those components which do occasionally fail without warning are often small and easily carried in the vehicle.

With any fault finding, the first step is to decide where to begin investigations. Sometimes this is obvious, but on other occasions a little detective work will be necessary. The owner who makes half a dozen haphazard adjustments or replacements may be successful in curing a fault (or its symptoms), but he will be none the wiser if the fault recurs and he may well have spent more time and money than was necessary. A calm and logical approach will be found to be more satisfactory in the long run. Always take into account any warning signs or abnormalities that may have been noticed in the period preceding the fault – power loss, high or low gauge readings, unusual noises or smells, etc – and remember that failure of components such as fuses or spark plugs may only be pointers to some underlying fault.

The pages which follow here are intended to help in cases of failure to start or breakdown on the road. There is also a Fault Diagnosis Section at the end of each Chapter which should be consulted if the preliminary checks prove unfruitful. Whatever the fault, certain basic principles apply. These are as follows:

Verify the fault. This is simply a matter of being sure that you know what the symptoms are before starting work. This is particularly important if you are investigating a fault for someone else who may not have described it very accurately.

Don't overlook the obvious. For example, if the vehicle won't start, is there petrol in the tank? (Don't take anyone else's word on this particular point, and don't trust the fuel gauge either!) If an electrical fault is indicated, look for loose or broken wires before digging out the test gear.

Cure the disease, not the symptom. Substituting a flat battery with a fully charged one will get you off the hard shoulder, but if the underlying cause is not attended to, the new battery will go the same way. Similarly, changing oil-fouled spark plugs for a new set will get you moving again, but remember that the reason for the fouling (if it wasn't simply an incorrect grade of plug) will have to be established and corrected.

Don't take anything for granted. Particularly, don't forget that a 'new' component may itself be defective (especially if it's been rattling round in the boot for months), and don't leave components out of a fault diagnosis sequence just because they are new or recently fitted. When you do finally diagnose a difficult fault, you'll probably realise that all the evidence was there from the start.

Electrical faults

Electrical faults can be more puzzling than straightforward mechanical failures, but they are no less susceptible to logical analysis if the basic principles of operation are understood. Vehicle electrical wiring exists in extremely unfavourable conditions – heat, vibration and chemical attack – and the first things to look for are loose or corroded connections and broken or chafed wires, especially where the wires

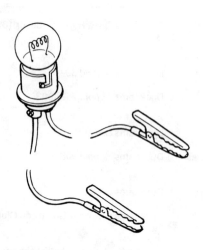

A simple test lamp is useful for tracing electrical faults

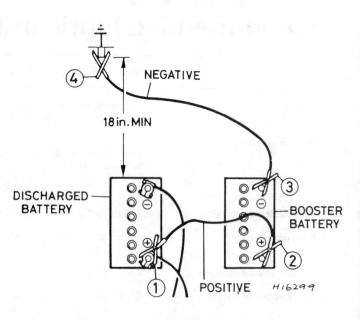

Jump start lead connections for negative earth vehicles – connect leads in order shown

Fault diagnosis

Carrying a few spares may save a long walk!

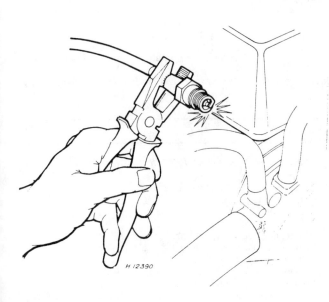

Crank engine and check for spark. Note use of insulated tool to hold plug lead

pass through holes in the bodywork or are subject to vibration.
All metal-bodied vehicles in current production have one pole of the battery 'earthed', ie connected to the vehicle bodywork, and in nearly all modern vehicles it is the negative (−) terminal. The various electrical components – motors, bulb holders etc – are also connected to earth, either by means of a lead or directly by their mountings. Electric current flows through the component and then back to the battery via the bodywork. If the component mounting is loose or corroded, or if a good path back to the battery is not available, the circuit will be incomplete and malfunction will result. The engine and/or gearbox are also earthed by means of flexible metal straps to the body or subframe; if these straps are loose or missing, starter motor, generator and ignition trouble may result.

Assuming the earth return to be satisfactory, electrical faults will be due either to component malfunction or to defects in the current supply. Individual components are dealt with in Chapter 12. If supply wires are broken or cracked internally this results in an open-circuit, and the easiest way to check for this is to bypass the suspect wire temporarily with a length of wire having a crocodile clip or suitable connector at each end. Alternatively, a 12V test lamp can be used to verify the presence of supply voltage at various points along the wire and the break can be thus isolated.

If a bare portion of a live wire touches the bodywork or other earthed metal part, the electricity will take the low-resistance path thus formed back to the battery: this is known as a short-circuit. Hopefully a short-circuit will blow a fuse, but otherwise it may cause burning of the insulation (and possibly further short-circuits) or even a fire. This is why it is inadvisable to bypass persistently blowing fuses with silver foil or wire.

Spares and tool kit

Most vehicles are supplied only with sufficient tools for wheel changing; the *Maintenance and minor repair* tool kit detailed in *Tools and working facilities,* with the addition of a hammer, is probably sufficient for those repairs that most motorists would consider attempting at the roadside. In addition a few items which can be fitted without too much trouble in the event of a breakdown should be carried. Experience and available space will modify the list below, but the following may save having to call on professional assistance:

Spark plugs, clean and correctly gapped
HT lead and plug cap – long enough to reach the plug furthest from the distributor
Drivebelt(s) – emergency type may suffice
Spare fuses
Set of principal light bulbs
Tin of radiator sealer and hose bandage
Exhaust bandage

Fault diagnosis

Roll of insulating tape
Length of soft iron wire
Length of electrical flex
Torch or inspection lamp (can double as test lamp)
Battery jump leads
Tow-rope
Ignition water dispersant aerosol
Litre of engine oil
Sealed can of hydraulic fluid
Emergency windscreen
Tube of filler paste

If spare fuel is carried, a can designed for the purpose should be used to minimise risks of leakage and collision damage. A first aid kit and a warning triangle, whilst not at present compulsory in the UK, are obviously sensible items to carry in addition to the above.

When touring abroad it may be advisable to carry additional spares which, even if you cannot fit them yourself, could save having to wait while parts are obtained. The items below may be worth considering:

Clutch and throttle cables
Cylinder head gasket
Alternator brushes
Tyre valve core

One of the motoring organisations will be able to advise on availability of fuel etc in foreign countries.

Engine will not start

Engine fails to turn when starter operated

Flat battery (recharge, use jump leads, or push start)
Battery terminals loose or corroded
Battery earth to body defective
Engine earth strap loose or broken
Starter motor (or solenoid) wiring loose or broken
Automatic transmission selector in wrong position, or inhibitor switch faulty
Ignition/starter switch faulty
Major mechanical failure (seizure)
Starter or solenoid internal fault (see Chapter 12)

Starter motor turns engine slowly

Partially discharged battery (recharge, use jump leads, or push start)
Battery terminals loose or corroded
Battery earth to body defective
Engine earth strap loose
Starter motor (or solenoid) wiring loose
Starter motor internal fault (see Chapter 12)

Starter motor spins without turning engine

Flat battery
Starter motor pinion sticking on sleeve
Flywheel gear teeth damaged or worn
Starter motor mounting bolts loose

Engine turns normally but fails to start

Damp or dirty HT leads and distributor cap (crank engine and check for spark) – try moisture dispersant such as Holts Wet Start
No fuel in tank (check for delivery at carburettor)
Excessive choke (hot engine) or insufficient choke (cold engine)
Fouled or incorrectly gapped spark plugs (remove, clean and regap)
Other ignition system fault (see Chapter 4)
Other fuel system fault (see Chapter 3)
Poor compression (see Chapter 1)
Major mechanical failure (eg camshaft drive)

Engine fires but will not run

Insufficient choke (cold engine)
Air leaks at carburettor or inlet manifold

Fuel starvation (see Chapter 3)
Other ignition fault (see Chapter 4)

Engine cuts out and will not restart

Engine cuts out suddenly – ignition fault

Loose or disconnected LT wires
Wet HT leads or distributor cap (after traversing water splash)
Coil failure (check for spark)
Other ignition fault (see Chapter 4)

Engine misfires before cutting out – fuel fault

Fuel tank empty
Fuel pump defective or filter blocked (check for delivery)
Fuel tank filler vent blocked (suction will be evident on releasing cap)
Carburettor needle valve sticking
Carburettor jets blocked (fuel contaminated)
Other fuel system fault (see Chapter 3)

Engine cuts out – other causes

Serious overheating
Major mechanical failure (eg camshaft drive)

Engine overheats

Ignition (no-charge) warning light illuminated

Slack or broken drivebelt – retension or renew (Chapter 2)

Ignition warning light not illuminated

Coolant loss due to internal or external leakage (see Chapter 2)
Thermostat defective
Low oil level
Brakes binding
Radiator clogged externally or internally
Electric cooling fan not operating correctly
Engine waterways clogged
Ignition timing incorrect or automatic advance malfunctioning
Mixture too weak
Note: *Do not add cold water to an overheated engine or damage may result*

Low engine oil pressure

Gauge reads low or warning light illuminated with engine running

Oil level low or incorrect grade
Defective gauge or sender unit
Wire to sender unit earthed
Engine overheating
Oil filter clogged or bypass valve defective
Oil pressure relief valve defective
Oil pick-up strainer clogged
Oil pump worn or mountings loose
Worn main or big-end bearings
Note: *Low oil pressure in a high-mileage engine at tickover is not necessarily a cause for concern. Sudden pressure loss at speed is far more significant. In any event, check the gauge or warning light sender before condemning the engine.*

Engine noises

Pre-ignition (pinking) on acceleration

Incorrect grade of fuel
Ignition timing incorrect
Distributor faulty or worn
Worn or maladjusted carburettor
Excessive carbon build-up in engine

Fault diagnosis

Whistling or wheezing noises

Leaking vacuum hose
Leaking carburettor or manifold gasket
Blowing head gasket

Tapping or rattling

Incorrect valve clearances
Worn valve gear
Worn timing belt
Broken piston ring (ticking noise)

Knocking or thumping

Unintentional mechanical contact (eg fan blades)
Worn drivebelt
Peripheral component fault (generator, water pump etc)
Worn big-end bearings (regular heavy knocking, perhaps less under load)
Worn main bearings (rumbling and knocking, perhaps worsening under load)
Piston slap (most noticeable when cold)

Chapter 1 Engine

Contents

Ancillary components – refitting	47	Engine reassembly – general	36
Ancillary components – removal	10	Engine (without transmission) – removal and refitting	6
Auxiliary shaft – examination	32	Examination and renovation – general	26
Auxiliary shaft – refitting	42	Fault diagnosis – engine	50
Auxiliary shaft – removal	15	Flywheel/driveplate – refitting	41
Camshaft – refitting	44	Flywheel/driveplate – removal	16
Camshaft – removal	12	Flywheel ring gear – examination and renovation	34
Camshaft and cam followers – examination and renovation	31	General description	1
Crankcase ventilation system – description and maintenance	25	Major operations possible – engine in vehicle	3
Crankshaft and bearings – examination and renovation	28	Major operations requiring engine removal	4
Crankshaft and main bearings – refitting	37	Methods of engine removal	5
Crankshaft and main bearings – removal	23	Oil filter – renewal	21
Crankshaft front oil seal – renewal	18	Oil pump – examination and renovation	27
Crankshaft rear oil seal – renewal	19	Oil pump – refitting	39
Cylinder block and bores – examination and renovation	29	Oil pump – removal	20
Cylinder head – decarbonising, valve grinding and renovation	35	Pistons and connecting rods – examination and renovation	30
Cylinder head – dismantling	13	Pistons and connecting rods – refitting	38
Cylinder head – reassembly	43	Pistons and connecting rods – removal	22
Cylinder head – refitting	45	Routine maintenance	2
Cylinder head – removal	11	Sump – refitting	40
Engine – initial start-up after overhaul or major repair	49	Sump – removal	17
Engine and transmission – removal and refitting	7	Timing belt – examination	33
Engine and transmission – separation and reconnection	8	Timing belt and sprockets – refitting	46
Engine dismantling – general	9	Timing belt and sprockets – removal	14
Engine mountings – renewal	24	Valve clearances – adjustment	48

Specifications

General

Type	Four cylinder, in-line, water-cooled, single overhead camshaft
Firing order	1-3-4-2 (No 1 at timing belt end)

	LAT (1.6)	**NAT (2.0)**
Engine code		
Bore	87.67 mm (3.454 in)	90.82 mm (3.576 in)
Stroke	66.00 mm (2.60 in)	76.95 mm (3.032 in)
Cubic capacity	1593 cc	1993 cc
Compression ratio	8.2 : 1	8.2 : 1
Compression pressure at starter speed	9 to 11 bar (131 to 160 lbf/in²)	10 to 12 bar (145 to 174 lbf/in²)
Maximum continuous engine speed	5800 rpm	5800 rpm

Cylinder block

Bore diameter mm (in):

	1.6 litre	**2.0 litre**
Standard 1	87.650 to 87.660 (3.4508 to 3.4512)	90.800 to 90.810 (3.5748 to 3.5752)
Standard 2	87.660 to 87.670 (3.4512 to 3.4516)	90.810 to 90.820 (3.5752 to 3.5756)
Standard 3	87.670 to 87.680 (3.4516 to 3.4520)	90.820 to 90.830 (3.5756 to 3.5760)
Standard 4	87.680 to 87.690 (3.4520 to 3.4524)	90.830 to 90.840 (3.5760 to 3.5764)
Oversize A	88.160 to 87.170 (3.4709 to 3.4713)	91.310 to 91.320 (3.5949 to 3.5953)
Oversize B	88.170 to 88.180 (3.4713 to 3.4717)	91.320 to 91.330 (3.5953 to 3.5957)

Cylinder block

	1.6 litre	2.0 litre
Oversize C	88.180 to 88.190 (3.4717 to 3.4720)	91.330 to 91.340 (3.5957 to 3.5961)
Standard service size	87.680 to 87.690 (3.4520 to 3.4524)	90.830 to 90.840 (3.5760 to 3.5764)
Oversize 0.5	88.180 to 88.190 (3.4717 to 3.4720)	91.330 to 91.340 (3.5957 to 3.5961)
Oversize 1.0	88.680 to 88.690 (3.4913 to 3.4917)	91.830 to 91.840 (3.6154 to 3.6157)

Crankshaft

Endfloat	0.08 to 0.28 mm (0.003 to 0.011 in)
Number of main bearings	5
Main bearing journal diameter:	
Standard	56.970 to 56.990 mm (2.2429 to 2.2437 in)
Undersize 0.25	56.720 to 56.740 mm (2.2331 to 2.2339 in)
Undersize 0.50	56.470 to 56.490 mm (2.2232 to 2.2240 in)
Undersize 0.75	56.220 to 56.240 mm (2.2134 to 2.2142 in)
Undersize 1.0	55.970 to 55.990 mm (2.2035 to 2.2043 in)
Main bearing running clearance	0.010 to 0.064 mm (0.0004 to 0.0025 in)
Crankpin journal diameter:	
Standard	51.980 to 52.000 mm (2.0465 to 2.0472 in)
Undersize 0.25	51.730 to 51.750 mm (2.0366 to 2.0374 in)
Undersize 0.50	51.480 to 51.500 mm (2.0268 to 2.0276 in)
Undersize 0.75	51.230 to 51.250 mm (2.0169 to 2.0177 in)
Undersize 1.00	50.980 to 51.000 mm (2.0071 to 2.0079 in)
Big-end bearing running clearance	0.006 to 0.060 mm (0.0002 to 0.0024 in)
Main bearing thrustwasher thickness:	
Standard	2.30 to 2.35 mm (0.0906 to 0.0925 in)
Oversize	2.50 to 2.55 mm (0.0984 to 0.1004 in)

Camshaft

Number of bearings	3
Drive	Toothed belt
Thrust plate thickness	3.98 to 4.01 mm (0.1567 to 0.1579 in)
Bearing journal diameter:	
Front	41.987 to 42.013 mm (1.6530 to 1.6541 in)
Centre	44.607 to 44.633 mm (1.7562 to 1.7572 in)
Rear	44.987 to 45.103 mm (1.7711 to 1.7722 in)
Endfloat	0.104 to 0.204 mm (0.004 to 0.008 in)

Auxiliary shaft

Endfloat	0.050 to 0.204 mm (0.0020 to 0.008 in)

Pistons

	1.6 litre	2.0 litre
Diameter mm (in):		
Standard 1	87.615 to 87.625 (3.4494 to 3.4498)	90.765 to 90.775 (3.5734 to 3.5738)
Standard 2	87.625 to 87.635 (3.4498 to 3.4502)	90.775 to 90.785 (3.5738 to 3.5742)
Standard 3	87.635 to 87.645 (3.4502 to 3.4506)	90.785 to 90.795 (3.5742 to 3.5746)
Standard 4	87.645 to 87.655 (3.4506 to 3.4510)	90.795 to 90.805 (3.5746 to 3.5750)
Standard service size	87.630 to 87.655 (3.4500 to 3.4510)	90.790 to 90.815 (3.5744 to 3.5753)
Oversize	88.130 to 88.155 (3.4696 to 3.4706)	91.290 to 91.315 (3.5940 to 3.5950)
Oversize 1.0	88.630 to 88.655 (3.4893 to 3.4903)	91.790 to 91.815 (3.6137 to 3.6147)
Clearance in bore (new)	0.015 to 0.050 (0.0006 to 0.0020)	0.015 to 0.050 (0.0006 to 0.0020)
Ring gap (fitted) mm (in):		
Top and centre	0.300 to 0.500 (0.012 to 0.019)	0.400 to 0.600 (0.016 to 0.024)
Bottom	0.40 to 1.40 (0.016 to 0.055)	0.40 to 1.40 (0.016 to 0.055)

Gudgeon pin

Diameter – Red	23.994 to 23.997 mm (0.9446 to 0.9448 in)
Blue	23.997 to 24.000 mm (0.9448 to 0.9449 in)
Yellow	24.000 to 24.003 mm (0.9449 to 0.9450 in)
Clearance in piston	0.008 to 0.014 mm (0.0003 to 0.0006 in)
Interference in connecting rod	0.018 to 0.039 mm (0.0007 to 0.0015 in)

Connecting rod

Big-end parent bore diameter	55.00 to 55.02 mm (2.1654 to 2.1661 in)
Small-end parent bore diameter	23.964 to 23.976 mm (0.9435 to 0.9439 in)

28   Chapter 1   Engine

Cylinder head

Cast marking:
  1.6 litre 6
  2.0 litre 0
Valve seat angle .... 44° 30′ to 45° 00′
Valve seat width .... 1.5 to 2.0 mm (0.059 to 0.079 in)
Valve stem bore:
  Standard .... 8.063 to 8.088 mm (0.3174 to 0.3184 in)
  Oversize 0.2 .... 8.263 to 8.288 mm (0.3253 to 0.3263 in)
  Oversize 0.4 .... 8.463 to 8.488 mm (0.3332 to 0.3342 in)

Valves

Valve clearance (cold):
  Inlet .... 0.20 ± 0.03 mm (0.008 ± 0.001 in)
  Exhaust .... 0.25 ± 0.03 mm (0.010 ± 0.001 in)
Valve timing:
  Inlet opens .... 18° BTDC
  Inlet closes .... 58° ABDC
  Exhaust opens .... 70° BBDC
  Exhaust closes .... 6° ATDC

Inlet valve

	1.6 litre	2.0 litre
Length	112.65 to 113.65 mm (4.435 to 4.474 in)	110.65 to 111.65 mm (4.356 to 4.396 in)
Head diameter	41.80 to 42.20 mm (1.646 to 1.661 in)	41.80 to 42.20 mm (1.646 to 1.661 in)

Stem diameter:
  Standard .... 8.025 to 8.043 mm (0.3159 to 0.3167 in)
  Oversize 0.2 .... 8.225 to 8.243 mm (0.3238 to 0.3245 in)
  Oversize 0.4 .... 8.425 to 8.443 mm (0.3317 to 0.3324 in)
  Oversize 0.6 .... 8.625 to 8.643 mm (0.3396 to 0.3403 in)
  Oversize 0.8 .... 8.825 to 8.843 mm (0.3474 to 0.3481 in)
Stem to guide clearance .... 0.020 to 0.063 mm (0.0008 to 0.0025 in)
Spring free length .... 47.0 mm (1.85 in)

Exhaust valve

	1.6 litre	2.0 litre
Length	112.05 to 113.05 mm (4.4114 to 4.4507 in)	110.05 to 111.05 mm (4.3326 to 4.3720 in)
Head diameter	34.00 to 34.40 mm (1.339 to 1.354 in)	35.80 to 36.20 mm (1.409 to 1.425 in)

Stem diameter:
  Standard .... 7.999 to 8.017 mm (0.3149 to 0.3156 in)
  Oversize 0.2 .... 8.199 to 8.217 mm (0.3228 to 0.3235 in)
  Oversize 0.4 .... 8.399 to 8.417 mm (0.3307 to 0.3314 in)
  Oversize 0.6 .... 8.599 to 8.617 mm (0.3385 to 0.3393 in)
  Oversize 0.8 .... 8.799 to 8.817 mm (0.3464 to 0.3471 in)
Stem to guide clearance .... 0.046 to 0.089 mm (0.0018 to 0.0035 in)
Spring free length .... 47.0 mm (1.85 in)

Lubrication system

Type .... Bi-rotor pump driven by auxiliary shaft
Oil type/specification .... Multigrade engine oil, viscosity range SAE 10W/30 to 20W/50, to API SF/CC or better (Duckhams QXR, Hypergrade, or 10W/40 Motor Oil)
Oil filter .... Champion C102
Oil capacity:
  Without filter .... 3.25 litres (5.7 pints)
  With filter .... 3.75 litres (6.6 pints)
Minimum oil pressure:
  At 750 rpm .... 2.1 bar (30.45 lbf/in²)
  At 2000 rpm .... 2.5 bar (36.25 lbf/in²)
Oil pressure relief valve opens at .... 4.0 to 4.7 bar (58.0 to 68.2 lbf/in²)
Warning light operates at .... 0.3 to 0.5 bar (4.4 to 7.25 lbf/in²)
Oil pump clearances:
  Rotor to body .... 0.153 to 0.304 mm (0.006 to 0.011 in)
  Inner rotor to outer rotor .... 0.05 to 0.20 mm (0.002 to 0.008 in)
  Rotor to cover .... 0.039 to 0.104 mm (0.001 to 0.004 in)

Torque wrench settings

	Nm	lbf ft
Main bearing cap	88 to 102	65 to 75
Big-end bearing cap	40 to 47	30 to 35
Crankshaft pulley	100 to 115	74 to 85
Camshaft sprocket	45 to 50	33 to 37
Auxiliary shaft sprocket	45 to 50	33 to 37
Flywheel	64 to 70	47 to 52
Oil pump to cylinder block	17 to 21	13 to 15

Chapter 1 Engine

Torque wrench settings (continued)

	Nm	lbf ft
Oil pump cover to body	9 to 13	7 to 10
Sump:		
Stage 1	1 to 2	0.7 to 1.5
Stage 2	6 to 8	4 to 6
Stage 3 (after 20 minutes running)	8 to 10	6 to 7
Sump drain plug	21 to 28	15 to 21
Oil pressure switch	12 to 15	9 to 11
Valve adjustment ball-pins	50 to 55	37 to 41
Cylinder head*:		
Stage 1	35 to 40	26 to 30
Stage 2	70 to 75	52 to 55
Stage 3 (after 5 minutes)	Tighten further 90°	Tighten further 90°
Front oil seal housing	13 to 17	10 to 13
Timing belt tensioner bolts	20 to 25	15 to 18
Valve (rocker) cover sequence (see Section 45):		
Bolts 1 to 6	6 to 8	4 to 6
Bolts 9 and 10	2 to 3	1.5 to 2.2
Bolts 7 and 8	6 to 8	4 to 6
Bolts 9 and 10 (further tighten)	6 to 8	4 to 6
Oil inlet pipe to pump	11 to 14	8 to 10
Oil inlet pipe to cylinder block	17 to 21	13 to 15
Cylinder block adapter plate	40.5 to 57.5	30 to 42
Clutch housing adapter plate	40.5 to 57.5	30 to 42
Rubber insulators to body	40.5 to 57.5	30 to 42

The manufacturers recommend that these bolts are renewed once they are removed as they must not be retorqued

1 General description

Two engine sizes are covered by this manual, the 1.6 litre (1593cc) and the 2.0 litre (1993cc). Both are similar in design and are shown in Fig. 1.2.

The cylinder head is of crossflow design with the inlet manifold one side and the exhaust manifold on the other. As flat top pistons are used the combustion chambers are contained in the cylinder head.

The combined crankcase and cylinder block is made of cast iron and houses the pistons and crankshaft. Attached to the underside of the crankcase is a pressed steel sump which acts as a reservoir for the engine oil.

The cast iron cylinder head is mounted on top of the cylinder block and acts as a support for the overhead camshaft. The slightly angled valves operate directly in the cylinder head and are controlled by the camshaft via cam followers. The camshaft is driven by a toothed reinforced composite rubber belt from the crankshaft. To eliminate backlash and prevent slackness of the belt, a spring-loaded tensioner in the form of a jockey wheel is in contact with the back of the belt. It serves two further functions, to keep the belt away from the water pump and also to increase the contact area of the camshaft and crankshaft sprocket.

The drivebelt also drives the auxiliary shaft sprocket and it is from this shaft that the oil pump, distributor and fuel pump operate.

The inlet manifold is mounted on the left-hand side of the cylinder head and to this the carburettor is fitted. A water jacket is incorporated in the inlet manifold so that the petrol/air charge may be preheated before entering the combustion chambers.

The exhaust manifold is mounted on the right-hand side of the cylinder head and connects to a single downpipe and twin silencer system.

Aluminium alloy pistons are connected to the crankshaft by forged steel connecting rods and gudgeon pins. The gudgeon pin is a press fit in the small end of the connecting rod but a floating fit in the piston boss. Two compression rings and one scraper ring, all located above the gudgeon pin, are fitted.

The forged crankshaft runs in five main bearings and endfloat is accommodated by thrustwashers either side of the centre main bearing.

Before commencing any overhaul work on the engine refer to Section 9 where information is given about special tools that are required for some overhaul operations.

2 Routine maintenance

The following maintenance tasks must be made at the intervals given in the Routine maintenance Section at the start of this manual.

Check the engine oil level

1 This check must be made with the vehicle standing on level ground, preferably when the engine is cold. Remove the engine oil dipstick, wipe it clean and then re-insert it fully into the location port in the side of the engine. Withdraw it and observe the oil level reading. The level should

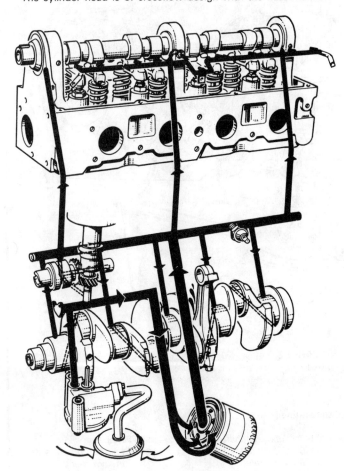

Fig. 1.1 Engine lubrication circuit (Sec 1)

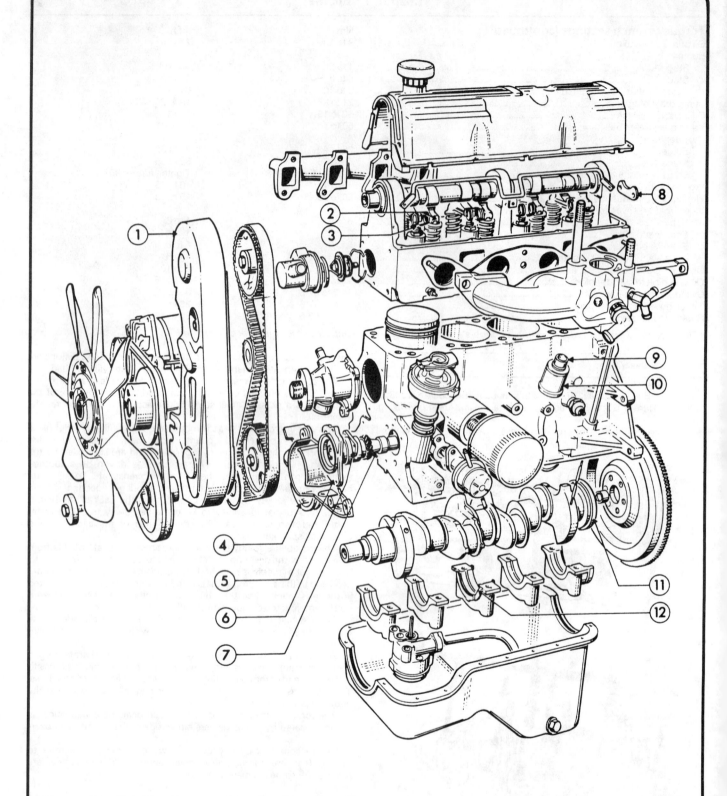

Fig. 1.2 Exploded view of the engine (Sec 1)

1 Timing cover	4 Crankshaft front oil seal housing	7 Auxiliary shaft	10 Oil separator
2 Cam follower	5 Auxiliary shaft front cover	8 Camshaft thrustplate	11 Crankshaft rear oil seal
3 Retaining spring clip	6 Thrustplate	9 Vent valve	12 Thrustwasher

Chapter 1 Engine

2.1 Topping up the engine oil level

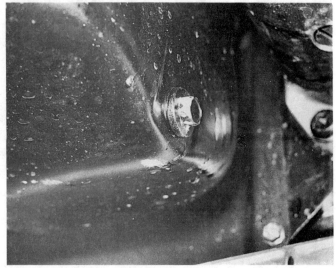

2.2 Engine sump drain plug

be between the MAX and MIN markings on the dipstick. If necessary, remove the oil filler cap on the top of the valve cover and top up the oil level through the filler neck. Do not overfill. Refit the cap (photo).

Oil renewal

2 With the vehicle standing on level ground, position a suitable container under the sump, unscrew the sump drain plug (photo), and drain the oil into the container for disposal. If the engine is hot, take care not to scald yourself as the plug is removed. When the oil is fully drained, refit the plug and tighten it to the specified torque setting. Refill with the specified grade of engine oil. When the engine is restarted, initially run it at idle speed for a couple of minutes to allow the oil to recirculate. Check for any signs of oil leaks. A further small adjustment to the oil level may be required after the engine had been initially run to replace the displacement caused by circulation.

3 Clean the oil filler cap filter using fuel or a suitable solvent and dry thoroughly.

4 Renew the oil filter as described in Section 21 of this Chapter.

5 Check and if necessary adjust the valve clearances as described in Section 48.

6 Renew the crankcase vent valve as described in Section 25.

7 Renew the timing belt as described in Section 14.

3 Major operations possible – engine in vehicle

The following major operations may be undertaken with the engine in the vehicle:

(a) Removal and refitting of the camshaft
(b) Removal and refitting of the cylinder head
(c) Removal and refitting of the timing belt
(d) Removal and refitting of the auxiliary shaft
(e) Removal and refitting of the engine front mountings
(f) Removal and refitting of the sump
(g) Removal and refitting of the oil pump
(h) Removal and refitting of the big-end bearings
(i) Removal and refitting of the pistons and connecting rods
(j) Removal and refitting of the flywheel/driveplate (after removal of the gearbox or automatic transmission)

4 Major operations requiring engine removal

The following operations can only be carried out with the engine removed from the vehicle:

(a) Removal and refitting of the main bearings
(b) Removal and refitting of the crankshaft

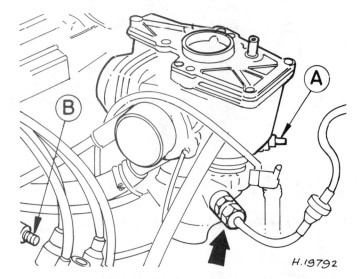

Fig. 1.3 Disconnect wiring from idle cut-off valve (A) temperature gauge sender (B) and the vacuum hose to the brake servo unit (arrowed) (Sec 6)

5 Methods of engine removal

1 The engine may be lifted out either on its own or in unit with the gearbox. On models fitted with automatic transmission it is recommended that the engine be lifted out on its own, unless a substantial crane or overhead hoist is available, because of the weight factor.

2 The engine or engine and gearbox (as applicable) must be removed through the front end of the vehicle, and therefore it is essential that the radiator and front panel are removed.

6 Engine (without transmission) – removal and refitting

1 Before starting work it is essential to have a good hoist which can be positioned over the engine, and a trolley jack if an inspection pit is not available.

6.14 Oil pressure switch location

6.17 Unbolting the engine mounting rubber insulators

7.8 Engine and manual gearbox removal

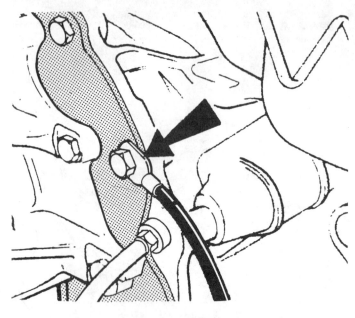

Fig. 1.4 Engine earth lead location (arrowed) (Sec 6)

2 Refer to Chapter 11 and remove the bonnet and front grille.
3 Disconnect the battery earth lead then drain the engine oil as described in Section 2, and the cooling system (Chapter 2).
4 Disconnect the radiator top and bottom hoses and the expansion tank hose. Disconnect the bonnet release cable at the bonnet lock then undo the two bolts each side securing the front upper body panel (bonnet lock platform). Undo the two lower bolts securing the support member and lift out the body panel complete with radiator.
5 Detach the small bore hose from the thermostat housing to the coolant expansion tank.
6 Remove the air cleaner unit as described in Chapter 3.
7 Detach the fuel hose to the fuel pump, plug the hose to prevent leakage, and position it out of the way.
8 Disconnect the heater hose from the automatic choke housing.
9 Disconnect the accelerator cable from the carburettor and position it out of the way.
10 Detach the ignition HT leads from the spark plugs and from the retainer on the valve cover. Detach the HT lead from the ignition coil, then remove the cap and leads from the engine.
11 Disconnect the brake servo vacuum hose at the inlet manifold.
12 Disconnect the alternator wiring plug from the rear of the alternator.
13 Disconnect the leads from the starter motor. On automatic transmission models, unbolt and remove the starter motor.
14 Disconnect the leads from the oil pressure sender/warning switch (photo), the carburettor idle cut-off, and the coolant temperature sender.

15 Undo the retaining bolts and detach the exhaust downpipe from the manifold.
16 On automatic transmission models, unbolt and detach the engine/transmission brace. Working through the starter motor aperture, unscrew the four torque converter to drive plate nuts. The engine can be turned by means of a spanner on the crankshaft pulley bolt to bring the torque converter nuts into an accessible position. Disconnect the transmission oil filler pipe from the engine.
17 Undo the nuts securing the two engine mounting rubber insulators to the body brackets (photo).
18 Attach a suitable crane or hoist to the engine using the lifting eyes located at the front and rear of the cylinder head.
19 Position the jack under the transmission, and just take the unit weight.
20 Unscrew the engine to transmission retaining bolts, and unbolt the clutch housing lower cover on manual gearbox models.
21 Check around the engine to ensure that all of the necessary items are detached and positioned out of the way.
22 With the aid of an assistant, lift the engine and simultaneously withdraw it through the front of the vehicle. On automatic transmission models, ensure that the torque converter remains in place on the transmission.
23 Refitting is a reversal of the removal procedure, but note the following points:

(a) Before refitting the engine, lightly smear the input shaft (manual gearbox models) with the Ford special grease (see Specifications – Chapter 6)
(b) As the engine is mated with the transmission, engage the input shaft with the clutch driven plate, align the engine/transmission locating dowels then insert and tighten the retaining bolts
(c) Remember to attach the engine earth lead when refitting the engine/transmission retaining bolts
(d) Tighten all fixings to the specified torque wrench settings (where applicable)
(e) Fill/top up the engine with the specified grade of oil
(f) Fill the cooling system as described in Chapter 2
(g) Adjust the tension of the drivebelt(s) as described in Chapter 2
(h) Adjust the accelerator cable with reference to Chapter 3

7 Engine and transmission – removal and refitting

1 Proceed as described in the previous Section, in paragraphs 1 to 15 inclusive, then continue as follows according to type.

Manual gearbox models
2 Working inside the vehicle, prise free the gearlever gaiter from the base, then detach the lever from the gearbox.
3 Remove the propeller shaft as described in Chapter 7.
4 Unscrew the retaining bolt and detach the speedometer cable. Detach the clutch cable as described in Chapter 5.
5 Detach the reversing lamp switch lead from the gearbox.

Chapter 1 Engine   33

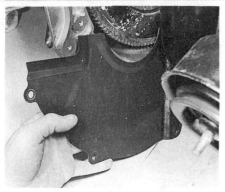
8.2 Clutch cover plate removal

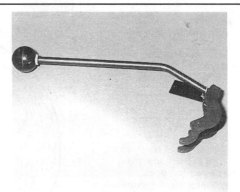
9.10 Special tool which can be used to compress the valve springs (tool number 21-005-B)

11.1 Remove the timing belt cover

11.3 Timing belt tensioner

11.6 Cylinder head Torx type bolt (arrowed)

11.7 Removing the cylinder head

6 Reaching up through the access aperture in the gearbox crossmember, unscrew the two retaining bolts.
7 Position a jack under the gearbox to support it. A trolley jack should be used if possible so that it can move forwards with the gearbox as it is withdrawn. Alternatively a length of strong wood or a metal bar positioned between the rear and forward crossmembers may suffice.
8 Fit the lift sling and hoist to the engine, then proceed as described in paragraphs 17, 18, 21 and 22 in the previous Section and withdraw the combined engine and gearbox from the vehicle. As they are withdrawn, have an assistant guide the gearbox through the tunnel and clear of the engine compartment components (photo).

Automatic transmission models
9 Jack up the front and rear of the vehicle and support it adequately on stands, or alternatively position it over an inspection pit; ensure that the vehicle is unladen.
10 Refer to Chapter 7 and disconnect the propeller shaft from the rear of the transmission. Tie the shaft to one side.
11 Disconnect the speedometer cable from the rear of the transmission.
12 Disconnect the selector rods from the transmission and relay lever.
13 Disconnect the wiring connector from the starter inhibitor switch on the left-hand side of the transmission.
14 Undo the bolt and remove the earth cable at the transmission housing.
15 Remove the engine and automatic transmission in the manner described in paragraph 8 for manual gearbox models, but allow for the considerably heavier weight of the automatic unit.

Manual gearbox and automatic transmission refitting
16 Refer to the refitting notes given for the engine in the previous Section, and for the transmission, refer to the appropriate Sections in Chapter 6.

8 Engine and transmission – separation and reconnection

Separation – manual gearbox models
1 Undo the bolts and remove the starter motor.
2 Remove the clutch cover plate from the lower part of the clutch housing (photo).
3 Unscrew and remove the engine to gearbox retaining bolts.
4 Withdraw the gearbox from the engine. Support the weight of the unit so that the input shaft is not distorted while still in engagement with the splines of the clutch driven plate.

Separation – automatic transmission models
5 Undo the bolts and remove the starter motor.
6 Remove the engine to transmission brace, reach through the starter motor aperture and undo the four torque converter to drive plate nuts. The engine can be turned by means of a spanner on the crankshaft pulley bolt to bring each torque converter nut into an accessible position.
7 Disconnect the transmission oil filler pipe from the engine.
8 Withdraw the transmission from the engine. As they are separated, retain the torque converter on the transmission to prevent fluid leakage.

Reconnection – all models
9 Reconnection of both transmission types is a reversal of the removal procedure, but note the following points according to type:

(a) On manual gearbox models, smear the splines of the input shaft with the Ford special grease before reconnection (see Specifications – Chapter 6)
(b) On automatic transmission models, ensure that the torque converter remains on its shaft as the two units are reconnected

(c) Tighten all fixings to the specified torque (where applicable) but leave tightening the earth strap bolt until after the units are refitted into the vehicle.

9 Engine dismantling – general

1 It is best to mount the engine on a dismantling stand but if one is not available, then stand the engine on a strong bench so as to be at a comfortable working height. Keep the engine upright at least until the sump is removed to avoid sludge running onto the engine components.
2 During the dismantling process the greatest care should be taken to keep the exposed parts free from dirt. As an aid to achieving this, it is a sound scheme to thoroughly clean the outside of the engine, removing all traces of oil and congealed dirt.
3 Use paraffin or a good grease solvent. The latter compound will make the job much easier, as, after the solvent has been applied and allowed to stand for a time, a vigorous jet of water will wash off the solvent and all the grease and filth. If the distributor and carburettor are still fitted, wrap them in polythene to prevent them getting wet. If they have already been removed, plug their apertures in the engine to prevent water entering.
4 Finally wipe the exterior of the engine with a rag and only then, when it is quite clean, should the dismantling process begin. As the engine is stripped, clean each part thoroughly.
5 Never immerse parts with oilways in paraffin or petrol, eg the crankshaft, but to clean, wipe down carefully with a petrol dampened rag. Oilways can be cleaned out with wire. If an air line is present all parts can be blown dry and the oilways blown through as an added precaution.
6 Re-use of old engine gaskets is a false economy and can give rise to oil and water leaks, if nothing worse. To avoid the possibility of trouble after the engine has been reassembled **always** use new gaskets throughout.
7 Do not throw away the old gaskets, as it sometimes happens that an immediate replacement cannot be found, and the old gasket is then very useful as a template. Hang up the old gaskets as they are removed on a suitable hook or nail.
8 To strip the engine it is best to work from the top down. The sump provides a firm base on which the engine can be supported with a wooden block in an upright position. When the stage where the sump must be removed is reached, the engine can be turned on its side and all other work carried out with it in this position.
9 Wherever possible, refit nuts, bolts and washers finger tight from wherever they were removed. This helps avoid later loss and muddle. If they cannot be refitted then lay them out in such a fashion that it is clean where they came from.
10 Multi-spline and Torx type sockets will be required for some operations. The types required are shown in photographs at the front of this manual. If working on the cylinder head valves with the camshaft in position, the special tool number 21.005-B will be required to compress the springs (photo).

10 Ancillary components – removal

1 Before basic engine dismantling begins the engine should be stripped of all its ancillary components. These items should also be removed if a factory exchange reconditioned unit is being purchased. The items comprise:

Alternator and brackets
Water pump and thermostat housing
Distributor and spark plugs
Inlet and exhaust manifold and carburettor
Fuel pump and fuel pipes
Oil filter and dipstick
Clutch assembly
Engine mountings
Oil pressure sender unit and coolant temperature sender unit
Oil separator unit and crankcase ventilation valve

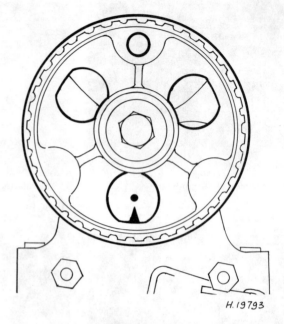

Fig. 1.5 Camshaft sprocket at TDC position (Sec 11)

2 Without exception all these items can be removed with the engine *in situ*, if it is merely an individual item which requires attention. (It is necessary to remove the gearbox if the clutch is to be renewed with the engine in position.)
3 Remove each of the listed items as described in the relevant Chapter of this manual.

11 Cylinder head – removal

If the engine is still in the vehicle, first carry out the following operations:

(a) Disconnect the battery earth lead
(b) Drain the cooling system (Chapter 2)
(c) Remove the air cleaner (Chapter 3)
(d) Detach the coolant hoses from the cylinder head
(e) Detach/remove the carburettor, inlet manifold and the exhaust manifold and associated components as required (Chapter 3)
(f) Disconnect the distributor cap, the HT leads from the spark plugs, the LT leads to the distributor and position them out of the way
(g) Detach any remaining wires or attachments from the cylinder head

1 Unscrew the bolts and withdraw the timing cover (photo).
2 Using a socket on the crankshaft pulley bolt, turn the engine clockwise until the TDC (top dead centre) notch on the pulley is aligned with the pointer on the crankshaft front oil seal housing, and the pointer on the camshaft sprocket is aligned with the indentation on the cylinder head (Fig. 1.5). Note the position of the distributor rotor arm.
3 Loosen the timing belt tensioner bolts and pivot it inwards to loosen off the belt tension (photo).
4 Remove the timing belt from the camshaft sprocket and position it to one side without damaging it or bending it.
5 Unscrew the bolts and remove the valve cover and gasket.
6 Using the special Torx socket unscrew the cylinder head bolts half a turn at a time in the reverse order to that shown in Fig. 1.11 (photo).
7 With the bolts removed, lift the cylinder head from the block (photo). If it is stuck, tap it free with a wooden mallet. *Do not insert a lever into the gasket joint otherwise the mating surfaces will be damaged.* Place the cylinder head on blocks of wood to prevent damage to the valves. Note that new cylinder head bolts will be required for refitting.
8 Remove the cylinder head gasket from the block.

Chapter 1 Engine 35

12.2 Camshaft retaining lug (arrowed)

12.3A Remove the camshaft sprocket ...

12.3B ... and backplate

12.4A Unscrew retaining bolts and remove ...

12.4B ... the camshaft oil supply tube

12.5 Note location of retaining spring clips ...

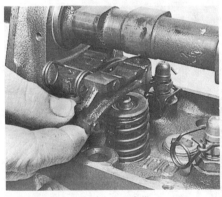
12.6 ... and remove the cam followers

12.7A Unscrew the bolts ...

12.7B ... and remove the camshaft thrustplate

12.8 Removing the camshaft

12.9 Prising out the camshaft front oil seal

13.2A Compressing the valve springs

36　　　　　　　　　　　　　　　　　　Chapter 1　Engine

13.2B Removing the valve springs and caps

13.3 Removing a valve

13.4 Removing a valve stem oil seal

14.3 Timing belt tensioner bolts (arrowed)

14.6A Remove the crankshaft pulley (using a puller if necessary) ...

14.6B ... followed by the guide washer

14.8 Removing the crankshaft sprocket

14.10 Removing the auxiliary shaft sprocket

12　Camshaft – removal

1　Remove the cylinder head as described in Section 11.
2　Unscrew the camshaft sprocket bolt while holding the camshaft stationary with a spanner on the camshaft special lug (photo).
3　Remove the camshaft sprocket using a puller if necessary. Remove the backplate (photos).
4　Unscrew the bolts and remove the camshaft oil supply tube (photos).
5　Note how the cam follower retaining spring clips are fitted then unhook them from the cam followers (photo).
6　If the special tool 21-005-B is available compress the valve springs in turn and remove the cam followers keeping them identified for location. Alternatively loosen the locknuts and back off the ball-pins until the cam followers can be removed (photo).
7　Unscrew the bolts and remove the camshaft thrustplate (photos).
8　Carefully withdraw the camshaft from the rear of the cylinder head taking care not to damage the bearings (photo).

9　Prise the oil seal from the front bearing (photo), taking care not to damage the bearing surface.

13　Cylinder head – dismantling

1　Remove the camshaft as described in Section 12, however if tool 21-005-B is available leave the camshaft in position while the valve springs are being compressed.
2　Using a valve spring compressor, compress each valve spring in turn until the split collets can be removed. Release the compressor and remove the cap and spring keeping them identified for location (photos). If the caps are difficult to release do not continue to tighten the compressor but gently tap the top of the tool with a hammer. Always make sure that the compressor is held firmly over the cap.
3　Remove each valve from the cylinder head keeping them identified for location (photo).
4　Prise the valve stem oil seals from the tops of the valve guides (photo).

Chapter 1 Engine

15.4A Auxiliary shaft front cover and oil seal (engine inverted)

15.4B Removing the auxiliary shaft front cover (engine inverted)

15.5A Auxiliary shaft thrustplate location (engine inverted)

15.5B Auxiliary shaft thrustplate removal (engine inverted)

15.5C Auxiliary shaft removal (engine inverted)

5 If necessary unscrew the cam follower ball-pins from the cylinder head keeping them identified for location.
6 If necessary unscrew the bolt and pivot and remove the timing belt tensioner.
7 Remove the thermostat housing with reference to Chapter 2.
8 If necessary, unscrew and remove the temperature gauge sender unit.

14 Timing belt and sprockets – removal

If the engine is still in the car, first carry out the following operations:

(a) Disconnect the battery negative lead
(b) Remove the radiator (Chapter 2) and disconnect the top hose from the thermostat housing
(c) Remove the alternator/power-steering pump drivebelt with reference to Chapter 2

1 Unscrew the three bolts and withdraw the timing cover. Note the location of the special bolt relative to the cover.
2 Using a socket on the crankshaft pulley bolt, turn the engine clockwise until the TDC (top dead centre) notch on the pulley is aligned with the pointer on the crankshaft front oil seal housing, and the pointer on the camshaft sprocket is aligned with the indentation on the cylinder head. Note the position of the distributor rotor arm.
3 Loosen the timing belt tensioner bolts and pivot the tensioner to release the pressure on the timing belt (photo).
4 Remove the timing belt from the camshaft sprocket.
5 On vehicles with manual gearbox, select 4th gear and have an assistant apply the brake pedal hard. On automatic transmission models, remove the starter motor (Chapter 12) and hold the crankshaft against rotation using a lever in the starter ring gear teeth. This method may be used as an alternative on manual gearbox versions.

6 Unscrew the crankshaft pulley bolt and remove the pulley and guide washer; if necessary, refit the bolt part way and use a puller (photos).
7 Remove the timing belt from the crankshaft and auxiliary shaft sprockets.
8 Remove the crankshaft sprocket using a puller if necessary (photo).
9 Unscrew the auxiliary shaft sprocket bolt while holding the sprocket stationary with a screwdriver inserted through one of the holes.
10 Remove the auxiliary shaft sprocket using a puller if necessary (photo).
11 Unscrew the camshaft sprocket bolt while holding the sprocket stationary with a screwdriver engaged in one of the grooves. Alternatively remove the valve cover and use a spanner on the camshaft special lug.
12 Remove the camshaft sprocket using a puller if necessary, then remove the backplate. Note that the oil seal can be removed using a special removal tool or by using self-tapping screws and a pair of grips.

15 Auxiliary shaft – removal

1 Remove the timing belt and auxiliary shaft sprocket only, as described in Section 14.
2 Remove the distributor as described in Chapter 4.
3 Remove the fuel pump and operating rod as described in Chapter 3.
4 Unscrew the bolts and remove the auxiliary shaft front cover (photos).
5 Unscrew the cross-head screws using an impact screwdriver if necessary, remove the thrustplate and withdraw the auxiliary shaft from the block (photos).
6 Cut the front cover gasket along the top of the crankshaft front oil seal housing and scrape off the gasket.

16.1 Flywheel retaining bolts. Also shown is the input shaft pilot bearing in the rear end of the crankshaft

16.2A Engine backplate and retaining bolts – starter motor side

16.2B Engine backplate and retaining bolts – exhaust side

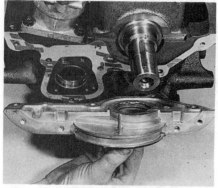
18.3A Crankshaft front oil seal housing removal

18.3B Driving out the crankshaft front oil seal

18.4 Using a socket to fit the new crankshaft front oil seal

18.5A Crankshaft front oil seal housing and auxiliary shaft cover gasket on the front face of the cylinder block

18.5B Checking the alignment of the crankshaft front oil seal housing

19.2 Crankshaft rear oil seal location

16 Flywheel/driveplate – removal

If the engine is still in the vehicle remove the clutch as described in Chapter 5 or the automatic transmission as described in Chapter 6.

1   Hold the flywheel/driveplate stationary using a piece of angle iron engaged with the ring gear teeth, then unscrew the bolts and withdraw the unit from the crankshaft (photo).
2   If necessary unbolt and remove the engine backplate from the dowels (photos).

17 Sump – removal

If the engine is still in the vehicle first carry out the following operations:

(a)   *Disconnect the battery negative lead*
(b)   *Jack up the front of the vehicle and support with axle stands. Apply the handbrake*
(c)   *Drain the engine oil into a suitable container*
(d)   *Remove the starter motor as described in Chapter 12*

1   Unscrew the bolts and remove the sump. If it is stuck tap the sump sideways to free it.
2   Remove the gaskets and sealing strips from the block.

18 Crankshaft front oil seal – renewal

1   Remove the timing belt and crankshaft sprocket only, as described in Section 14.
2   If an oil seal removal tool is available, the oil seal can be removed at this stage. It may also be possible to remove the oil seal by drilling the outer face and using self-tapping screws and a pair of grips.

Chapter 1 Engine 39

20.2 Unscrewing the oil pump strainer bracket

20.3 Removing the retaining bolts using multi-spline socket

20.4 Removing the oil pump driveshaft

21.1 Oil filter removal using a strap wrench

22.2 Big-end cap and connecting rod identification numbers

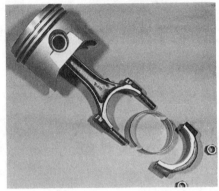
22.4 Views of piston, connecting rod, big-end cap and bearing shells

3   If the oil seal cannot be removed as described in paragraph 2, remove the sump as described in Section 17, then unbolt the oil seal housing and auxiliary shaft front cover and remove the gasket. The oil seal can then be driven out from the inside (photos).
4   Clean the oil seal seating then drive in a new seal using metal tubing or a suitable socket (photo). Make sure that the sealing lip faces into the engine, and lightly oil the lip.
5   If applicable fit the oil seal housing and auxiliary shaft front cover to the block together with a new gasket and tighten the bolts. Make sure that the bottom face of the housing is aligned with the bottom face of the block (photos). Fit the sump as described in Section 40.
6   Refit the timing belt and crankshaft sprocket (Section 46).

19   Crankshaft rear oil seal – renewal

1   Remove the flywheel/driveplate and engine backplate as described in Section 16.
2   Using a special removal tool extract the oil seal. However it may be possible to remove the oil seal by drilling the outer face and using self-tapping screws and a pair of grips (photo).
3   Clean the oil seal seating then drive in a new seal using a suitable metal tube. Make sure that the sealing lip faces into the engine, and lightly oil the lip.
4   Fit the flywheel/driveplate and engine backplate as described in Section 41.

20   Oil pump – removal

1   Remove the sump as described in Section 17.
2   Unscrew the bolt securing the pick-up tube and strainer to the block (photo).

3   Using the special multi-spline socket, unscrew the bolts and withdraw the oil pump and strainer (photo).
4   Withdraw the hexagon-shaped driveshaft which engages the bottom of the distributor, noting which way round it is fitted (photo).

21   Oil filter – renewal

1   The oil filter should be renewed at the specified intervals. Place a container directly beneath the oil filter, then using a strap wrench, unscrew and remove the filter (photo). If a strap wrench is not available it may be possible to unscrew the filter by driving a screwdriver through the filter canister and using it as a lever.
2   Wipe clean the filter face on the block.
3   Smear a little oil on the new filter seal and screw on the filter until it just contacts the block, then tighten it a further three quarters of a turn.

22   Pistons and connecting rods – removal

1   Remove the sump as described in Section 17, and the cylinder head as described in Section 11.
2   Check the big-end caps for identification marks and if necessary use a centre-punch to identify the caps and connecting rods (photo).
3   Turn the crankshaft so that No 1 crankpin is at its lowest point, then unscrew the nuts and tap off the cap. Keep the bearing shells in the cap and connecting rod.
4   Using the handle of a hammer, push the piston and connecting rod up the bore and withdraw from the top of the cylinder block. Loosely refit the cap to the connecting rod (photo).

Chapter 1 Engine

23.5 Main bearing cap identification marks – the arrow must face the front of the engine

23.6 Checking the crankshaft endfloat using a feeler gauge

23.7A Removing the centre main bearing cap

23.7B Removing the rear main bearing cap

23.8A Removing the crankshaft

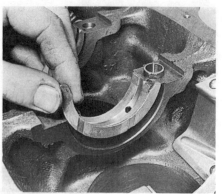

23.8B Removing the thrustwashers from the centre main bearing

23.9 Remove the centre main bearing shell

25.2 Vent valve removal from the separator

5   Repeat the procedure in paragraphs 3 and 4 on No 4 piston and connecting rod, then turn the crankshaft through half a turn and repeat the procedure on Nos 2 and 3 pistons and connecting rods.

23 Crankshaft and main bearings – removal

1   With the engine removed from the vehicle, remove the pistons and connecting rods as described in Section 22, however unless work is required on the pistons or bores it is not necessary to remove the pistons completely from the cylinder block.
2   Remove the timing belt and crankshaft sprocket with reference to Section 14 and remove the flywheel/driveplate with reference to Section 16.
3   Unbolt the crankshaft front oil seal housing and auxiliary shaft front cover and remove the gasket.
4   Remove the oil pump and strainer as described in Section 20.

5   Check the main bearing caps for identification marks and if necessary use a centre-punch to identify them (photo).
6   Before removing the crankshaft check that the endfloat is within the specified limits by inserting a feeler blade between the centre crankshaft web and the thrustwashers (photo). This will indicate whether new thrustwashers are required or not.
7   Unscrew the bolts and tap off the main bearing caps complete with bearing shells (photos). If the thrustwashers are to be re-used identify them for location.
8   Lift the crankshaft from the crankcase and remove the rear oil seal. Remove the remaining thrustwashers (photos).
9   Extract the bearing shells keeping them identified for location (photo).

24 Engine mountings – renewal

1   The engine mountings incorporate rubber insulators and must be renewed if excessive engine movement is evident.

Chapter 1 Engine 41

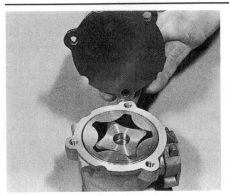
27.1 Removing the oil pump cover

27.2A Checking the oil pump rotor to body clearance

27.2B Oil pump inner to outer rotor clearance check

27.2C Oil pump rotor to cover clearance check

27.2D Removing the oil pump pick-up tube and strainer

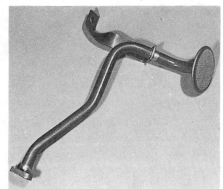
27.2E Oil pump pick-up tube and strainer

2 If the engine is in the vehicle, support its weight with a jack but take care not to damage the sump. A piece of flat timber positioned between the jack and the sump should protect it from damage.
3 Unscrew the engine mounting bolts/nuts each side, raise the jack a fraction to take the weight of the engine and remove the mountings.
4 Fit the new mounting(s) using a reversal of the removal procedure. Tighten the rubber insulators to the specified torque wrench setting.

25 Crankcase ventilation system – description and maintenance

1 The crankcase ventilation system consists of the special oil filler cap, containing a steel wool filter, and an oil separator and vent valve on the left-hand side of the engine. This is connected by hose to the inlet manifold. The system operates according to the vacuum in the inlet manifold. Air is drawn through the filler cap, through the crankcase, and then together with piston blow-by gases through the oil separator and vent valve to the inlet manifold. The blow-by gases are then drawn into the engine together with the fuel/air mixture.
2 Renew the vent valve at the specified intervals, shown in the *Routine maintenance* Section at the start of this manual, by pulling it from the oil separator and loosening the hose clip (photo). Fit the new valve, tighten the clip, and insert it into the oil separator grommet.

26 Examination and renovation – general

With the engine completely stripped, clean all the components and examine them for wear. Each part should be checked, and where necessary renewed or renovated as described in the following Sections. Renew main, big-end shell bearings and oil seals as a matter of course, unless you know that they have had little wear and are in perfect condition.

27 Oil pump – examination and renovation

1 Unscrew the bolts and remove the oil pump cover (photo).
2 Using feeler gauges check that the rotor clearances are within the limits given in Specifications. If not, unbolt the pick-up tube and strainer and obtain a new unit (photos). Fit the pick-up tube and strainer to the new pump using a new gasket, and tighten the bolts.
3 If the oil pump is serviceable prime the unit using clean engine oil then refit the cover and tighten the bolts.

28 Crankshaft and bearings – examination and renovation

1 Examine the bearing surfaces of the crankshaft for scratches or scoring and, using a micrometer, check each journal and crankpin for ovality. Where this is found to be in excess of 0.001 in (0.0254 mm) the crankshaft will have to be reground and undersize bearings fitted.
2 Crankshaft regrinding should be carried out by a suitable engineering works, who will normally supply the matching undersize main and big-end shell bearings.
3 Note that undersize bearings may already have been fitted, either in production or by a previous repairer. Check the markings on the backs of the old bearing shells, and if in doubt take them along when buying new ones. Production undersizes are also indicated by paint marks as follows:

White line on main bearing cap – parent bore 0.40 mm oversize
Green line on crankshaft front counterweight – main bearing journals 0.25 mm undersize
Green spot on counterweight – big-end bearing journals 0.25 mm undersize

4 If the crankshaft endfloat is more than the maximum specified amount, new thrustwashers should be fitted to the centre main bearings. These are usually supplied together with the main and big-end bearings on a reground crankshaft.

rate accompanied by blue smoke from the exhaust.
2 Use an internal micrometer to measure the bore diameter just below the ridge and compare it with the diameter at the bottom of the bore, which is not subject to wear. If the difference is more than 0.006 in (0.152 mm), the cylinders will normally require reboring with new oversize pistons fitted.
3 Provided the cylinder bore wear does not exceed 0.008 in (0.203 mm), special oil control rings and pistons can be fitted to restore compression and stop the engine burning oil.
4 If new pistons are being fitted to old bores, it is essential to roughen the bore walls slightly with fine glasspaper to enable the new piston rings to bed in properly.
5 Thoroughly examine the crankcase and cylinder block for cracks and damage and use a piece of wire to probe all oilways and waterways to ensure they are unobstructed.

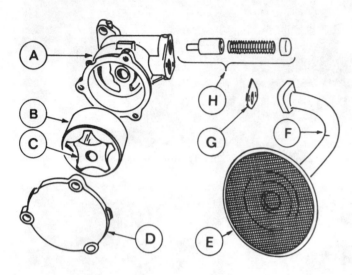

Fig. 1.6 Exploded diagram of the oil pump (Sec 27)

A Body	E Strainer
B Outer rotor	F Pick-up tube
C Inner rotor	G Gasket
D Cover	H Relief valve

5 An accurate method of determining bearing wear is by the use of Plastigage. The crankshaft is located in the main bearings (and big-end bearings if necessary) and the Plastigage filament located across the journal which must be dry. The cap is then fitted and the bolts/nuts tightened to the specified torque. On removal of the cap the width of the filaments is checked against a scale which shows the bearing running clearance. This clearance is then compared with that given in the Specifications (photos).
6 If the spigot bearing in the rear of the crankshaft requires renewal extract it with a suitable puller. Alternatively fill it with heavy grease and use a close fitting metal dowel driven into the centre of the bearing. Drive the new bearing into the crankshaft with a soft metal drift.

29 Cylinder block and bores – examination and renovation

1 The cylinder bores must be examined for taper, ovality, scoring and scratches. Start by examining the top of the bores; if these are worn, a slight ridge will be found which marks the top of the piston ring travel. If the wear is excessive, the engine will have had a high oil consumption

30 Pistons and connecting rods – examination and renovation

1 Examine the pistons for ovality, scoring, and scratches. Check the connecting rods for wear and damage.
2 The gudgeon pins are an interference fit in the connecting rods, and if new pistons are to be fitted to the existing connecting rods the work should be carried out by a Ford garage who will have the necessary tooling. Note that the oil splash hole on the connecting rod must be located on the right-hand side of the piston (the arrow on the piston crown faces forwards) (Fig. 1.7).
3 If new rings are to be fitted to the existing pistons, expand the old rings over the top of the pistons. The use of two or three old feeler blades will be helpful in preventing the rings dropping into empty grooves. Note that the oil control ring is in three sections.
4 Before fitting the new rings to the pistons, insert them into the cylinder bore and use a feeler gauge to check that the end gaps are within the specified limits (photos). Note that the figures given in the Specifications apply to gauge rings used in production. In service it is allowable for the measured gaps to exceed these figures by 0.15 mm (0.006 in).

31 Camshaft and cam followers – examination

1 Examine the surface of the camshaft journals and lobes, and the cam followers for wear. If excessive, considerable noise would have been noticed from the top of the engine and a new camshaft and followers must be fitted.
2 Check the camshaft bearings for wear and if necessary have them renewed by a Ford garage.
3 Check the camshaft lubrication tube for obstructions and make sure that the jet holes are clear.

28.4A Showing flattened Plastigage filament (arrowed)

28.4B Checking the bearing running clearance with the special gauge

30.4A Checking the piston ring gap at the top end of the cylinder (underneath the wear ridge if engine is not rebored)

Chapter 1 Engine 43

30.4B Checking the piston ring gap at the lower end of the cylinder

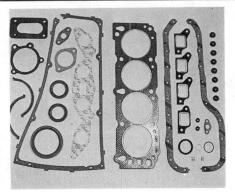
36.3 Complete engine gasket set

37.9 Lubricate the crankshaft main bearings

37.12 Applying sealing compound to the rear main bearing cap

37.14 Locate the thrustwashers on the side faces of the main bearing cap

37.18 Fitting the wedges to the rear main bearing cap

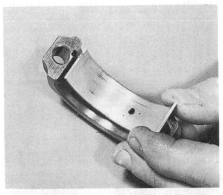
38.2A Fitting the big-end bearing shells to their caps

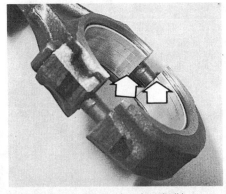
38.2B Showing big-end bearing shell lugs adjacent to each other

38.3 Lubricate the cylinder bores ...

38.4A ... insert the piston assembly and fit the ring compressor over the piston

38.4B Arrow mark on piston crown indicates front

38.5A Lubricate the crankpins ...

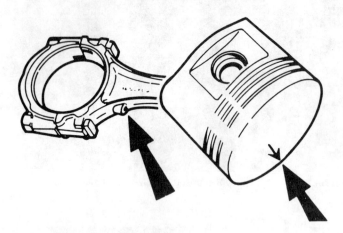

Fig. 1.7 Showing correct relationship of piston and connecting rod (Sec 30)

32 Auxiliary shaft – examination

1 Examine the auxiliary shaft for wear and damage and renew it if necessary.
2 If the auxiliary shaft endfloat is outside the limits given in the Specifications fit a new thrustplate and renew the shaft if necessary.

33 Timing belt – examination

1 Whenever the timing belt is removed it is worthwhile renewing it especially if it has covered a high mileage.
2 Do not allow oil, water or grease to come into contact with the timing belt.
3 If you are considering re-using the timing belt that was removed, inspect it thoroughly for cracking and deterioration. Remember, even though these faults may not be apparent, the belt may have stretched beyond an acceptable amount, and renewal is strongly recommended.

34 Flywheel ring gear – examination and renovation

1 If the ring gear is badly worn or has missing teeth it should be renewed. The old ring can be removed from the flywheel by cutting a notch between two teeth with a hacksaw and then splitting it with a cold chisel.
2 To fit a new ring gear requires heating the ring to 204°C (400°F). This can be done by polishing four equally spaced sections of the gear, laying it on a suitable heat-resistant surface (such as fire bricks) and heating it evenly with a blow lamp or torch until the polished areas turn a light yellow tinge. Do not overheat or the hard wearing properties will be lost. The gear has a chamfered inner edge which should go against the shoulder when put on the flywheel. When hot enough, place the gear in position quickly, tapping it home if necessary and let it cool naturally without quenching in any way.

35 Cylinder head – decarbonising, valve grinding and renovation

1 This operation will normally only be required at comparatively high mileages. However, if persistent pinking occurs and performance has deteriorated even though the engine adjustments are correct, decarbonising and valve grinding may be required.
2 With the cylinder head removed, use a scraper to remove the carbon from the combustion chambers and ports. Remove all traces of gasket from the cylinder head surface, then wash it thoroughly with paraffin.
3 Use a straight edge and feeler blade to check that the cylinder head surface is not distorted. If it is, it must be resurfaced by a suitably equipped engineering works.
4 If the engine is still in the car, clean the piston crowns and cylinder bore upper edges, but make sure that no carbon drops between the pistons and bores. To do this, locate two of the pistons at the top of their bores and seal off the remaining bores with paper and masking tape. Press a little grease between the two pistons and their bores to collect any carbon dust; this can be wiped away then the piston is lowered. To prevent carbon build-up, polish the piston crown with metal polish, but remove all traces of the polish afterwards.
5 Before examining and renovating the valves, first check the valve guides for excessive wear. If the guides are worn they will need reboring for oversize valves or for fitting guide inserts. The valve seats will also need recutting to ensure they are concentric with the stems. This work should be given to your Ford dealer or local engineering works.
6 To check the valve guides for wear, insert a new valve into the guide (inlet to inlet/exhaust to exhaust as applicable). Using a dial indicator as shown (Fig. 1.8), check the amount of valve side play with the valve flush with the end of the guide. If the play is excessive, have the guides reamed out to suit oversize valves.
7 Examine the heads of the valves for pitting and burning, especially the exhaust valve heads. Renew any valve which is badly burnt. Examine the valve seats at the same time. If the pitting is very slight, it can be removed by grinding the valve heads and seats together with coarse, then fine, grinding paste.
8 Where excessive pitting has occurred, the valve seats must be recut or renewed by a suitably equipped engineering works.
9 Valve grinding is carried out as follows. Place the cylinder head upside down on a bench on blocks of wood.
10 Smear a trace of coarse carborundum paste on the seat face and press a suction grinding tool onto the valve head. With a semi-rotary action, grind the valve head to its seat, lifting the valve occasionally to redistribute the grinding paste. Where a dull matt even surface is produced on both the valve seat and the valve, wipe off the paste and repeat the process with fine carborundum paste as before. A light spring balanced under the valve head will greatly ease this operation. When a smooth unbroken ring of light grey matt finish is produced on both the

Fig. 1.8 Method used to measure the lateral play of a valve in its guide (Sec 35)

Maximum allowable play – inlet 0.50 mm
Maximum allowable play – exhaust 0.60 mm

valve and seat, the grinding operation is complete.
11 Scrape away all carbon from the valve head and stem, and clean away all traces of grinding compound. Clean the valves and seats with a paraffin-soaked rag, then wipe with a clean rag.
12 If the free length of any of the valve springs is outside the specified dimension, the springs should be renewed as a set. Always renew the valve stem oil seals when the valves are removed.

36 Engine reassembly – general

1 To ensure maximum life with minimum trouble from a rebuilt engine, not only must everything be correctly assembled, but it must also be spotlessly clean. All oilways must be clear, and locking washers and spring washers must be fitted where indicated. Oil all bearings and other working surfaces thoroughly with engine oil during assembly.
2 Before assembly begins, renew any bolts or studs with damaged threads.
3 Gather together a torque wrench, oil can, clean rag, and a set of engine gaskets and oil seals, together with a new oil filter (photo).

37 Crankshaft and main bearings – refitting

1 Wipe the bearing shell locations in the crankcase with a soft, non-fluffy rag.
2 Wipe the crankshaft journals with a soft, non-fluffy rag.
3 If the old main bearing shells are to be renewed (not to do so is a false economy, unless they are virtually new) fit the five upper halves of the main bearing shells to their location in the crankcase.
4 Identify each main bearing cap and place in order. The number is cast on to the cap and on intermediate caps an arrow is also marked which should point towards the front of the engine.
5 Wipe the cap bearing shell location with a soft non-fluffy rag.
6 Fit the bearing half shell onto each main bearing cap.
7 Apply a little grease to each side of the centre main bearing so as to retain the thrustwasher.
8 Fit the upper halves of the thrustwashers into their grooves either side of the main bearing. The slots must face outwards.
9 Lubricate the crankshaft journals and the upper and lower main bearing shells with engine oil (photo) and locate the rear oil seal (with lip lubricated) on the rear of the crankshaft.
10 Carefully lower the crankshaft into the crankcase.
11 Lubricate the crankshaft main bearing journals again and then fit No. 1 bearing cap. Fit the two securing bolts but do not tighten yet.
12 Apply a little sealing compound to the crankshaft rear main bearing cap (photo).
13 Fit the rear main bearing cap. Fit the two securing bolts but as before do not tighten yet.
14 Apply a little grease to either side of the centre main bearing cap so as to retain the thrustwashers. Fit the thrustwashers with the tag located in the groove and the slots facing outwards (photo).
15 Fit the centre main bearing cap and the two securing bolts. Then refit the intermediate main bearing caps. Make sure that the arrows point towards the front of the engine.
16 Lightly tighten all main cap securing bolts and then fully tighten in a progressive manner to the specified torque wrench setting.
17 Check that the crankshaft rotates freely, then check that the endfloat is within the specified limits by inserting a feeler blade between the centre crankshaft web and the thrustwashers.
18 Make sure that the rear oil seal is fully entered onto its seating. Coat the rear main bearing cap wedges with sealing compound then press them into position with the rounded red face towards the cap (photo).
19 Refit the oil pump and strainer as described in Section 39.
20 Refit the crankshaft front oil seal housing, and auxiliary shaft front cover, if applicable, together with a new gasket and tighten the bolts. Make sure that the bottom face of the housing is aligned with the bottom face of the block.
21 Refit the flywheel/driveplate (Section 41) and timing belt and crankshaft sprocket (Section 46).
22 Refit the pistons and connecting rods as described in Section 38.

38 Pistons and connecting rods – refitting

1 Clean the backs of the bearing shells and the recesses in the connecting rods and big-end caps.
2 Press the bearing shells into the connecting rods and caps in their correct positions and oil them liberally. Note that the lugs must be adjacent to each other (photos).
3 Lubricate the cylinder bores with engine oil (photo).
4 Fit a ring compressor to No 1 piston then insert the piston and connecting rod into No 1 cylinder. With No 1 crankpin at its lowest point, drive the piston carefully into the cylinder with the wooden handle of a hammer, and at the same time guide the connecting rod onto the crankpin. Make sure that the arrow on the piston crown is facing the front of the engine (photos).
5 Oil the crankpin then fit the big-end bearing cap in its previously noted position. Lubricate the threads and contact faces of the cap nuts with engine oil, then fit and tighten the nuts to the specified torque (photos).
6 Check that the crankshaft turns freely.
7 Repeat the procedure given in paragraphs 4 to 6 inclusive on the remaining pistons.
8 Refit the cylinder head as described in Section 45 and the sump as described in Section 40.

39 Oil pump – refitting

1 Insert the oil pump driveshaft into the cylinder block in its previously noted position.
2 Prime the oil pump by injecting oil into the pump outlet, then locate the pump on the cylinder block, insert the bolts, and tighten them with the special multi-spline socket (photo).
3 Insert the pick-up tube securing bolt and tighten it.

38.5B ... then fit and tighten the big-end bearing caps and nuts

39.2 Priming the oil pump with oil

46 Chapter 1 Engine

40.1A Apply sealing compound ...

40.1B ... then fit the rubber sealing strips

40.2 Locate the sump gaskets beneath the sealing strips

40.3 Refitting the sump

41.3 Apply liquid locking agent to the flywheel retaining bolts

4 Where applicable refit the crankshaft front oil seal housing together with a new gasket and tighten the bolts. Make sure that the bottom face of the housing is aligned with the bottom face of the block.
5 Refit the sump as described in Section 40.

40 Sump – refitting

1 Apply sealing compound to the corners of the front and rear rubber sealing strip locations then press the strips into the grooves of the rear main bearing cap and crankshaft front oil seal housing (photos).
2 Apply a little sealing compound to the bottom face of the cylinder block. Then fit the sump gaskets in position and locate the end tabs beneath the rubber sealing strips (photo).
3 Locate the sump on the gaskets and insert the bolts loosely (photo).
4 Tighten the bolts to the specified torques in the three stages given in Specifications. Refer to Fig. 1.9 and tighten to the first stage in circular sequence starting at point A, then tighten to the second stage starting at point B. Tighten to the third stage starting at point A after the engine has been running for twenty minutes.
5 If the engine is in the car, reverse the introductory procedure given in Section 17.

41 Flywheel/driveplate – refitting

1 If applicable, locate the engine backplate on the rear of the cylinder block and fit the retaining bolts.
2 Wipe the mating faces then locate the flywheel/driveplate on the rear of the crankshaft.
3 Coat the threads of the bolts with a liquid locking agent then insert them (photo). Note that the manufacturers recommend using new bolts.
4 Using a piece of angle iron to hold the flywheel/driveplate stationary, tighten the bolts evenly to the specified torque in diagonal sequence.
5 If the engine is in the vehicle refit the automatic transmission as described in Chapter 6 or the clutch as described in Chapter 5.

42 Auxiliary shaft – refitting

1 Oil the auxiliary shaft journals then insert the shaft into the cylinder block.

Fig. 1.9 Sump bolt tightening sequence – refer to text (Sec 40)

Chapter 1 Engine

42.3A Drive out the auxiliary shaft cover oil seal

42.3B Using a socket to fit the new auxiliary shaft oil seal into the cover

43.3 Showing ball-pin locations and cam follower retaining spring clips

44.2 Lubricate the camshaft bearings

44.7 Fit the camshaft sprocket backplate

2 Locate the thrustplate in the shaft groove, then insert the crosshead screws and tighten them with an impact screwdriver. Check that the auxiliary shaft endfloat is as specified using feeler gauges.
3 Support the front cover on blocks of wood and drive out the old oil seal. Drive in the new seal using a suitable metal tube or socket (photos). Make sure that the sealing lip faces toward the engine. Smear a little oil on the lip.
4 If applicable cut the unwanted top half of a new gasket and locate it on the cylinder block, then fit the front cover and tighten the bolts.
5 Refit the fuel pump and operating rod as described in Chapter 3.
6 Refit the distributor as described in Chapter 4.
7 Refit the auxiliary shaft sprocket and timing belt as described in Section 46.

43 Cylinder head – reassembly

1 Refit the thermostat housing with reference to Chapter 2.
2 If applicable, locate the timing belt tensioner on the front of the cylinder head and screw in the bolt and pivot.
3 If applicable, screw the cam follower ball-pins in their correct locations (photo).
4 Oil the valve stems and insert the valves in their correct guides.
5 Wrap some adhesive tape over the collet groove of each valve, then oil the oil seals and slide them over the valves onto the guides. Use a suitable metal tube if necessary to press them onto the guides. Remove the adhesive tape.

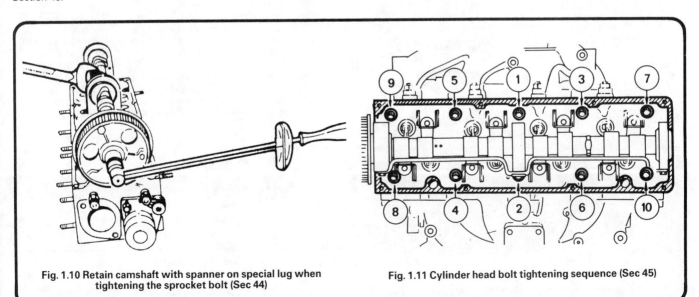

Fig. 1.10 Retain camshaft with spanner on special lug when tightening the sprocket bolt (Sec 44)

Fig. 1.11 Cylinder head bolt tightening sequence (Sec 45)

Chapter 1 Engine

45.3 Locate the cylinder head gasket

46.7 Fitting the timing belt

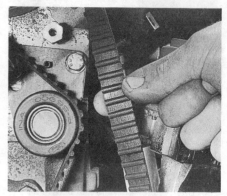
46.11 Twisting the timing belt to assess the tension

6 Working on each valve in turn, fit the valve spring and cap then compress the spring with the compressor and insert the split collets. Release the compressor and remove it. Tap the end of the valve stem with a non-metallic mallet to settle the collets. If tool 21-005-B is being used, first locate the camshaft in its bearings.
7 Refit the camshaft as described in Section 44.

44 Camshaft – refitting

1 Drive the new oil seal into the camshaft front bearing location on the cylinder head using a suitable metal tube or socket. Smear the lip with engine oil.
2 Lubricate the bearings with hypoid SAE 80/90 oil then carefully insert the camshaft (photo).
3 Locate the thrustplate in the camshaft groove then insert and tighten the bolts.

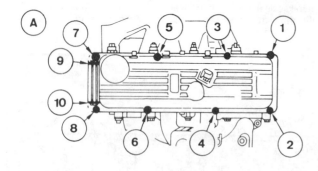
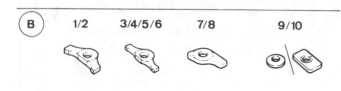

Fig. 1.12 Valve cover bolts (A) and reinforcing plates (B). For tightening sequence see text (Sec 45)

4 Using feeler gauges check that the endfloat is as given in the Specifications.
5 Lubricate the ball-pins with hypoid SAE 80/90 oil, then fit the cam followers in their correct locations and retain with the spring clips. It will be necessary to rotate the camshaft during this operation.
6 Fit the oil supply tube and tighten the bolts.
7 Fit the camshaft sprocket backplate and sprocket. Insert and tighten the bolt while holding the camshaft stationary with a spanner on the special lug (photo).
8 Refit the cylinder head as described in Section 45.

45 Cylinder head – refitting

1 Adjust the valve clearances as described in Section 48. This work is easier to carry out on the bench rather than in the vehicle.
2 Turn the engine so that No 1 is approximately 2 cm (0.8 in) before top dead centre. This precaution will prevent any damage to open valves.
3 Make sure that the faces of the cylinder block and cylinder head are perfectly clean. Then locate the new gasket on the block making sure that all the internal holes are aligned (photo). *Do not use jointing compound.*
4 Turn the camshaft so that the TDC pointer is aligned with the indentation on the front of the cylinder head.
5 Lower the cylinder head onto the gasket. The help of an assistant will ensure that the gasket is not dislodged.
6 Lightly oil the new cylinder head bolts completely, then insert them.
7 Using the special Torx type socket tighten the bolts in the order given in Fig. 1.11 to the three stages given in Specifications. Stage three will be completed after the engine has been running for 5 minutes.
8 Fit the valve cover together with new gaskets and tighten the bolts to the specified torque in the order given below. When fitting the valve cover, arrange the reinforcing plates as detailed in Fig. 1.12.

Stage 1 – Bolts 1 to 6
Stage 2 – Bolts 9 and 10
Stage 3 – Bolts 7 and 8
Stage 4 – Bolts 9 and 10 (again)

46 Timing belt and sprockets – refitting

1 If applicable fit the camshaft sprocket backplate and sprocket. Then insert and tighten the bolt while holding the camshaft stationary either

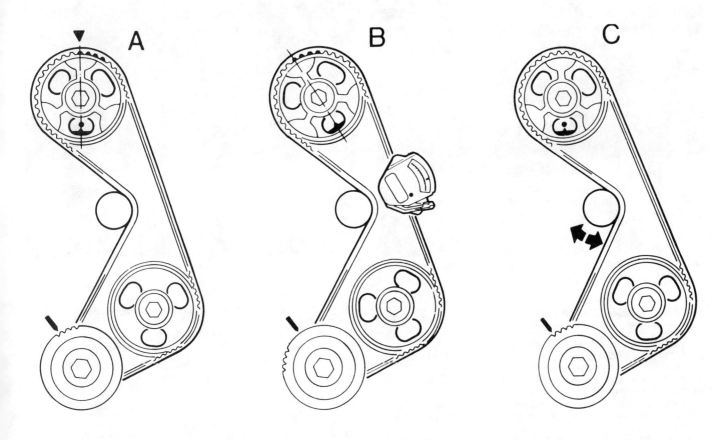

Fig. 1.13 Timing belt tension checking sequence (Sec 46)

    A No 1 at TDC
    B 60° BTDC for checking of belt adjustment
    C Return to TDC for adjustment of timing belt tension

with a screwdriver engaged in one of the grooves or with a spanner on the camshaft special lug (see Fig. 1.10).
2 Fit the auxiliary shaft sprocket, insert the bolt, and tighten it while holding the sprocket stationary with a screwdriver through one of the holes. Note that the reinforcement ribs on the sprocket must be toward the engine.
3 Fit the crankshaft sprocket to the front of the crankshaft with the chamfered side innermost.
4 Make sure that the TDC pointer on the camshaft sprocket backplate is aligned with the indentation on the front of the cylinder head.
5 Temporarily locate the crankshaft pulley and guide washer (convex side towards the toothed belt sprocket) onto the front end of the crankshaft and turn the crankshaft by the shortest route until the TDC notch in the pulley is aligned with the pointer on the crankshaft front oil seal housing. As the crankshaft is turned, take care not to allow the pistons to come in contact with the valves. Remove the pulley and guide washer.
6 Fit the distributor (Chapter 4) and then turn the auxiliary shaft sprocket until the contact end of the rotor arm is aligned with No 1 HT segment position in the distributor cap.
7 Locate the timing belt over the toothed sprockets and around the tensioner. Position, and set the tensioner so that it initially takes up any slack in the timing belt. The timing belt tension must now be checked and if necessary, adjusted. If possible use Ford tension gauge No 21-113 to make this check. Note that it is essential that the engine is cold whenever this check/adjustment is made (photo).
8 Turn the crankshaft clockwise, using the crankshaft pulley bolt, temporarily refitted through at least two complete revolutions. Finish with No 1 piston at TDC firing.
9 From this position, turn the crankshaft 60° anti-clockwise. This distance corresponds to three teeth on the camshaft sprocket.

10 The belt tension should now ideally be checked by applying Ford tension gauge 21-113 to the longest run. Desired gauge readings are:

Used belt – 4 to 6
New belt – 10 to 11

11 If the tension gauge is not available, a rough guide is that belt tension is correct when the belt can be twisted 90° in the middle of the longest run with the fingers (photo).
12 If adjustment is necessary, turn the crankshaft clockwise to bring No 1 cylinder to TDC firing.
13 Slacken the belt tensioner fastenings and move the tensioner to increase or decrease the tension as required. Tighten the fastening.
14 Turn the crankshaft 90° clockwise, then turn it anti-clockwise to the position previously used when checking the tension (60° BTDC).
15 Repeat the procedure as necessary until the tension is correct.
16 Fit the timing cover and tighten the bolts.
17 Refit the crankshaft pulley and guide washer as previously described and tighten the retaining bolt to the specified torque wrench setting.
18 If the engine is in the vehicle, reverse the introductory procedures given in Section 14.

47 Ancillary components – refitting

1 Refer to Section 10 and refit the listed components with reference to the Chapters indicated. Apply a liquid locking agent to the crankcase ventilation oil separator tube before pressing it into the cylinder block.

48.5 Valve clearance check with feeler gauge inserted between the cam lobe base and the cam follower

48.7 Valve clearance adjustment

48 Valve clearances – adjustment

1 Valve clearances are checked with the engine cold.
2 Remove the air cleaner as described in Chapter 3 and disconnect the HT leads from the spark plugs and valve cover.
3 Although not essential, it will make the engine easier to turn if the spark plugs are removed (Chapter 4).
4 Remove the ten bolts which secure the valve cover, noting the location of the different shapes of reinforcing plates. Remove the cover and gasket.
5 One of the cam lobes will be seen to be pointing upwards. Measure the clearance between the base of this cam and the cam follower, finding the thickness of feeler blade which gives a firm sliding fit (photo).
6 The desired valve clearances are given in the Specifications. Note that the clearances for inlet and exhaust valves are different. Numbering from the front (sprocket) end of the camshaft, the exhaust valves are 1, 3, 5 and 7, and the inlet valves 2, 4, 6 and 8.
7 If adjustment is necessary, slacken the ball-pin locknut and screw the ball-pin up or down until the clearance is correct. Hold the ball-pin stationary and tighten the locknut (photo). Recheck the clearance after tightening the locknut in case the ball-pin is moved.
8 Turn the engine to bring another cam lobe to the vertical position and repeat the above procedure. Carry on until all eight valves have been checked.
9 Access to some of the ball-pins is made difficult by the carburettor. To avoid having to remove the offending components, double cranked spanners or cutaway socket spanners can be used.
10 When adjustment is complete, refit the valve cover using a new gasket. Make sure that the dovetail sections of the gasket fit together correctly.
11 Fit the valve cover bolts and reinforcing plates. Tighten the bolts as described in Section 45.
12 Refit the other disturbed components.
13 Run the engine and check that there are no oil leaks from the valve cover.

49 Engine – initial start-up after overhaul or major repair

1 Make a final check to ensure that everything has been reconnected to the engine and that no rags or tools have been left in the engine bay.
2 Check that oil and coolant levels are correct.
3 Start the engine. This may take a little longer than usual as fuel is pumped up to the engine.
4 Check that the oil pressure light goes out when the engine starts.
5 Run the engine at a fast tickover and check for leaks of oil, fuel and coolant. Also check power steering and transmission fluid cooler unions, when applicable. Some smoke and odd smells may be experienced as assembly lubricant burns off the exhaust manifold and other components.
6 Bring the engine to operating temperature. Check the ignition timing (Chapter 4) then adjust the idle speed.
7 Stop the engine and allow it to cool, then re-check the oil and coolant levels.
8 If new bearings, pistons etc have been fitted, the engine should be run in at reduced speeds and loads for the first 500 miles (800 km) or so. It is beneficial to change the engine oil and filter after this mileage.

50 Fault diagnosis – engine

Symptom	Reason(s)
Engine fails to start	Discharged battery Loose battery connection Loose or broken ignition leads Moisture on spark plugs, distributor cap or HT leads Incorrect spark plug gap Cracked distributor cap or rotor Dirt or water in carburettor Empty fuel tank Faulty fuel pump Faulty starter motor Low cylinder compression Faulty electronic ignition system

Chapter 1 Engine 51

Symptom	Reason(s)
Engine idles erratically	Inlet manifold air leak Leaking cylinder head gasket Worn camshaft lobes Incorrect valve clearances Loose crankcase ventilation hoses Incorrect slow running adjustment Uneven cylinder compressions Incorrect ignition timing
Engine misfires	Incorrect spark plug gap Faulty coil or electronic ignition Dirt or water in carburettor Slow running adjustment incorrect Leaking cylinder head gasket Distributor cap cracked Incorrect valve clearances Uneven cylinder compressions Moisture on spark plugs, distributor cap or HT leads
Engine stalls	Slow running adjustment incorrect Inlet manifold air leak Ignition timing incorrect
Excessive oil consumption	Worn pistons and cylinder bores Valve guides and valve stem seals worn Oil leaking from gasket or oil seal
Engine runs on after switching off	Anti-dieseling valve (if applicable) faulty Excessive carbon build-up in combustion chambers

Chapter 2
Cooling, heating and ventilation systems

Contents

Coolant mixture – general	3	Heater unit – dismantling and reassembly	14
Cooling fan – removal and refitting	8	Heater unit – removal and refitting	13
Cooling system – draining	4	Radiator – removal and refitting	7
Cooling system – filling	6	Routine maintenance	2
Cooling system – flushing	5	Temperature gauge and sender unit – testing, removal and	
Demister and air vent hoses – removal and refitting	15	refitting	12
Drivebelt – adjustment and renewal	11	Thermostat – removal, testing and refitting	9
Fault diagnosis – cooling, heating and ventilation systems	16	Water pump – removal and refitting	10
General description	1		

Specifications

General

Type	Pressurised, with belt-driven pump, crossflow radiator, cooling fan, thermostat, and expansion (degas) tank
Expansion tank filler cap pressure	1 bar (14.5 lbf/in²)
Drivebelt tension	10.0 mm (0.4 in) deflection at mid-point of longest span of belt under firm thumb pressure

Thermostat

Type	Wax
Opening temperature	88° C (190° F)
Fully open temperature	102° C (216° F)

System coolant capacity

With heater	8.4 litre (14.8 pints)
Without heater	7.2 litre (12.6 pints)

Coolant type/specification

	Soft water and antifreeze to Ford spec SSM-97B 9103-A (Duckhams Universal Antifreeze and Summer Coolant)
Antifreeze/corrosion inhibitor concentration	50% water/50% antifreeze
Antifreeze and water specific gravity	1065

Torque wrench settings

	Nm	lbf ft
Radiator to body panel	6.8 to 9.5	5 to 7
Thermostat housing	16.3 to 20.3	12 to 15
Water pump:		
M8 bolts	17 to 21	12.5 to 15.5
M10 bolts	35 to 42	26 to 31
Fan to water pump (1.6 litre)	20.5 to 25.5	15 to 19
Viscous drive to water pump (2.0 litre)	40 to 50	29.5 to 37
Fan to viscous drive (2.0 litre)	8.5 to 10.6	6 to 8

1 General description

The cooling system is of pressurised type and includes a front mounted crossflow radiator, belt driven water pump, belt driven fan (1.6 litre models) or temperature conscious thermo-viscous fan (2.0 litre models), wax type thermostat, and an expansion and degas tank. The radiator matrix is of aluminium construction and the end tanks are of plastic.

The thermostat is located behind the water outlet elbow at the front of the cylinder head, and its purpose is to ensure rapid engine warm-up by restricting the flow of coolant in the engine when cold, and also to assist in regulating the normal operating temperature of the engine.

The expansion tank incorporates a pressure cap which effectively pressurises the cooling system as the coolant temperature rises thereby increasing the boiling point temperature of the coolant. The tank also has a further degas function. Any accumulation of air bubbles in the coolant, in particular in the thermostat housing and the radiator, is

Chapter 2 Cooling, heating and ventilation systems

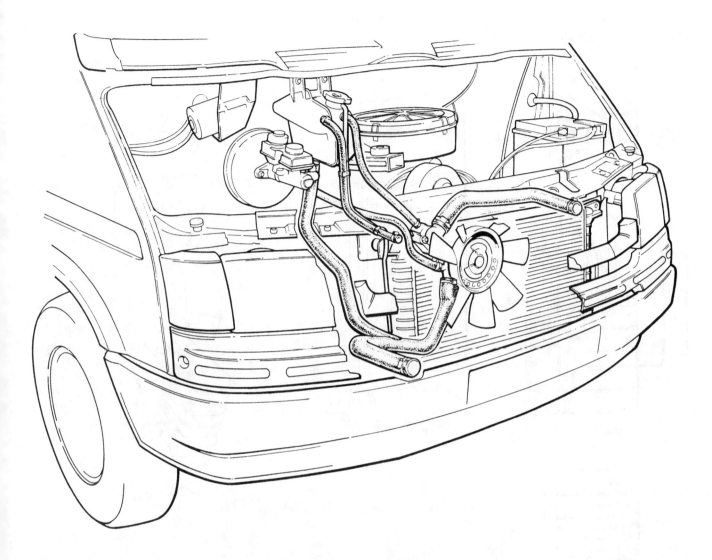

Fig. 2.1 Cooling system layout (Sec 1)

returned to the tank and released in the air space thus maintaining the efficiency of the coolant.

The system functions as follows. Cold water in the bottom of the radiator circulates through the bottom hose to the water pump where the pump impeller pushes the water through the passages within the cylinder block, cylinder head and inlet manifold. After cooling the cylinder bores, combustion surfaces, and valve seats, the water reaches the underside of the thermostat which is initially closed. A small proportion of water passes through the small hose from the thermostat housing to the expansion tank, but because the thermostat is closed the main circulation is through the inlet manifold, automatic choke, and heater matrix, returning to the water pump. When the coolant reaches the predetermined temperature the thermostat opens and hot water passes through the top hose to the top of the radiator. As the water circulates down through the radiator, it is cooled by the passage of air past the radiator when the vehicle is in forward motion, supplemented by the action of the cooling fan. Having reached the bottom of the radiator, the water is now cooled and the cycle is repeated. Circulation of water continues through the expansion tank, inlet manifold, automatic choke, and heater at all times, the heater temperature control being by an air flap.

The thermo-viscous fan fitted to 2.0 litre models is controlled by the temperature of air behind the radiator. When the air temperature reaches a predetermined level, a bi-metallic coil commences to open a valve within the unit and silicon fluid is fed through a system of vanes.

Half of the vanes are driven directly by the water pump and the remaining half are connected to the fan blades. The vanes are arranged so that drive is transmitted to the fan blades in relation to the drag or viscosity of the fluid, and this in turn depends on ambient temperature and engine speed. The fan is therefore only operated when required, and compared with direct drive type fans represents a considerable improvement in fuel economy, drivebelt wear and fan noise.

The heating and ventilation system operates in conjunction with the cooling system, coolant being circulated through the heater matrix (housed within the heater unit) then flowing back to the engine cooling circuit.

An electric blower motor, also housed in the heater unit, assists when required in supplying hot or cool air as necessary to the vehicle interior and windscreen demister vents.

2 Routine maintenance

Carry out the following maintenance procedures at the intervals given in the *Routine maintenance* Section at the start of this manual.

1 Check the coolant level: this check can be made by simply observ-

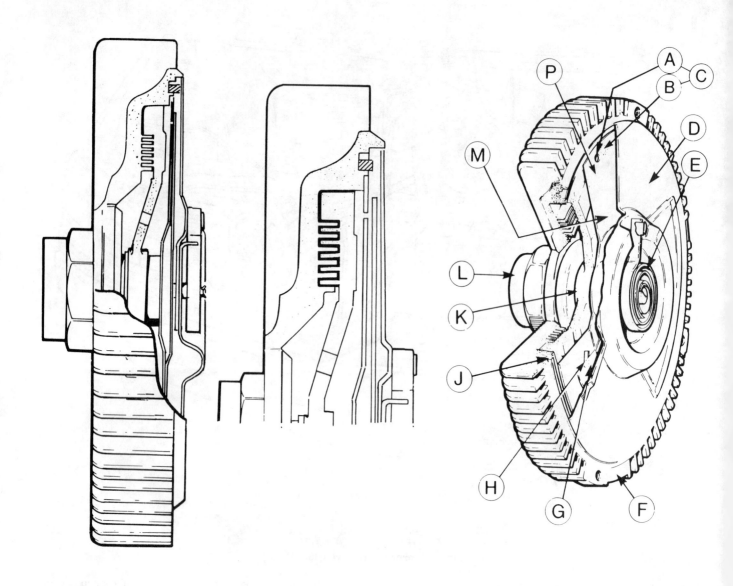

Fig. 2.2 Cut-away diagrams of the thermo-viscous type fan unit (Sec 1)

Left-hand diagram – fluid location at rest

A Discharge port
B Weir
C Ram pump
D Front casing
E Bi-metallic coil
F Main casing
G Control valve

Centre diagram – fluid circuit for drive

H Intake port
J Seal
K Rotor
L Drive shaft
M Fluid reservoir
P Pump plate

ing the coolant level through the translucent expansion tank. The level must be kept between the MAX and MIN level markings on the side of the tank. The level should normally be checked when the engine is cold. If the coolant is hot, the coolant may well rise above the MAX mark.
2   To top up the coolant level, remove the filler/pressure cap, then top up the coolant level through the filler neck in the top of the expansion tank. If the engine has been run and is still hot (or warm), care must be taken to release the pressure from the cooling system by first covering the cap with a cloth, then slowly release the cap to its first position.

Leave in this position until the pressure in the system is fully released, then remove the cap.
3   If possible, top up the coolant level using coolant containing the correct antifreeze mixture to avoid weakening the mixture. Do not overfill the system and, wherever possible, do not add cold coolant to a hot engine. Ensure that the filler cap is fully fitted to complete.
4   Check the cooling system hoses, thermostat housing and water pump for any signs of coolant leakage and rectify as necessary.
5   Drain the coolant, flush the system and fill the system with fresh

Chapter 2 Cooling, heating and ventilation systems

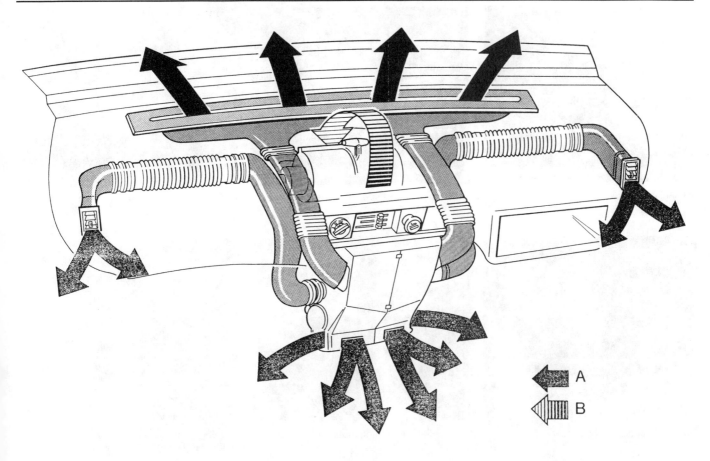

Fig. 2.3 Heating and fresh air ventilation system layout. Cold/warm air (A) Recirculated air (B) (Sec 1)

coolant as described in Sections 4, 5 and 6.
6 Check the drivebelt for correct adjustment and general condition. Adjust or renew if necessary as described in Section 11.

3 Coolant mixture – general

1 Plain water should never be used in the cooling system. Apart from giving protection against freezing, an antifreeze mixture protects the engine internal surfaces and components against corrosion. This is very important in an engine with alloy components.
2 Always use a top quality glycol-based antifreeze which is recommended for alloy engines.
3 The cooling system is initially filled with a solution of 50% antifreeze and it is recommended that this percentage is maintained throughout the year, as the solution supplied by Ford contains a rust and corrosion inhibitor. A suitably equipped Ford garage will have the hydrometer necessary to check the antifreeze strength.
4 After a period of two years the antifreeze solution should be renewed by draining and flushing the cooling system as described in Sections 4 and 5. Check all the hose connections for security and then mix the correct quantity of antifreeze solution in a separate container, which should be clean.
5 Fill the cooling system as described in Section 6.

4 Cooling system – draining

1 If the engine is cold, remove the filler cap from the expansion tank by turning the cap anti-clockwise. If the engine is hot, then turn the filler cap very slightly until pressure in the system has had time to be

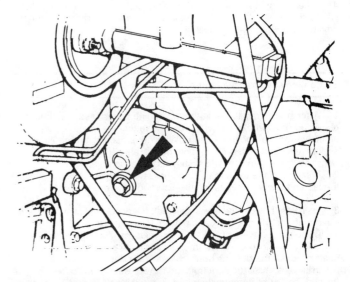

Fig. 2.4 Cylinder block drain plug (arrowed) (Sec 4)

released. Use a rag over the cap to protect your hand from escaping steam. *If with the engine very hot the cap is released suddenly, the drop in pressure can result in the water boiling. With the pressure released the cap can be removed.*
2 Position a bowl of suitable capacity beneath the radiator.
3 Unscrew the radiator drain tap a quarter of a turn and drain the

Chapter 2 Cooling, heating and ventilation systems

4.3 Radiator drain tap

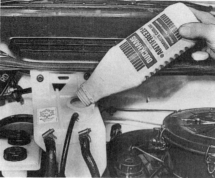
6.2 Topping up the coolant level in the expansion tank

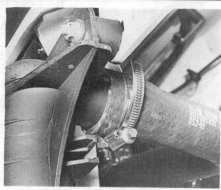
7.3A Radiator top hose connection

7.3B Radiator bottom hose connection

7.9A Radiator mounting rubber bush location

7.9B Engage foot of radiator over the mounting rubber bush each side

coolant from the radiator (photo). A drain plug is also fitted to the side of the cylinder block and this should be unscrewed to drain the engine (see Fig. 2.4).
4 When the water has finished running, probe the drain plug orifices with a short piece of wire to dislodge any particles of rust or sediment which may be causing a blockage.

5 Cooling system – flushing

1 After some time the radiator and engine waterways may become restricted or even blocked with scale or sediment, which reduces the efficiency of the cooling system. When this occurs, the coolant will appear rusty and dark in colour and the system should then be flushed. Begin by draining the cooling system as just described.
2 Disconnect the top and bottom hoses from the radiator, then insert a hose and allow water to circulate through the radiator until it runs clear.
3 Insert the hose in the expansion tank filler neck and allow water to run out of the cylinder block and bottom hose until clear.
4 Disconnect the inlet hose from the inlet manifold, insert the hose, and allow water to circulate through the manifold, automatic choke, heater and out through the bottom hose until clear.
5 In severe cases of contamination the system should be reverse flushed. To do this, remove the radiator, invert it, and insert a hose in the outlet. Continue flushing until clear water runs from the inlet.
6 The engine should also be reverse flushed. To do this, remove the thermostat and insert the hose into the cylinder head. Continue flushing until clear water runs from the cylinder block drain plug and bottom hose.
7 If, after a reasonable period the water still does not run clear, the radiator can be flushed with a good proprietary cleaning agent such as Holts Radflush or Holts Speedflush. Regular renewal of the antifreeze should prevent contamination of the system recurring.

6 Cooling system – filling

1 Check that the radiator drain tap and the cylinder block drain plug are refitted and tightened.
2 Pour the coolant mixture (see Sec 3) into the expansion tank filler neck until it reaches the maximum level mark, .nen refit the cap (photo). Add the coolant slowly to allow air to bubble through.
3 Run the engine at a fast idling speed for several minutes, then stop the engine and check the level in the expansion tank. Top up the level as necessary, being careful to release pressure from the system before removing the filler cap.

7 Radiator – removal and refitting

1 Drain the cooling system as described in Section 4.
2 On 2.0 litre models remove the finger guard.
3 Slacken the hose clips then detach the top and bottom hoses at the radiator (photos).
4 Slacken the clip and detach the expansion tank hose from the radiator.
5 Remove the front grille together with the centre brace by undoing the two screws (one at each end), and the five fasteners (three along the top and two in the aperture above the number plate).
6 Undo the two radiator upper retaining nuts, one each side, and lift the unit forwards and upwards out of the lower mounting bushes.
7 The radiator should be flushed through whilst removed as it can be inverted and reverse flushed. Clean the exterior of dead flies and dirt with a water jet from a garden hose.
8 Due to the nature of its construction, radiator repairs should be left to a specialist. However in an emergency, minor leaks from the radiator

Chapter 2 Cooling, heating and ventilation systems   57

9.4 Thermostat hose connections (arrowed)

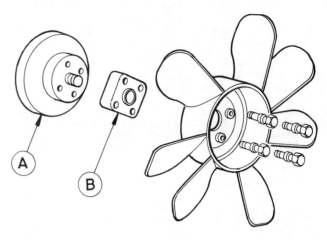

Fig. 2.5 Fixed type cooling fan water pump pulley (A) and spacer (B) (Sec 8)

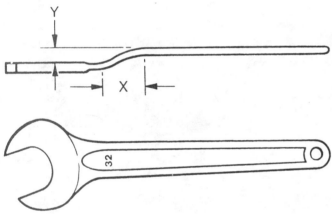

Fig. 2.6 Modified spanner required for removing the thermo-viscous coupling fan (Sec 8)

X = 12 mm (0.5 in)   Y = 12 mm (0.5 in)

may be cured by using a radiator sealant such as Holts Radweld with the radiator *in situ*.
9 Refitting the radiator is a reversal of the removal procedure, but the following additional points should be noted:

(a) Examine and renew any clips and hoses which have deteriorated
(b) Renew the radiator lower mounting bushes if they are in poor condition (photo). Ensure that the radiator engages with the bushes fully as it is fitted (photo)
(c) Refill the cooling system as described in Section 6

8 Cooling fan – removal and refitting

Fixed type fan (1.6 litre engine)
1 Loosen off the alternator retaining bolts, pivot the alternator in towards the engine to allow the drivebelt to be removed from the fan pulley, then undo the four fan retaining bolts. Remove the fan, spacer and pulley from the water pump hub.

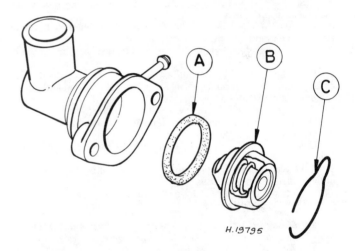

Fig. 2.7 Seal ring (A), thermostat (B), retaining clip (C) and housing (Sec 9)

2 Refit in the reverse order of removal. Adjust the drivebelt as described in Section 11.

Thermo-viscous type fan (2.0 litre engine)
3 Detach and remove the fan finger guard, then use a 32 mm open-end spanner to unscrew the fan clutch unit from the water pump pulley hub. The spanner will need to be 5 mm in section and cranked as shown to allow it to engage over the flats of the retaining nut (Fig. 2.6). Note that the fan clutch has a left-hand thread and to stop the pulley from rotating as it is loosened, hold it with the aid of a chain wrench or similar. If necessary, tap the spanner to free the clutch unit.
4 On removal, the fan can be separated from the clutch by undoing the four retaining bolts.
5 Refit in the reverse order of removal. Tighten the retaining bolts to their specified torque wrench settings.

9 Thermostat – removal, testing and refitting

1 Disconnect the battery negative lead.
2 Drain the cooling system as described in Section 4.

3 On 2.0 litre models, detach and remove the finger guard from above the fan.
4 Disconnect the expansion tank and top hoses from the thermostat housing at the front of the cylinder head (photo).
5 Unscrew the bolts and remove the housing and gasket.
6 Using a screwdriver prise the retaining clip from the housing, and extract the thermostat and sealing ring.
7 To test the thermostat suspend it with a piece of string in a container of water. Gradually heat the water and note the temperature at which the thermostat starts to open. Remove the thermostat from the water and check that it is fully closed when cold (Fig. 2.8).
8 Renew the thermostat if the opening temperature is not as given in the Specifications or if the unit does not fully close when cold.
9 Clean the housing and the mating face of the cylinder head. Check the thermostat sealing ring for condition and renew it if necessary.
10 Refitting is a reversal of removal, but use a new gasket. Note that the thermostat wax capsule must face into the cylinder head with the flow direction arrow facing forward. Refill the cooling system as described in Section 6.

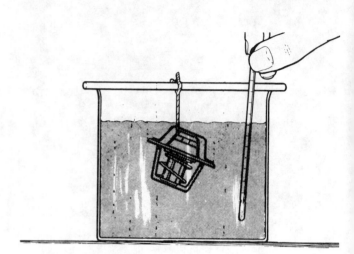

Fig. 2.8 Method of checking thermostat opening temperature (Sec 9)

10 Water pump – removal and refitting

1 An impeller type water pump is fitted on the front face of the cylinder block.
2 Drain the cooling system as described in Section 4. Remove the radiator as given in Section 7.
3 Slacken the alternator mounting bolts and push the alternator towards the cylinder block. Lift away the drivebelt.
4 Undo the bolt securing the alternator adjustment strap to the water pump.
5 Undo and remove the bolts and washers that secure the fan assembly to the water pump spindle hub. Lift away the fan and pulley. (Refer to Section 8 for further details).
6 Detach the timing belt cover which is secured by three bolts.
7 Disconnect and remove the heater hose and bottom hose from the water pump.
8 Unscrew the water pump retaining bolts and withdraw the pump unit.
9 If the water pump is faulty renew it, as it is not possible to readily obtain individual components.
10 Refitting the water pump is a reversal of the removal procedure, but note the following points:

(a) Make certain that the mating surfaces of the pump and cylinder block are perfectly clean, and fit a new gasket
(b) Tighten the pump retaining bolts and fan pulley bolts to the specified torque
(c) Adjust the drivebelt tension as described in Section 11
(d) On completion, run the engine and check for leaks around the pump mating surface and hose connections

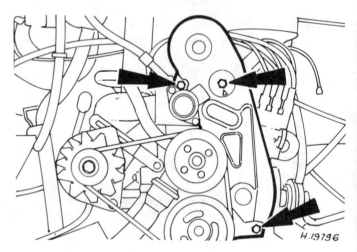

Fig. 2.9 Timing belt cover retaining bolts (arrowed) (Sec 10)

11 Drivebelt – adjustment and renewal

Water pump/alternator drivebelt
1 The belt tension is correct when there is 10.0 mm (0.4 in) of lateral movement under firm thumb pressure at a point midway between the alternator and crankshaft pulleys (Fig. 2.11). On 2.0 litre models, remove the fan finger guard for access.
2 To adjust the belt, loosen the alternator pivot bolt(s) fully but only slightly loosen the alternator adjustment strap bolt. Carefully move the alternator towards or away from the engine until the correct tension is obtained, then tighten the adjusting strap bolt and the alternator pivot bolt(s) in that order (Fig. 2.12).
3 If the belt is worn or stretched unduly it should be renewed. However, the most common reason for renewing a belt is that the original has broken, and it is therefore advisable to carry a replacement on the vehicle for such an occurrence.
4 To remove the belt first remove the power steering drivebelt as described below (where fitted), then loosen the alternator mounting and adjustment bolts and swivel the unit towards the engine, thus releasing the tension.
5 Slip the old belt over the crankshaft, alternator and fan pulleys and lift it over the fan blades. It may be necessary to turn the engine by hand in order to assist the belt over the alternator pulley.
6 Place the new belt onto the pulleys and adjust its tension as described in paragraphs 1 and 2.
7 On completion refit the power steering drivebelt (where applicable), then check that the alternator mounting and adjuster bolts are securely tightened. After fitting a new belt, its tension should be rechecked after about 600 miles (1000 km) and if necessary readjusted to take up the initial stretch.
8 If belt renewal is a frequent occurrence, check the pulleys for misalignment and/or damage.

Power steering drivebelt
9 The procedures for adjustment and replacement are similar to those given above for the water pump/alternator drivebelt. Instead of adjusting the position of the alternator, loosen the retaining bolts and move

Chapter 2 Cooling, heating and ventilation systems

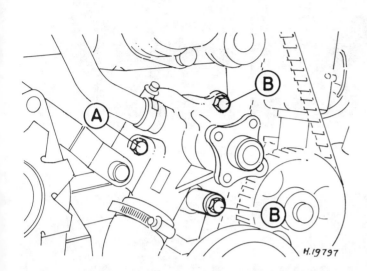

Fig. 2.10 Water pump retaining bolts (Sec 10)

A M10 bolts   B M8 bolts

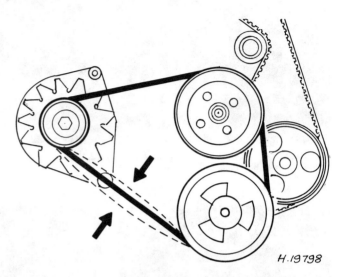

Fig. 2.11 Drivebelt tension check point (arrowed) (Sec 11)

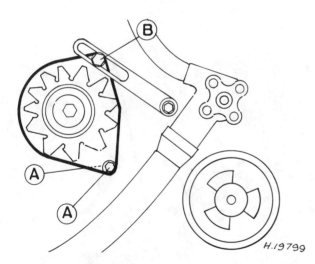

Fig. 2.12 Alternator showing lower (A) and upper (B) mounting bolts (Sec 11)

the power steering idler pulley as required to loosen or take up the drivebelt adjustment (Fig. 2.13).

12 Temperature gauge and sender unit – testing, removal and refitting

1 If the temperature gauge is faulty and gives an incorrect reading, either the gauge, sender unit, wiring or connections are responsible.
2 First check that all the wiring and connections are clean and secure. The gauge and sender unit cannot be repaired by the home mechanic,

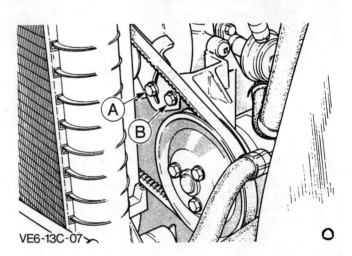

Fig. 2.13 Power steering pump drivebelt and adjuster bolt (A) and pivot bolt (B) (Sec 11)

and therefore they must be renewed if faulty. Further information will be found in Chapter 12.
3 The wiring can be checked by connecting a substitute wire between the sender unit and the temperature gauge and observing the result.
4 Before removing the sender unit the cooling system must be partially drained of about 2.27 litres (4 pints) of coolant, then detach the supply lead and unscrew the unit.
5 Refit the sender unit in the reverse order and top up the cooling system.
6 Details of the removal and refitting of the temperature gauge are given in Chapter 12.

13 Heater unit – removal and refitting

1 Disconnect the battery earth lead.
2 Drain the engine coolant as described in Section 4.
3 Loosen off the retaining clips and detach the heater flow and return hoses at the bulkhead connections (photo).

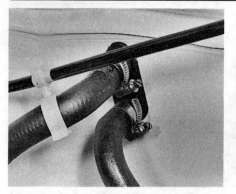
13.3 Heater coolant flow and return hose connections at the bulkhead

13.4 Heater control panel removal

13.5A Undo the retaining nut each side (arrowed)...

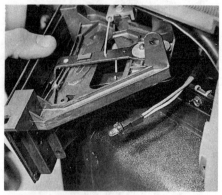
13.5B ... and withdraw the heater flap control panel

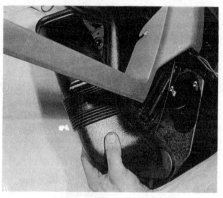
13.6 Disconnect the heater/ventilation ducts and hoses from the heater unit

13.7 Heater unit with lower retaining screw on left-hand side (arrowed)

4   Working inside the vehicle, undo the four retaining screws and withdraw the heater control panel. As it is withdrawn, detach the wiring connectors and bulb holders from the fan motor switch and the cigar lighter units (photo). Remove the ashtray.
5   Disconnect the heater flap control panel by undoing the two retaining nuts from the positions shown (photos).
6   Detach the heater unit demister and ventilation hoses (photos).
7   Unscrew and remove the four heater unit to bulkhead retaining screws, and then carefully withdraw the heater assembly from the vehicle (photo).As it is withdrawn, disconnect the fan motor wiring connections.
8   Refitting is a reversal of the removal procedure, but note the following special points:

  (a) When refitting the heater flap control panel, position the control levers 2 mm (0.08 in) from their end stops, and fit the control cable sheath securing clips. Each flap should be at the end of its travel
  (b) Ensure that the wiring connectors are securely attached to the fan motor switch and the cigar lighter
  (c) Check that the heater hose connections are securely made, then refill the cooling system as described in Section 6
  (d) When the engine is restarted, operate the heater and ventilation system to check for satisfactory operation and also check for any signs of coolant leaks from the system

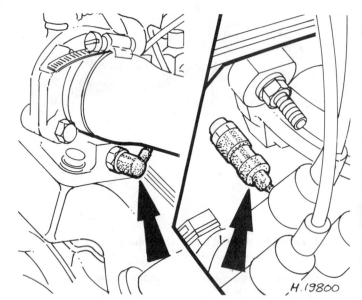

Fig. 2.14 Coolant temperature sender unit location (Sec 12)

14 Heater unit – dismantling and reassembly

1   Detach the heater control cables from the heater unit and remove the control cables and panel.
2   Undo the eight retaining screws and separate the upper casing from the lower casing (Fig. 2.15).
3   To remove the fan motor, unscrew the three securing screws, and then withdraw the unit from the casing (Fig. 2.16).
4   Remove the coolant pipe support screw and the foam seal. Remove the seal from the end of the matrix, then carefully withdraw the matrix from the heater unit case.
5   Further dismantling of the casings is unlikely to be necessary, but if

Chapter 2 Cooling, heating and ventilation systems   61

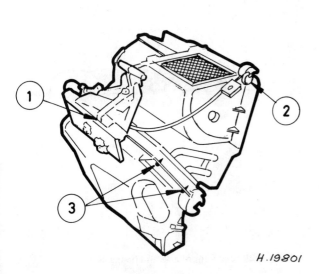

Fig. 2.15 Heater unit showing control (1) air flap assembly (2) and retaining screw locations (3) (Sec 14)

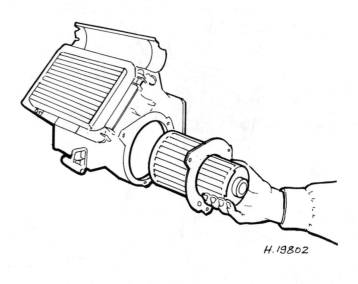

Fig. 2.16 Withdrawing the fan motor (Sec 14)

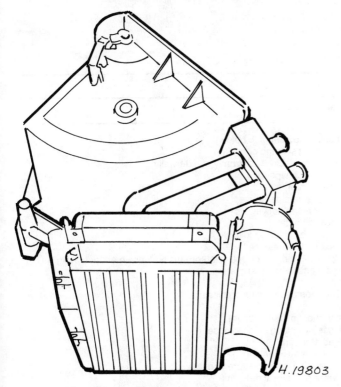

Fig. 2.17 Withdrawing the heater matrix (Sec 14)

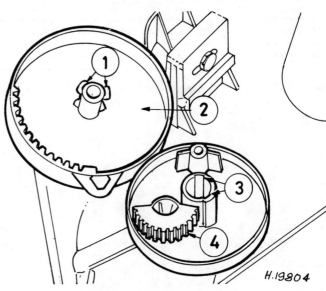

Fig. 2.18 Air distributor flap unit (Sec 14)

1 Actuator spigot shaft lugs
2 Actuator (with internal teeth)
3 Housing slots
4 Toothed quadrant

required can be achieved by releasing the retaining clips.
6 Renew any defective components as necessary. Repairs to individual items are either not possible or not practical.
7 Reassemble in the reverse order of dismantling. If the flap valve control quadrant was dismantled, fit the quadrant as shown in Fig. 2.18, then locate the actuator and align the spigot shaft lugs with the casing slots (rotating slightly to secure in position).
8 When reassembling the casing halves, locate a new foam seal. Also fit a new seal to the matrix when installing it.

15 Demister and air vent hoses – removal and refitting

Demister hose
1 Detach the hose from the nozzle and the heater casing by simply pulling it free, then withdraw it.
2 Refit the demister hose by fitting it to the demister nozzle first.

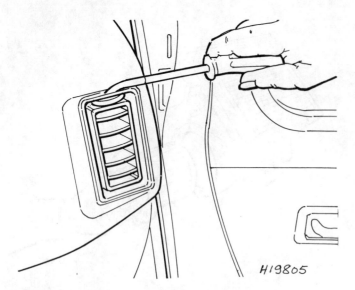 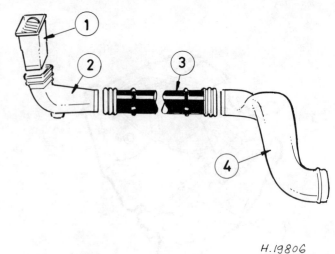

Fig. 2.19 Face vent nozzle grille removal (Sec 15)

Fig. 2.20 Face vent nozzle and hoses (Sec 15)

1 Nozzle
2 Adaptor (to nozzle)
3 Hose
4 Adaptor (to heater)

Face vent nozzle

3 Carefully prise free the nozzle grill using a screwdriver as a lever. Prise at the top and bottom edges to ease it free, then withdraw the nozzle (Fig. 2.19).

4 To refit the nozzle unit, align it correctly in its aperture then carefully press it into position so that it is felt to engage with the adaptor.

16 Fault diagnosis – cooling, heating and ventilation systems

Symptom	Reason(s)
Overheating	Loss of coolant (see below)
	Drivebelt slipping or broken
	Radiator blocked (internally or externally)
	Hose(s) collapsed
	Thermostat faulty
	Water pump defective
	New engine not yet run-in
	Brakes binding
	Engine oil level too low
	Mixture too weak
	Ignition timing incorrect, or automatic advance malfunctioning
	Blockage in exhaust system
Overcooling	Thermostat defective or missing
Coolant loss – external	Overheating (see above)
	Hose(s) leaking or clips loose
	Radiator core leaking
	Heater matrix leaking
	Expansion tank pressure cap leaking
	Water pump seal defective
Coolant loss – internal	Blown cylinder head gasket (oil in coolant and/or coolant in oil)
	Cylinder head or block cracked

Chapter 3 Fuel and exhaust systems

Contents

Accelerator cable – removal, refitting and adjustment	9
Accelerator pedal – removal and refitting	10
Air cleaner and element – removal and refitting	3
Air cleaner temperature control – testing	4
Carburettor – general description	11
Carburettor – overhaul	14
Carburettor – removal and refitting	13
Carburettor – slow running adjustment	12
Exhaust manifold – removal and refitting	16
Exhaust system – checking, removal and refitting	17
Fault diagnosis – fuel and exhaust systems	19
Fuel gauge sender unit – removal and refitting	8
Fuel pump – testing, removal, servicing and refitting	5
Fuel tank – removal, servicing and refitting	6
Fuel tank filler pipe – removal and refitting	7
General description	1
Inlet manifold – removal and refitting	15
Routine maintenance	2
Unleaded fuel – general	18

Specifications

General

Air cleaner	Automatic air intake temperature control. Renewable paper type element
Air cleaner element	Champion W184
Air intake heat sensor temperature rating	28° C (82° F)
Fuel tank capacity	68 litres (15 gallons)
Fuel pump type	Mechanical diaphragm operated by pushrod and eccentric cam on auxiliary shaft
Fuel pump delivery pressure	0.26 to 0.34 bar (3.77 to 4.93 lbf/in^2)

Carburettor

Type	Ford variable venturi (VV) downdraught
Identification number:	
1.6 litre	86 HF-9510-KHA
2.0 litre	80 HF-9510-KJA
Idle speed	750 to 850 rpm
Idle mixture (CO)	1.0 ± 0.5%

Fuel requirement

Fuel octane rating (minimum)	97 RON (4-star) leaded or 91 RON/82.5 MON unleaded*

*See Section 18 for unleaded fuel suitability

Torque wrench settings

	Nm	lbf ft
Exhaust manifold-to-downpipe nuts	40.5 to 57.5	30 to 42
Exhaust system U-clamp bolts	30 to 42	22 to 31
Inlet manifold-to-cylinder head	17 to 21	13 to 16
Exhaust manifold-to-cylinder head	21 to 25	16 to 18
Fuel pump bolts	14 to 18	10 to 13

1 General description

The fuel system consists of a centre mounted fuel tank of polyethylene construction, a mechanical diaphragm fuel pump, a downdraught variable venturi (VV) carburettor and a thermostatically controlled air cleaner.

The fuel pump is mounted on the side of the cylinder block just in front of the oil filter, and is driven by a cam on the auxiliary shaft.

The air cleaner element is a renewable cartridge type and it is important that this is replaced at the specified intervals. The air cleaner intake duct contains a thermostatically controlled air inlet flap valve which regulates the temperature of the air entering the inlet system.

When the engine is cold, the valve diverts hot air from the area around the exhaust manifold into the intake duct. As the engine warms up, the valve pivots and shuts off the hot air entry and simultaneously opens the ambient air port to allow cooler air to be drawn into the system.

The carburettor is of Ford manufacture and is described fully in Section 11.

The exhaust system comprises two sections, front and rear, each having its own silencer unit.

Certain Transit models are suitable for continuous use on unleaded fuel and this can be determined by referring to code letters stamped on the cylinder head. Details are contained in Section 18.

Warning: *Many of the procedures in this Chapter entail the removal*

Chapter 3 Fuel and exhaust systems

3.1 Removing an air filter retaining screw

3.2 Air filter cover alignment arrow

3.3 Air filter unit removal

of fuel pipes and connections which may result in some fuel spillage. Before carrying out any operation on the fuel system refer to the precautions given in Safety First! at the beginning of this manual and follow them implicitly. Petrol is a highly dangerous and volatile liquid and the precautions necessary when handling it cannot be overstressed.

2 Routine maintenance

Carry out the following procedures at the intervals given in Routine maintenance at the start of this manual.

1 Check all fuel lines and hoses for damage and general condition, (including those on the underbody). Ensure that all hoses and lines are secure. Make any repairs (or renew) as necessary.
2 Renew the air filter element (Section 3).
3 Check the accelerator pedal and operating cable for satisfactory operation. Lubricate the linkages.
4 Check the engine idle speed and carburettor mixture adjustments (Section 12).
5 Inspect the exhaust system for signs of damage and excessive corrosion. Check the joints for security and any signs of leakages. Check the condition and security of the mountings (Section 17).

3 Air cleaner and element – removal and refitting

1 Undo the top cover retaining screws, release the securing clips, and then lift off the top cover from the air cleaner unit body (photo).
2 Lift out the element. Wipe out the cleaner unit body with a cloth, then insert the new element into position and refit the cover. As the cover is fitted, align the arrow on the cover with the centre line on top of the intake duct (photo).
3 If the air cleaner unit is to be removed, remove the top cover screws as described above, detach the vacuum hoses and the intake duct and remove the air cleaner as a complete unit (photo).
4 Refitting is a reversal of removal.

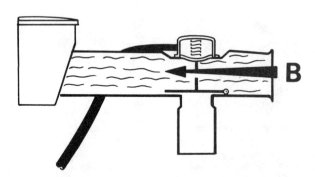

Fig. 3.1 Air cleaner control valve operation (Sec 4)

A Valve open allowing warm air into air cleaner
B Valve closed allowing cool air into air cleaner

4 Air cleaner temperature control – testing

1 The following tests must be carried out with the engine cold.
2 Look into the inlet tube and check that the control flap is fully shut onto the warm air port.
3 Start the engine and allow it to idle. Check that the flap is now fully open to admit warm air from the exhaust manifold shroud. If the flap does not fully open check the diaphragm unit and heat sensor as follows.
4 Disconnect the diaphragm vacuum pipe at the heat sensor and using a vacuum pump apply a vacuum of 100 mm (4.0 in) Hg. If the flap now opens, the heat sensor or vacuum line must be faulty. If the flap remains shut, the diaphragm unit or control flap is faulty (photo).

Chapter 3 Fuel and exhaust systems

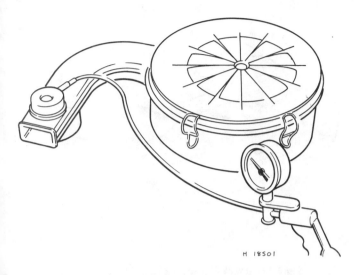

Fig. 3.2 Air cleaner diaphragm unit check using a vacuum pump (Sec 4)

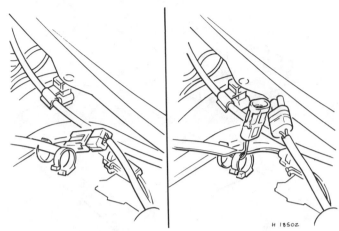

Fig. 3.3 Fuel gauge sender wires and in-line connection (Sec 6)

5 Fuel pump – testing, removal, servicing and refitting

Note: *Refer to the warning note in Section 1 before proceeding*
1 The fuel pump is located on the left-hand side of the engine next to the oil filter. To test its operation disconnect the outlet hose and hold a wad of rag near the pump. Disconnect the low tension negative wire from the ignition coil.
2 Have an assistant spin the engine on the starter and check that well defined spurts of fuel are ejected from the fuel pump outlet.
3 If a pressure gauge is available the fuel pump can be checked more accurately. Twist connect it to the pump outlet. Do not disconnect the coil low tension wire. Start the engine and allow it to idle, then check that the delivery pressure is as given in the Specifications. Note that for the engine to run there must be sufficient fuel in the carburettor.
4 To remove the fuel pump disconnect and plug the inlet and outlet hoses. If crimped type clips secure the hoses, cut them free and renew them with screw type clips on refitting (photo).
5 Unscrew the two bolts and withdraw the pump from the cylinder block. Remove the gasket. If necessary extract the pushrod.
6 Clean the exterior of the pump in paraffin and wipe dry. Clean all traces of gasket from the cylinder block and pump flange.
7 Refitting is a reversal of removal, but fit a new gasket and tighten the bolts to the specified torque. If necessary discard the crimped type hose clips and fit screw type clips.

6 Fuel tank – removal, servicing and refitting

Note: *Refer to the warning note in Section 1 before proceeding*
1 Disconnect the battery earth lead.
2 Disconnect the fuel supply hose to the fuel pump (at the pump). Connect up an auxiliary pump to the hose to syphon any remaining fuel from the tank into a suitable sealed container for storage. It is not possible to syphon the fuel from the tank via the filler pipe. If a suitable auxiliary pump is not readily available, allow the fuel level to run down as much as possible prior to removal.
3 Raise and support the vehicle on axle stands, to allow sufficient working clearance underneath, for detaching and removing the tank.
4 Disconnect the ventilation pipes from the filler pipe.
5 Disconnect the fuel gauge sender unit wires at the in-line connector, and detach the wires from the chassis member.
6 Detach the fuel filler pipe at the tank end (photo). The fuel outlet pipe-to-tank ties should be cut free at this stage also.
7 Locate a trolley jack or stands under the tank and set at a height which will allow the tank to be partially lowered and supported whilst the fuel feed pipe is detached. Note the position of the nuts on the strap hook bolts before removal to ensure correct refitting.
8 Undo the fuel tank support strap retaining nuts, then lower the tank onto the jack or stands. Detach the fuel feed pipe, then carefully lower the tank and withdraw it from under the vehicle. As it is removed, feed the ventilation pipes through the chassis member (photo).
9 If the fuel tank is to be renewed, detach the ventilation pipes and remove the fuel gauge sender unit as described in Section 8.

4.4 The air cleaner heat sensor

5.4 Fuel pump showing connecting hoses and clips

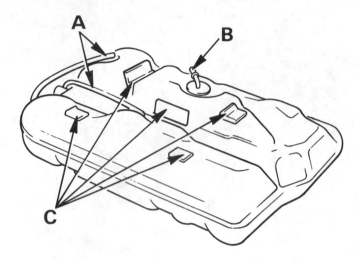

Fig. 3.4 Fuel tank unit and fittings – LCX type (Sec 6)

(A) Vent pipes
(B) Fuel supply outlet
(C) Pads

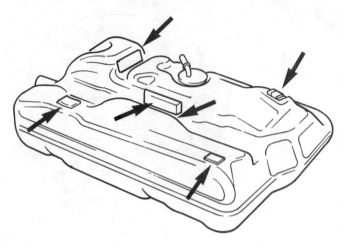

Fig. 3.5 Fuel tank unit and fittings – LCY type (Sec 6)

Arrows indicate pad locations

10 If the tank is contaminated with sediment or water, swill it out with clean fuel. If the tank is damaged or leaks, it must be renewed.

11 Refitting is a reversal of the removal procedure, but note the following special points:

(a) Fit a new sender unit seal (if the sender was removed)
(b) Replace any crimped type hose clips with screw types
(c) Fit new tank insulation pads into position as shown (Figs. 3.4 and 3.5)
(d) Tighten the tank retaining strap hook bolt nuts to the positions noted during removal to ensure that they are not overtightened
(e) Fit new tie wraps when securing the fuel line(s)
(f) On LCY models, ensure that the filler pipes are secured by the chassis retaining clips
(g) Ensure that the fuel ventilation pipes are routed correctly according to model (Fig. 3.6)

5 Release the retaining clips and detach the filler hose and the ventilation pipes from the tank.
6 Withdraw the filler hose, detaching the sealing grommet at the bottom end and the gasket at the top end.
7 Refit in the reverse order of removal. Renew the grommet and gasket, and ensure that there are two clips fitted to the filler hose to pipe connection.

8 Fuel gauge sender unit – removal and refitting

1 Remove the fuel tank as described in Section 6.
2 Unscrew the sender unit from the tank. There is a Ford tool (No. 23-014) which engages with the lugs on the unit, but with patience a pair of crossed screwdrivers or similar items can be used instead.
3 Remove the sender unit, taking care not to damage the float or bend the float arm. Recover the seal.
4 A defective sender unit must be renewed; spares are not available. Renew the seal in any case.
5 Refit by reversing the removal operations.

7 Fuel tank filler pipe – removal and refitting

1 Disconnect the battery earth lead.
2 Drain the fuel tank as described in paragraph 2 of the previous Section.
3 Unscrew and remove the two filler pipe to body retaining screws.
4 Raise and support the vehicle on axle stands.

9 Accelerator cable – removal, refitting and adjustment

1 Disconnect the battery negative lead.
2 Prise off the clip retaining the cable to the accelerator pedal, and unhook the cable (photo).

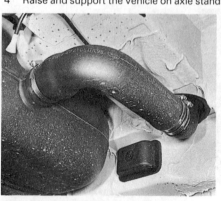

6.6 Fuel tank fuel filler pipe connections

6.8 Fuel tank support strap and retaining nut

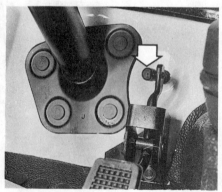

9.2 Accelerator cable connection to the accelerator pedal (arrowed)

Chapter 3 Fuel and exhaust systems

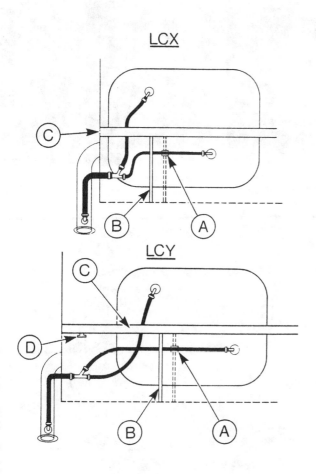

Fig. 3.6 Fuel vent pipes to be routed according to model (Sec 6)

(A) Chassis cab variants – over outrigger (secure with clip)
(B) Vans and bus variants – through outrigger
(C) All models – through crossmember
(D) Spare chassis clip (not used)

3 Working in the engine compartment release the cable from the bulkhead and pull it through.
4 Remove the air cleaner as described in Section 3.
5 Rotate the throttle lever segment to open the throttle then disconnect the inner cable from the segment.
6 Prise the spring clip from the cable bracket using a screwdriver.
7 Depress the four plastic legs and withdraw the cable from the bracket (photo). If difficulty is experienced make up a tool as shown in Fig. 3.9 and push it onto the plastic fitting to depress the legs.
8 Refitting is a reversal of removal, but before refitting the air cleaner adjust the cable as follows. Using a broom or length of wood fully depress the accelerator pedal and retain it in this position. On automatic transmission models make sure that the downshift cable does not restrict the accelerator pedal movement. Unscrew the cable ferrule at the carburettor end until the throttle lever segment is fully open. Release the accelerator pedal then fully depress it again and check that the throttle lever segment is fully open.

10 Accelerator pedal – removal and refitting

1 Disconnect the battery earth lead.

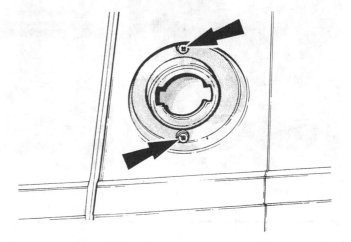

Fig. 3.7 Fuel filler pipe neck to body screws (Sec 7)

2 Prise free the retaining clip, and detach the accelerator cable from the pedal.
3 Pivot the two square head pedal bushes through 45° and detach them. Withdraw the pedal and shaft unit.
4 Refit in the reverse order of removal and then check the accelerator cable adjustment as described in Section 9.

11 Carburettor – general description

The Ford variable venturi carburettor is theoretically more efficient than fixed jet types due mainly to the improved fuel atomisation especially at low engine speeds and loads. The carburettor operates as follows.

Fuel is supplied to the carburettor via a needle valve which is actuated by the float. When the fuel level is low in the float chamber in the carburettor, the float drops and opens the needle valve. When the correct fuel level is reached the float will close the valve and shut off the fuel supply.

The float level on this type of carburettor is not adjustable since minor variations in the fuel level do not affect the performance of the carburettor. The valve needle is prevented from vibrating by means of a ball and light spring and to further ensure that the needle seals correctly it is coated in a rubber-like coating of Viton.

The float chamber is vented internally via the main jet body and carburettor air inlet, thus avoiding the possibility of petrol vapour escaping into the atmosphere.

The air/fuel mixture intake is controlled by the air valve which is opened or closed according to the operating demands of the engine. The valve is actuated by a diaphragm which opens or closes according to the vacuum supplied through the venturi between the air valve and the throttle butterfly. As the air valve and diaphragm are connected they open or close correspondingly.

When the engine is idling the air intake requirement is low and therefore the valve is closed, causing a high air speed over the main jet exit. However as the throttle plate is opened, the control vacuum (depression within the venturi) increases and is channelled to the diaphragm which then opens the air valve to balance the control spring and control vacuum.

When the throttle is opened further this equality of balance is maintained as the air valve is progressively opened to equalise the control spring and control vacuum forces throughout the speed range.

Fuel from the float chamber is drawn up the pick-up tube and then regulated through two jets and the tapered needle and into the engine. The vacuum within the venturi draws the fuel. This is shown in Fig. 3.11. At low engine speeds the needle taper enters the main jet to restrict the fuel demand. On acceleration and at high engine speeds the needle is withdrawn through the main jet by the action of the air valve to which it is attached. As the needle is tapered, the amount by which it is moved regulates the amount of fuel passing through the main jet.

The sonic idle system as used on other Ford fixed jet carburettors is

68 Chapter 3 Fuel and exhaust systems

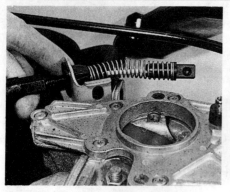

9.7 Disconnecting the accelerator cable from the support bracket

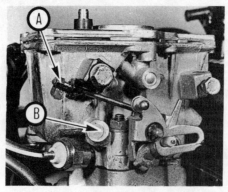

12.4 Ford VV carburettor showing the idle speed adjustment screw (A) and mixture adjusting screw (B)

13.4 Detach the automatic choke hoses

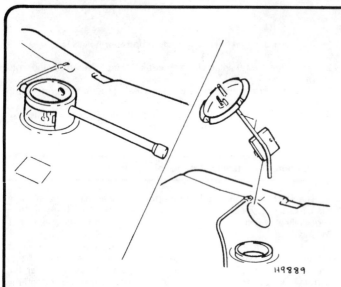

Fig. 3.8 Fuel gauge sender removal from fuel tank (Sec 8)

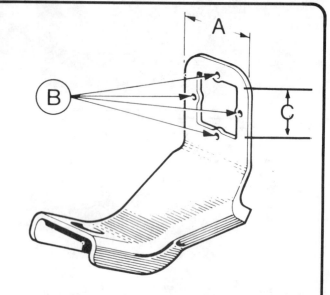

Fig. 3.9 Tool for removing the accelerator cable (Sec 9)

A 25.4 mm (1.0 in)
B Centre punch holes
C 16.0 mm ($\frac{5}{8}$ in) square hole

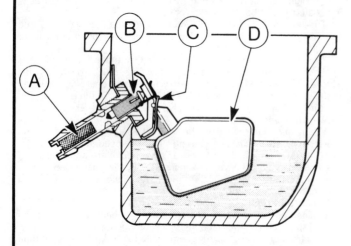

Fig. 3.10 Cross section of the float chamber on the Ford VV carburettor (Sec 11)

A Filter
B Needle valve
C Pivot
D Float

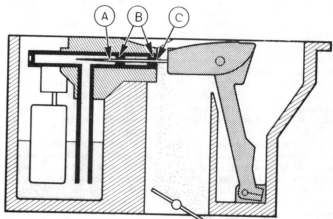

Fig. 3.11 Main jet system on the Ford VV carburettor (Sec 11)

A Tapered metering rod
B Main and secondary jets
C Main fuel outlets

also employed in the VV type, with 70% of the idle fuel mixture supplied via the sonic idle system and 30% from the main system. When idling, fuel is drawn through the main pick-up tube (Fig. 3.12) passes through the idle jet and then mixes with the air stream being supplied from the air bleed in the main jet body. The air/fuel mixture then passes on through the inner galleries at the mixture control screw which regulates the fuel supply at idle. This mixture then mixes with the air from the by-pass idle channel and finally enters the inlet manifold via the sonic discharge tube at an accelerated rate of flow.

Throttle actuation is via a progressive linkage which has a cam and roller mechanism. The advantage of this system is that a large initial throttle movement allows only a small throttle plate opening. As the throttle is opened up and approaches its maximum travel the throttle plate movement accelerates accordingly. This system aids economy, gives a good engine response through the range on smaller engines, and enables the same size of carburettor to be employed on other models in the range.

To counterbalance the drop in vacuum when initially accelerating, a restrictor is fitted into the air passage located between the control vacuum areas and the control diaphragm. This restrictor causes the valve to open slowly when an increase in air flow is made which in turn causes a higher vacuum for a brief moment in the main jet, caused by the increase in air velocity. This increase in vacuum causes the fuel flow to increase thus preventing a 'flat spot'. The large amounts of fuel required under heavy acceleration are supplied by the accelerator pump.

The accelerator pump injects fuel into the venturi direct when acceleration causes a drop in manifold vacuum. This richening of the mixture prevents engine hesitation under heavy acceleration. The accelerator pump is a diaphragm type and is actuated from vacuum obtained from under the throttle plate. During acceleration the vacuum under the throttle plate drops, the diaphragm return spring pushes the diaphragm and the fuel in the pump is fed via the inner galleries through the one-way valve and into the venturi. The system incorporates a back bleeder and vacuum break air hole. Briefly explained, the back bleed allows any excess fuel vapour to return to the float chamber when prolonged idling causes the carburettor temperature to rise and the fuel in the accelerator pump reservoir to become overheated. The vacuum break air hole allows air into the pump outlet pipe to reduce the vacuum at the accelerator pump jet at high speed. Fuel would otherwise be drawn out of the accelerator pump system.

A fully automatic choke system is fitted incorporating a coolant operated bi-metallic spring. According to the temperature of the coolant, the spring in the unit opens or closes. This in turn actuates the choke mechanism, which consists of a variable needle jet and a variable supply of air. Fuel to the choke jet is fed from the main pick-up tube via the internal galleries within the main jet body. When the bi-metal spring is contracted (engine cold), it pulls the tapered needle from the jet to increase the fuel delivery rate. The spring expands as the engine warms up and the needle reduces the fuel supply as it re-enters the jet. The choke air supply is supplied via the venturi just above the throttle plate. The fuel mixes with the air in the choke air valve and is then delivered to the engine.

A choke pull-down system is employed whereby if the engine is under choke but is only cruising, ie not under heavy loading, the choke is released. This is operated by the vacuum piston which is connected to the choke spindle by levers.

Last but not least, an anti-dieselling valve is fitted on the outside of the body of the carburettor. This valve shuts off the fuel supply to the idle system when the engine is turned off and so prevents the engine running on or 'dieselling'. The solenoid valve is actuated electrically. When the ignition is turned off, it allows a plunger to enter and block the sonic discharge tube to stop the supply of fuel into the idle system. When the ignition is switched on the solenoid is actuated and the plunger is withdrawn from the tube.

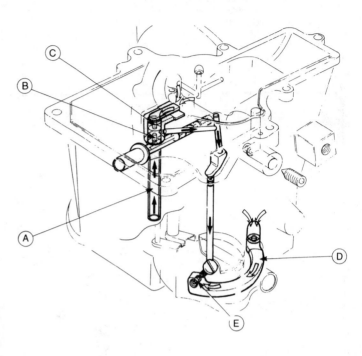

Fig. 3.12 Sonic idle system on the Ford VV carburettor (Sec 11)

A Main pick up tube
B Idle fuel jet
C Idle air jet
D By-pass gallery
E Sonic discharge tube

the engine. The air cleaner must be fitted during the checks/adjustments.
3 Run the engine at 3000 rpm for 30 seconds, then allow it to idle and note the idle speed and CO content.
4 Adjust the idle speed screw to give the specified idle speed (photo).
5 Adjustment of the CO content (mixture) is not normally required during routine maintenance, but if the reading noted in paragraph 3 is not as given in the Specifications first remove the tamperproof plug using a thin screwdriver.
6 Run the engine at 3000 rpm for 30 seconds then allow it to idle. Adjust the mixture screw within 10 to 30 seconds. If more time is required run the engine at 3000 rpm again for 30 seconds.
7 Adjust the idle speed if necessary and recheck the CO content.
8 Fit a new tamperproof plug, if required. Note that it is not possible to adjust the idling mixture accurately without an exhaust gas analyser.

12 Carburettor – slow running adjustment

1 Run the engine to normal operating temperature then stop it.
2 Connect a tachometer and, if available, an exhaust gas analyser to

13 Carburettor – removal and refitting

1 Disconnect the battery negative lead.
2 Remove the air cleaner as described in Section 3.
3 Remove the filler cap from the cooling system expansion tank. If the engine is warm refer to Chapter 2 for safety precautions. After releasing the pressure refit the cap.
4 Identify the hoses for position then loosen the clips and disconnect the hoses from the automatic choke. Either plug the hoses or secure them with their ends facing upwards to prevent loss of coolant (photo).
5 Disconnect the wire from the anti-dieselling valve.
6 Disconnect the vacuum and fuel pipes noting their various locations.

7 Disconnect the accelerator cable.
8 Unscrew the nuts and withdraw the carburettor from the inlet manifold. Remove the gasket.
9 Refitting is a reversal of removal, but make sure that the mating faces are clean and always fit a new gasket. If necessary discard the crimped type hose clips and fit screw type clips, but make sure that they do not obstruct any surrounding component. Finally check and adjust the idling speed and mixture as described in Section 12.

14 Carburettor – overhaul

Note: *Before attempting to overhaul a well worn carburettor, ensure that spares are available and reasonably priced. It may be both quicker and cheaper to obtain a complete carburettor on an exchange basis.*

1 Before dismantling the carburettor clean it off externally and prepare a suitable work space on the bench to lay out the respective

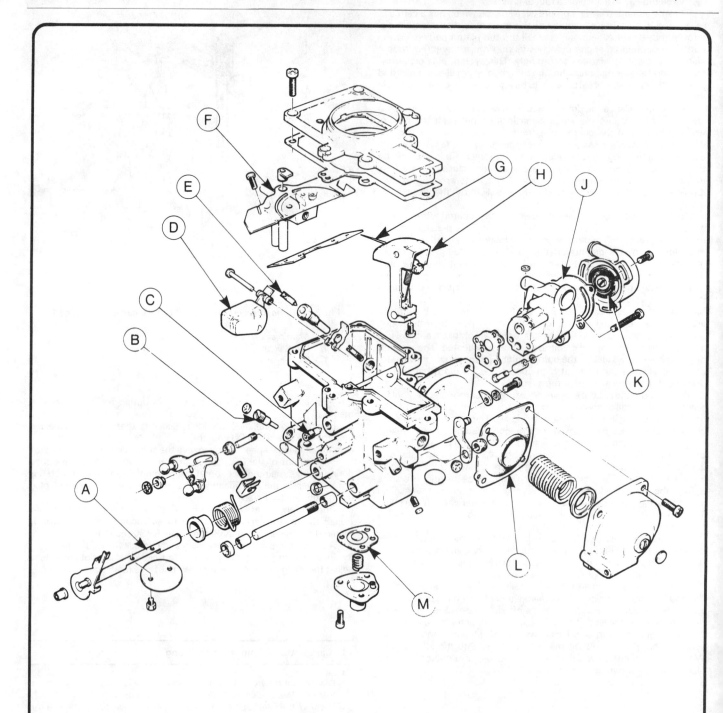

Fig. 3.13 Exploded view of the Ford VV carburettor (Sec 11)

A Throttle spindle
B Mixture screw
C By-pass leak adjuster
D Float
E Needle valve
F Main jet body
G Metering rod
H Air valve
J Choke assembly
K Bi-metal coil
L Vacuum diaphragm
M Accelerator pump diaphragm

Chapter 3 Fuel and exhaust systems

14.2 Lift the cover from the carburettor

14.5A Main jet body retaining screw locations

14.5B Main jet body removal

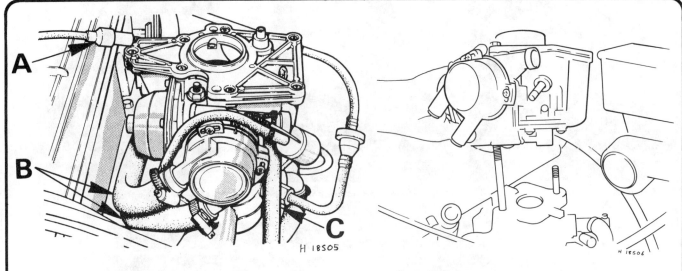

Fig. 3.14 VV carburettor removal (Sec 13)

A Accelerator cable
B Automatic choke hoses
C Fuel pipe

Fig. 3.15 Removing the VV carburettor from the manifold (Sec 13)

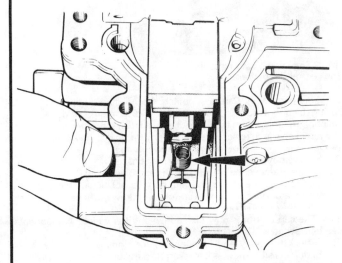

Fig. 3.16 Correct position at the metering rod spring (Sec 14)

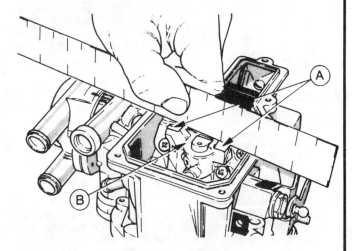

Fig. 3.17 Using a straight edge to check the alignment of the main jet body (Sec 14)

A Alignment flanges
B Main jet body

Chapter 3 Fuel and exhaust systems

14.7 Float and needle valve

14.8A Remove the circlip ...

14.8B ... and remove the diaphragm

14.10A Note alignment marks ...

14.10B ... then remove the automatic choke housing

14.14 Align the diaphragm holes correctly

14.18 Fit a new gasket before fitting the main jet body

14.19 Engage the automatic choke coil with the central slot of the lever

components in order of appearance.
2   Unscrew and remove the seven carburettor cover retaining screws. Carefully lift the cover clear trying not to break the gasket. Remove the gasket (photo).
3   Drain any remaining fuel from the float chamber.
4   If the variable choke metering rod (or needle) is to be removed, prise the tamperproof plug from the body and insert a suitable screwdriver through the hole. Unscrew the metering rod and withdraw it. However note that the manufacturers do not recommend removing the rod. If the rod is damaged, the carburettor should be renewed.
5   To remove the main jet body, unscrew the four retaining screws (photo), and carefully lift the body clear, noting the gasket. If the metering rod is still in position, retract it as far as possible from the jet, press the float down and carefully pull the jet body clear of the metering rod (photo). Great care must be taken here not to bend or distort the rod in any way.
6   The accelerator pump outlet one-way valve ball and weight can now be extracted by inverting the carburettor body.
7   Withdraw the float pivot pin followed by the float and needle valve (photo).
8   Unscrew and remove the four screws retaining the control

diaphragm housing. Carefully detach the housing, spring, and seat, taking care not so split or distort the diaphragm. Fold back the diaphragm rubber from the flange. Using a small screwdriver, prise free the retaining clip to release the diaphragm. Put the clip in a safe place to prevent it getting lost before reassembly (photos).
9   Now remove the accelerator pump by unscrewing the three retaining screws. Remove the housing, spring and diaphragm.
10   To remove the choke housing, note its positional markings, unscrew the retaining screws and carefully withdraw the housing (photos). Unscrew the solenoid unit.
11   The carburettor is now dismantled and the various components can be cleaned and inspected.
12   Check the body and components for signs of excessive wear and/or defects and renew as necessary. In particular inspect the main jet in the body. Excessive wear is present if the body is oval. Also pay particular attention to the air valve and linkage, the throttle plate (butterfly), its spindle and the throttle linkages for wear. The diaphragm rubber must be in good condition and not split or perished. Check also that the metering rod spring is correctly fitted to the air valve (Fig. 3.16). Renew all gaskets and seals during assembly and ensure that the mating surfaces are perfectly clean.

Chapter 3 Fuel and exhaust systems

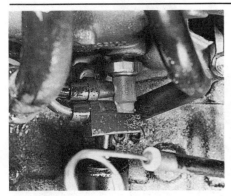
15.6A Disconnect the spark control system hoses and ...

15.6B ... the servo vacuum hose from the inlet manifold

15.9 Inlet manifold and rear engine lifting bracket

16.3 Exhaust manifold and hot air shroud

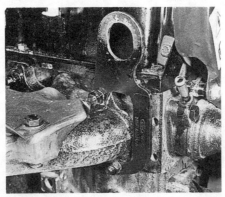
16.5 Exhaust manifold and front engine lifting bracket

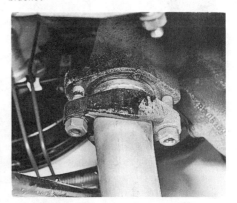
17.3 Exhaust downpipe and manifold flange

17.4 Exhaust system mounting bracket and rubber

17.6 Exhaust system flange joint

13 Commence assembly by refitting the accelerator pump. Locate the gasket face of the diaphragm towards the pump cover and when in position it must not be distorted at all. Fit the spring and cover and tighten the screws evenly.

14 Reconnect the diaphragm to the control linkage and retain by fitting the circlip. This is fiddly and requires a steady hand and a little patience. Check that the clip is fully engaged when in position. As the diaphragm is fitted ensure that the double holes on one corner align with the corresponding holes in the carburettor body (photo). With the diaphragm in position, relocate the housing and spring and insert the retaining screws to secure. Take care not to distort the diaphragm as the housing is tightened.

15 If removed, refit the mixture adjustment screw, but don't relocate the tamperproof plug yet as the mixture must be adjusted when the engine is restarted. Do not overtighten the screw! Back off the screw three full turns.

16 Insert the float needle, the float and the pivot pin. When installing the needle valve the spring-loaded ball must face towards the float.

17 Locate the accelerator pump ball and weight into the discharge gallery, fit a new gasket into position and then refit the main jet body. If the metering rod is already in position, retract and raise it to re-engage the main jet housing over the rod and then lower it into position. Do not force or bend the rod in any way during this operation. Tighten the jet body retaining screws. If the metering rod is still to be fitted, do not fully tighten the jet body retaining screws until after the jet is fitted and known to be centralised. If still to be fitted, slide the metering rod into position and screw it in until the rod shoulder aligns with the main body vertical face. Do not overtighten the rod. Should it bend during assembly, try re-centralising the main jet body, then tighten the retaining screws. Using a straight edge check that the main jet body alignment flanges are flush with the top face of the carburettor (Fig. 3.17). Where applicable fit the plug to the metering rod extraction hole in the carburettor body.

18 Position the new top cover gasket in position and refit the top cover. Tighten the retaining screws progressively and evenly (photo).

19 The auto choke housing can be refitted to complete assembly. Ensure that the body alignment marks correspond and as it is fitted engage the bi-metal coil with the middle choke lever slot (photo). Use a new gasket. Refit the three retaining screws (bottom screw first) and before tightening check that the body alignment markings correspond.

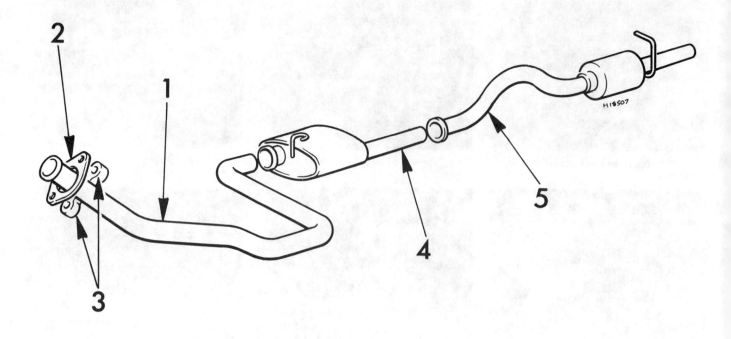

Fig. 3.18 Exhaust system (Sec 17)

1 Downpipe
2 Flange
3 Nuts
4 Front silencer and pipe
5 Rear silencer and pipe

15 Inlet manifold – removal and refitting

1 Disconnect the battery negative lead.
2 Drain the cooling system as described in Chapter 2.
3 Remove the air cleaner as described in Section 3.
4 Loosen the clips and disconnect the coolant hoses from the automatic choke cover and inlet manifold.
5 Note their position then disconnect the vacuum and fuel pipes from the carburettor and inlet manifold as necessary.
6 Disconnect the crankcase ventilation, spark control system, and brake servo vacuum pipes from the inlet manifold. Note that the brake servo pipe is attached with a union nut (photos).
7 Disconnect the accelerator cable with reference to Section 9.
8 Disconnect the wire from the anti-dieselling valve on the carburettor.
9 Unscrew the retaining nuts and bolts, noting the location of the rear engine lifting bracket (photo).
10 Withdraw the inlet manifold from the cylinder head and remove the gasket.
11 Unscrew the nuts and withdraw the carburettor from the inlet manifold. Remove the gasket.
12 Refitting is a reversal of removal, but first make sure that all mating faces are perfectly clean. Fit new gaskets and apply sealing compound either side of the water aperture. Fill the cooling system as described in Chapter 2 and refit the air cleaner as described in Section 3. Adjust the accelerator cable as described in Section 9. Adjust the slow running as described in Section 12.

16 Exhaust manifold – removal and refitting

1 Remove the air cleaner as described in Section 3.
2 Unscrew the nuts securing the downpipe to the manifold then lower the downpipe onto an axle stand and remove the gasket.
3 Unbolt the hot air shroud from the exhaust manifold noting that the heater hose bracket is located on the front bolt (photo).
4 Disconnect the HT leads from the spark plugs.
5 Unscrew the retaining nuts noting the location of the front engine lifting bracket (photo).
6 Withdraw the exhaust manifold from the cylinder head and remove the gaskets.
7 Refitting is a reversal of removal, but first make sure that all mating faces are perfectly clean. Fit new gaskets and tighten the nuts to the specified torque.

17 Exhaust system – checking, removal and refitting

1 Examine the exhaust system for leaks, damage and security at the intervals specified in the *Routine maintenance* Section at the start of this manual. To do this, apply the handbrake and allow the engine to idle. Lie down on each side of the vehicle in turn and check the full length of the exhaust system for leaks while an assistant temporarily places a wad of cloth over the tailpipe. If a leak is evident, stop the engine and use a proprietary repair kit such as Holts Flexiwrap and Holts Gun Gum exhaust repair systems. These can be used for effective repairs to exhaust pipes and silencer bores, including ends and bends. Holts Flexiwrap is an MOT-approved permanent exhaust repair. If the leak is excessive or damage is evident, renew the section. Check the rubber mountings for deterioration, and renew them if necessary.
2 To remove the exhaust system jack up the front and rear of the vehicle and support it on axle stands.
3 Unscrew the nuts from the manifold flange, lower the exhaust downpipe and remove the gasket (photo).
4 Disconnect the mounting rubbers and lower the complete system from the vehicle (photo).
5 Refitting is a reversal of removal, but clean the flange mating faces and fit a new gasket. Tighten the flange nuts to the specified torque.
6 The exhaust system sections are secured by U-bolt clamps or by flange joints (photo). To separate the sections, undo the flange retaining

Chapter 3 Fuel and exhaust systems

nuts or the U-bolt nuts as applicable then detach the joint. With the U-bolt coupling, the sections are sleeve jointed and these will separate by either soaking in penetrating oil, then gripping and pulling them apart, or by cutting them free with a hacksaw, or possibly by prising them free using a suitable screwdriver.

7 When reconnecting the pipe sections, apply a suitable jointing compound to the joint. Where U-bolt fixings are used, leave fully tightening the joints until after the pipe/system is fully relocated so that any adjustments to the front and/or rear sections can be made.

18 Unleaded fuel – general

1 All Transit vehicles covered by this manual may be operated on unleaded fuel provided that the following precautions are observed where applicable.
2 It is generally believed that continuous use of unleaded fuel can cause rapid wear of conventional cylinder head valve seats. Valve seat inserts which can tolerate unleaded fuel are fitted to some engines. These engines are identified as follows:

1.6 litre   'A', 'K', 'M', or 'MM' stamped on the cylinder head adjacent to No 4 spark plug
2.0 litre   'A', 'L', 'P', or 'PP' stamped on the cylinder head adjacent to No 4 spark plug

3 Engines which are marked as above can run entirely on unleaded fuel.
4 Engines which are not fitted with the special valve seat inserts can still be run on unleaded fuel, but one tankful of leaded fuel should be used for every four tankfuls of unleaded. This will protect the valve seat inserts.

Note: *On all vehicles, the ignition timing will have to be retarded when unleaded fuel is used. See Chapter 4, Specifications.*

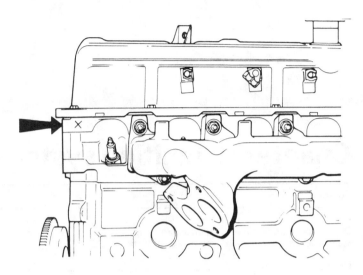

Fig. 3.19 Identification letter position (arrowed) for cylinder heads with valve seat inserts (Sec 18)

19 Fault diagnosis – fuel and exhaust systems

Symptom	Reason(s)
Excessive fuel consumption	Air cleaner element choked Leaks in carburettor, fuel tank or fuel lines Float level incorrect Mixture adjustment incorrect Excessively worn carburettor
Insufficient fuel supply or weak mixture	Faulty fuel pump Mixture adjustment incorrect Leaking inlet manifold or carburettor gasket Leaking fuel line
Difficulty starting	Faulty automatic choke or out of adjustment

Chapter 4 Ignition system

Contents

Coil – description and testing	5	Ignition timing – adjustment	4
Distributor – removal, examination and refitting	3	Routine maintenance	2
Electronic amplifier module – removal and refitting	7	Spark control system – general description	8
Fault diagnosis – ignition system	10	Spark control system components – removal and refitting	9
General description	1	Spark plugs and HT leads – general	6

Specifications

General

System type .... Electronic, breakerless distributor, with mechanical and vacuum control. Separate electronic amplifier module

Firing order .... 1–3–4–2 (No 1 at timing belt end)

Distributor

Type .... Bosch electronic breakerless
Drive/rotation .... From skew gear on auxiliary shaft/clockwise (viewed from above)
Dwell angle .... Automatic control via electronic module
Trigger coil resistance .... 1000 to 1200 ohm

Ignition timing

Initial, static, or at idling speed with vacuum hose(s) detached and plugged:

    1.6 and 2.0 litre .... 6° BTDC (or 2° BTDC for use with unleaded fuel)

Additional no-load advance (at 2000 rpm, vacuum hose(s) attached):

Engine/distributor	Mechanical	Vacuum	Total
1.6/Bosch 84HF-12100-AB	9.5° to 17.8°	13.0° to 21.0°	22.5° to 38.8°
1.6/Bosch 86HF-12100-AB	9.5° to 18.0°	13.0° to 21.0°	22.5° to 39.0°
2.0/Bosch 84HF-12100-CA	6.0° to 15.0°	13.0° to 21.0°	19.0° to 36.0°
2.0/Bosch 86HF-12100-EA	5.5° to 14.5°	9.0° to 17.0°	14.5° to 31.5°
2.0/Bosch 86HF-12100-CA	5.5° to 14.5°	13.0° to 21.0°	18.5° to 35.3°

Note that the figures quoted are in crankshaft degrees and do not include the initial static advance

Coil

Output (minimum) .... 25.0 kilovolt
Primary winding resistance .... 0.72 to 0.88 ohm
Secondary winding resistance .... 4500 to 7000 ohm

Spark plugs

Make and type (all models) .... Champion F7YCC or F7YC
Electrode gap:
    Champion F7YCC .... 0.8 mm (0.032 in)
    Champion F7YC .... 0.7 mm (0.028 in)

HT leads

Type .... Champion CLS 4, boxed set
Maximum resistance per lead .... 30 000 ohm

Torque wrench settings

	Nm	lbf ft
Spark plugs	20 to 28	15 to 21

Chapter 4 Ignition system

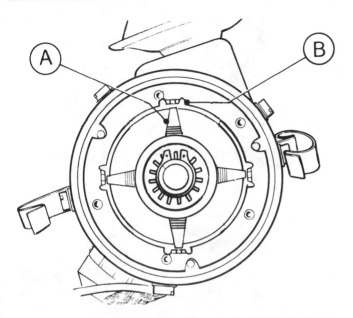

Fig. 4.1 Magnetic reluctance components in the Bosch distributor (Sec 1)

A Trigger rotor   B Stator pick-up post

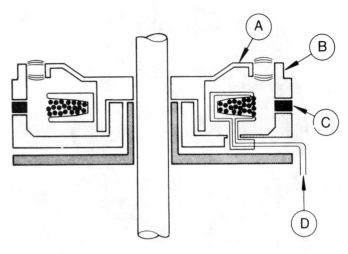

Fig. 4.2 Cross section diagram of the Bosch magnetic reluctance system (Sec 1)

A Trigger rotor   C Permanent magnet
B Stator pick-up post   D Wires

1 General description

The ignition system is divided into two circuits, low tension (primary) and high tension (secondary). The low tension circuit consists of the battery, ignition switch, primary coil windings, electronic amplifier module and the signal generating system inside the distributor. The signal generating system comprises the trigger coil, trigger wheel, stator, permanent magnets and stator pick-up. The high tension circuit consists of the secondary coil windings, the heavy ignition lead from the centre of the distributor cap to the coil, the rotor arm and the spark plug leads and spark plugs.

When the system is in operation, low tension voltage is changed in the coil into high tension voltage by the action of the electronic amplifier module in conjunction with the signal generating system. As each of the trigger wheel teeth pass through the magnetic field created around the trigger coil in the distributor, a change in the magnetic field force (flux) is created which induces a voltage in the trigger coil. This voltage is passed to the electronic amplifier module which switches off the ignition coil primary circuit. This results in the collapse of the magnetic field in the coil which generates the high tension voltage. The high tension voltage is then fed via the carbon brush in the centre of the distributor cap to the rotor arm. The voltage passes across to the appropriate metal segment in the cap and via the spark plug lead to the spark plug where it finally jumps the spark plug gap to earth.

The ignition advance is a function of the distributor and is controlled both mechanically and by a vacuum operated system. The mechanical governor mechanism consists of two weights which move out from the distributor shaft as the engine speed rises due to centrifugal force. As they move outwards, they rotate the trigger wheel relative to the distributor shaft and so advance the spark. The weights are held in position by two light springs and it is the tension of the springs which is largely responsible for correct spark advancement.

The vacuum control consists of a diaphragm, one side of which is indirectly connected via a small bore hose to the carburettor inlet manifold, and the other side to the distributor. Depression in the inlet manifold and/or carburettor, which varies with engine speed and throttle position, causes the diaphragm to move, so moving the baseplate and advancing or retarding the spark. A fine degree of control is achieved by a spring in the diaphragm assembly. Additionally a spark control system utilizing vacuum valves and water temperature sensitive control valves is used to modify the spark advance according to engine temperature. Further details of this system will be found in Section 8.

2.1 Distributor cap showing HT segments and spring tensioned carbon brush (in centre)

Warning: *When working on the electronic ignition system remember that the high tension voltage can be considerably higher than on a conventional system and in certain circumstances could prove fatal. Depending on the position of the distributor trigger components it is also possible for a single high tension spark to be generated simply by knocking the distributor with the ignition switched on. It is therefore important to keep the ignition system clean and dry at all times, and to make sure that the ignition switch is off when working on the engine.*

2 Routine maintenance

1 At the intervals specified in *Routine maintenance* at the beginning of this manual, remove the distributor cap and thoroughly clean it inside and out with a dry lint-free cloth. Examine the four HT lead segments inside the cap. If the segments appear badly burnt or pitted, renew the

Chapter 4 Ignition system

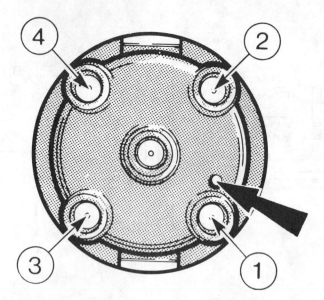

Fig. 4.3 Distributor cap and lead terminals, with No 1 terminal identification mark arrowed (Sec 3)

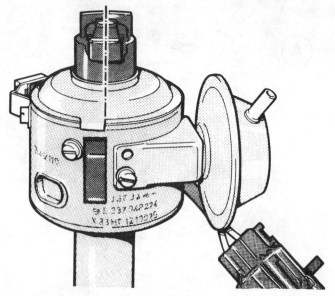

Fig. 4.4 Alignment of distributor rotor arm prior to installation (Sec 3)

cap. Make sure that the carbon brush in the centre of the cap is free to move and that it protrudes significantly from its holder (photo).
2 Remove, clean, regap and test the spark plugs as described in Section 6. Renew the plugs in sets of four if any show signs of failure when tested, or if under visual inspection the end of the centre electrode is rounded off, or the earth electrode is eroded. The plugs must be renewed, regardless of visual appearance, if they have exceeded their recommended service life. Also check the spark plug HT leads for any signs of corrosion of the end fittings, which if evident should be carefully cleaned away. Wipe the leads clean (including the coil lead) over their entire length before refitting.
3 Check the condition and security of all leads and wiring associated with the ignition system. Make sure that no chafing is occurring on any of the wires and that all connections are secure, clean and free from corrosion.

3 Distributor – removal, examination and refitting

Note: *During manufacture the engine ignition timing is set using a microwave process, and sealant applied to the distributor clamp bolt. Removal of the distributor should be avoided except where excessive bearing wear has occurred due to high mileage or during major engine overhaul.*

1 Disconnect the battery negative lead.
2 Disconnect the HT leads from the spark plugs by pulling on the connectors, not the leads. Note that the location of No 1 HT lead is marked on the distributor cap by a small indentation. The remaining HT lead locations are shown in Fig. 4.3.
3 Prise away the spring clips with a screwdriver and remove the distributor cap (photo).
4 Pull the HT lead from the coil, slide the rubber HT lead holder from the valve cover clip, and withdraw the distributor cap (photo).
5 Disconnect the vacuum advance pipes (photo).
6 Turn the engine so that the specified initial static timing notch on the crankshaft pulley is aligned with the timing pointer on the timing cover. Use a socket on the crankshaft pulley bolt to turn the engine.
7 Pull off the rotor arm and remove the dust cover. Then refit the rotor arm and check that it is pointing towards the scribed line on the body rim (photo).
8 Disconnect the distributor multi-plug by depressing the spring tensioned plates (photo).
9 Mark the distributor body and cylinder block in relation to each other.
10 Scrape the sealant from the distributor clamp bolt then unscrew and remove the bolt and clamp (photo).

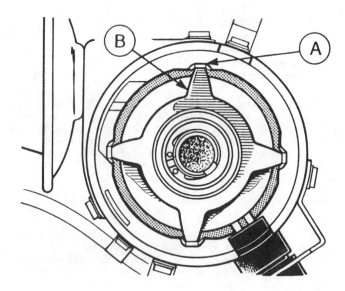

Fig. 4.5 Align the trigger wheel arms (B) with the stator arms (A) (Sec 3)

11 Withdraw the distributor from the engine (photo). As it is removed it will be noticed that the rotor arm will turn clockwise due to the skew gear drive. Mark the new position of the rotor arm on the body rim.
12 Check the distributor spindle for excessive side-to-side movement. If evident, the distributor must be renewed as it is not possible to obtain individual components.
13 To refit the distributor, first check that the timing marks on the crankshaft pulley and timing cover are still aligned.
14 Turn the rotor arm to the position shown in Fig. 4.4. This position is approximately 40° clockwise from the scribed line on the body rim and should coincide with the mark made in paragraph 11 (photo).
15 Hold the distributor directly over the aperture in the cylinder block with the previously made marks aligned, then lower it into position. As the skew gear drive meshes, the rotor arm will turn anti-clockwise.
16 With the distributor fully entered and the body mark aligned with the mark on the block, check that the rotor arm is pointing towards the scribed line on the body rim.
17 The rotor arm should point away from the engine at 90° to the centre line of the crankshaft. Also with the rotor arm and dust cover removed, the trigger rotor arms should be aligned with the stator

Chapter 4 Ignition system

3.3 Distributor cap retaining spring clip (Bosch)

3.4 Removing the HT lead rubber holder

3.5 Disconnecting the vacuum advance pipe

3.7 Rotor arm aligned with the BTDC mark (Bosch)

3.8 Wiring multi-plug

3.10 Unscrewing the distributor clamp bolt

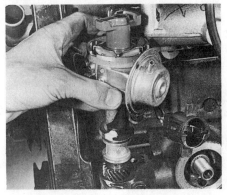
3.11 Removing the distributor (engine removed for clarity)

3.14 Upper view of Bosch distributor rotor arm alignment prior to installation

4.6 Showing crankshaft pulley timing marks (at TDC) and timing cover pointer

pick-up posts when the engine is turned to align the initial static advance timing marks (Fig. 4.5).
18 Refit the clamp, then insert and tighten the bolt.
19 Refit the multi-plug, reconnect the vacuum advance pipes and refit the rotor arm and dust cover.
20 Refit the distributor cap and reconnect the HT leads to the spark plugs and coil.
21 Reconnect the battery negative lead.
22 Check and adjust the ignition timing as described in Section 4.

4 Ignition timing – adjustment

Note: *During manufacture the engine ignition timing is set using a microwave process, and sealant applied to the distributor clamp bolt. Because the electronic components require no maintenance, checking the ignition timing does not constitute part of the routine maintenance, and the procedure is therefore only necessary after removal and refitting of the distributor.*

1 Disconnect and plug the vacuum pipe(s) at the distributor.
2 Wipe clean the crankshaft pulley notches and timing cover pointer. If necessary, use white paint or chalk to highlight the marks.
3 Connect a stroboscopic timing light to the engine in accordance with the manufacturer's instructions (usually to No 1 spark plug HT lead).
4 Connect a tachometer to the engine in accordance with the manufacturer's instructions.
5 Start the engine and run it to normal operating temperature, then allow it to idle at the recommended idling speed (see Chapter 3).
6 Point the timing light at the timing marks on the crankshaft pulley and check that the appropriate initial static timing notch is in alignment with the pointer on the timing cover. Note that in relation to pulley rotation the last heavy notch indicates TDC (top dead centre) and the preceding heavy notches are at 4° (also 4 mm) intervals (photo).
7 If adjustment is necessary (ie the appropriate notch is not aligned with the pointer) loosen the distributor clamp bolt and turn the distributor body anti-clockwise to advance the timing or clockwise to retard the timing. Tighten the clamp bolt when the setting is correct.
8 Run the engine at 2000 rpm and check that the *additional* centrifugal advance is within the limits given in Specifications.

Chapter 4 Ignition system

5.1 Ignition coil

6.3 Spark plug removal with correct type of socket is essential

6.10 Screw spark plugs into position initially by hand to avoid possibility of cross-threading

7.2 Ignition system electronic module (located under coil)

8.1 Typical arrangement of the spark control system showing PVS, spark delay/sustain valve and fuel traps

9.5 Vacuum line connections at the ported vacuum switch

9   Reconnect the vacuum pipe(s) then run the engine at 2000 rpm and check that the *total* advance is within the limits given in Specifications.
10   If the results obtained from paragraphs 8 and 9 are correct, it can be assumed that the vacuum advance is also correct. If the result of paragraph 8 is correct, but the result of paragraph 9 is below limits, check the vacuum hoses and vacuum unit for condition and security.
11   Switch off the engine and remove the tachometer and timing light.

5 Coil – description and testing

1   The coil is located on the left-hand side of the engine compartment and is retained by a metal strap (photo). It is of high output type and the HT tower should be kept clean at all times to prevent possible arcing. Bosch and Femsa coils are fitted with protective plastic covers and Polmot coils are fitted with an internal fusible link.
2   To ensure the correct HT polarity at the spark plugs, the LT coil leads must always be connected correctly. The black lead must always be connected to the terminal marked +/15, and the green lead to the terminal marked –/1. Incorrect connections can cause bad starting, misfiring, and short spark plug life.
3   To test the coil first disconnect the LT and HT leads. Connect an ohmmeter between both LT terminals and check that the primary winding resistance is as given in the Specifications. Connect the ohmmeter between the HT terminal (terminal 4) and either LT terminal and check that the secondary winding resistance is as given in the Specifications. Reconnect the leads after making the test.

6 Spark plugs and HT leads – general

1   The correct functioning of the spark plugs is vital for the correct running and efficiency of the engine. It is essential that the plugs fitted are appropriate for the engine, and the suitable type is specified at the beginning of this chapter. If this type is used and the engine is in good condition, the spark plugs should not need attention between scheduled replacement intervals. Spark plug cleaning is rarely necessary and should not be attempted unless specialised equipment is available as damage can easily be caused to the firing ends.
2   To remove the spark plugs, disconnect the HT leads by pulling on the connectors, not the leads. If necessary identify the leads for position. For better access, remove the air cleaner with reference to Chapter 3.
3   Clean the area around each spark plug using a small brush, then using a plug spanner (preferably with a rubber insert) unscrew and remove the plugs (photo). Cover the spark plug holes with a clean rag to prevent the ingress of any foreign matter.
4   The condition of the spark plugs will tell much about the overall condition of the engine.
5   If the insulator nose of the spark plug is clean and white, with no deposits, this is indicative of a weak mixture, or too hot a plug. (A hot plug transfers heat away from the electrode slowly – a cold plug transfers it away quickly).
6   If the tip and insulator nose is covered with hard black-looking deposits, then this is indicative that the mixture is too rich. Should the plug be black and oily, then it is likely that the engine is fairly worn, as well as the mixture being too rich.
7   If the insulator nose is covered with light tan to greyish brown deposits, then the mixture is correct and it is likely that the engine is in good condition.
8   The spark plug gap is of considerable importance, as, if it is too large or too small, the size of the spark and its efficiency will be seriously impaired. The spark plug gap should be set to the figure given in the Specifications at the beginning of this Chapter. To set it, measure the gap with a feeler gauge, and then bend open, or close the *outer* plug electrode until the correct gap is achieved. The centre electrode should *never* be bent as this may crack the insulation and cause plug failure, if nothing worse.
9   Before fitting the spark plugs check that the threaded connector sleeves are tight and that the plug exterior surfaces are clean. As the plugs incorporate taper seats also make sure that the 18 mm threads and seats are clean.

Chapter 4 Ignition system

10 Screw in the spark plugs by hand then tighten them to the specified torque. *Do not exceed the torque figure* (photo).
11 Push the HT leads firmly onto the spark plugs and where applicable refit the air cleaner.
12 The HT leads and distributor cap should be cleaned and checked at the intervals given in the *Routine maintenance* Section at the start of this manual. To test the HT leads remove them together with the distributor cap, then connect an ohmmeter to each end of the leads and the appropriate terminal within the cap in turn. If the resistance is greater than the maximum given in the Specifications, check that the lead connection in the cap is good before renewing the lead (Fig. 4.6).

7 Electronic amplifier module – removal and refitting

Note: *Do not run the engine with the module detached from the body panel as the body acts as an effective heat sink and therefore damage may occur through internal overheating.*
1 Disconnect the battery negative lead.
2 Disconnect the multi-plug from the module. Pull on the multi-plug and not on the wiring (photo).
3 Remove the plastic retainers and withdraw the module.
4 Refitting is a reversal of removal, but make sure that the underside of the module and the corresponding area of the body panel are clean.

8 Spark control system – general description

1 To improve driveability during warm-up conditions and to keep exhaust emission levels to a minimum, a vacuum-operated, temperature-sensitive spark control system is fitted to certain Transit models (photo). The system is designed to ensure that the rate of distributor vacuum advance is compatible with the change in fuel/air mixture flow under all throttle conditions, thus resulting in more complete combustion and reduced exhaust emissions.
2 Under part throttle cruising conditions, distributor vacuum advance is required to allow time for the fuel/air mixture in the cylinders to burn. When returning to a part throttle opening after accelerating or decelerating, the distributor vacuum increases before the fuel/air mixture has stabilised. On certain engines this can lead to short periods of incomplete combustion and increased exhaust emission. To overcome this condition a spark delay valve is incorporated in the vacuum line between the carburettor and distributor to reduce the rate at which the distributor advances. Under certain conditions, particularly during the period of engine warm-up, some engines may suffer from a lack of throttle response. This problem is overcome by the incorporation of a spark sustain valve either individually or in conjunction with the spark delay valve. This valve is used to maintain distributor vacuum under transient throttle conditions, thus stabilising the combustion process.
3 The operation of the valves is controlled by a ported vacuum switch (PVS) which has the vacuum lines connected to it. The PVS operates in a similar manner to that of the thermostat in the cooling system. A wax filled sensor is attached to a plunger which operates a valve. The PVS is actuated by the engine coolant and is sensitive to changes in engine operating temperature. As the engine warms up and coolant temperature increases, the wax expands causing the plunger to move within the switch to open or close the vacuum and outlet ports accordingly. In this way the vacuum supply to the sustain/delay valves can be applied or shut off according to engine temperature.
4 The actual arrangement, and number of valves and switches varies considerably according to model, equipment fitted and operating territory. Additionally, one or more fuel traps and one-way valves may be included to prevent fuel vapour from being drawn into the vacuum lines and to further control the application of vacuum.

9 Spark control system components – removal and refitting

Spark delay/sustain valve
1 Disconnect the vacuum lines at the valve and remove the valve from the engine.
2 When refitting a spark delay valve it must be positioned with the black side (marked CARB) towards the carburettor and the coloured

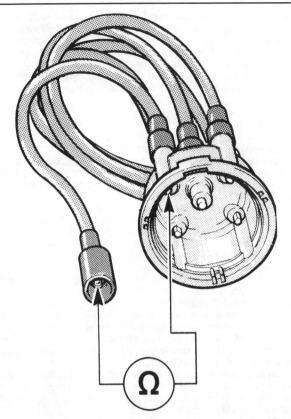

Fig. 4.6 Method of testing an HT lead with an ohmmeter (Sec 6)

side (marked DIST) towards the distributor.
3 When refitting a spark sustain valve the side marked VAC must be toward the carburettor and the side marked DIST towards the distributor.

Ported vacuum switch
4 Remove the cap from the expansion tank to reduce pressure in the cooling system. If the engine is hot, remove the cap slowly using a rag to prevent scalding.
5 Disconnect the vacuum lines after marking them for location (photo).
6 Unscrew the valve from its adaptor and remove it from the engine.
7 Refitting is the reversal of removal. On completion check, and if necessary top up the cooling system.

Fuel trap
8 Disconnect the vacuum lines and remove the fuel trap from the engine.
9 When refitting, make sure that the fuel trap is positioned with the black side (marked CARB) towards the carburettor and the white side (marked DIST) towards the PVS.

Spark control system additional components
10 According to model, engine and equipment, additional components such as check valves or solenoids may also be fitted as part of the spark control system.
11 The removal and refitting of these components is basically the same as previously described and providing that all attachments are marked for position prior to removal, no problems should be encountered.

10 Fault diagnosis – ignition system

The electronic ignition fitted is far less likely to cause trouble than the contact breaker type fitted to many cars, largely because the low tension circuit is electronically controlled. However the high tension

Chapter 4 Ignition system

circuit remains identical and therefore the associated faults are the same. There are two main symptoms indicating ignition faults. Either the engine will not start or fire, or the engine is difficult to start and misfires. If it is a regular misfire, ie, the engine is only running on two or three cylinders, the fault is almost sure to be in the secondary, or high tension circuit. Loss of power and overheating, apart from incorrect carburettor settings, are normally due to incorrect ignition timing.

Engine fails to start

1   If the starter motor fails to turn the engine check the battery and starter motor with reference to Chapter 12.

2   Disconnect an HT lead from any spark plug and hold the end of the cable approximately 5 mm (0.2 in) away from the cylinder head using *well insulated pliers*. While an assistant spins the engine on the starter motor, check that a regular blue spark occurs. If so, remove, clean, and re-gap the spark plugs as described in Section 6.

3   If no spark occurs, disconnect the main feed HT lead from the distributor cap and check for a spark as in paragraph 2. If sparks now occur, check the distributor cap, rotor arm, and HT leads as described in Sections 2 and 6, and renew them as necessary.

4   If no sparks occur check the resistance of the main feed HT lead as described in Section 6 and renew as necessary. Should the lead be serviceable check that all wiring and multi-plugs are secure on the electronic module and distributor.

5   Disconnect the wire to the coil terminal marked + /15 and connect a voltmeter or 12 volt test lamp between the wire and an earth point. With the ignition switched on (position II), battery voltage should be registered or the testlamp should light. If not, check the ignition switch and wiring with reference to Chapter 12.

6   Disconnect both wiring connectors from the coil LT terminals and connect a 12 volt, 21 watt test bulb between the connectors. Spin the engine on the starter and check that the bulb flashes, If so, the coil is proved faulty and should be renewed.

7   If the bulb does not flash, disconnect the distributor multi-plug and connect an ohmmeter between the two parallel pins. If the trigger coil resistance is not as given in the Specifications, renew the distributor complete.

Engine misfires

8   If the engine misfires regularly, run it at a fast idling speed. Pull off each of the plug HT leads in turn and listen to the note of the engine. *Hold the plug leads with a well insulated pair of pliers as protection against a shock from the HT supply.*

9   No difference in engine running will be noticed when the lead from the defective circuit is removed. Removing the lead from one of the good cylinders will accentuate the misfire.

10   Remove the plug lead from the end of the defective plug and hold it about 5 mm (0.2 in) away from the cylinder head. Restart the engine. If the sparking is fairly strong and regular, the fault must lie in the spark plug.

11   The plug may be loose, the insulation may be cracked, or the points may have burnt away, giving too wide a gap for the spark to jump. Worse still, one of the points may have broken off. Either renew the plug, or clean it, reset the gap, and then test it.

12   If there is no spark at the end of the plug lead, or if it is weak and intermittent check the HT lead from the distributor to the plug. If the insulation is cracked or perished of if its resistance is incorrect, renew the lead. Check the connections at the distributor cap.

13   If there is still no spark, examine the distributor cap carefully for tracking. This can be recognised by a very thin black line running between two or more electrodes, or between an electrode and some other part of the distributor. These lines are paths which now conduct electricity across the cap, thus letting it run to earth. The only answer in this case is a new distributor cap. Tracking will also occur if the inside or outside of the distributor cap is damp. If this is evident use a proprietary water repellant spray such as Holts Wet Start. To prevent the problem recurring, Holts Damp Start may be used to supply a sealing coat, so excluding any further moisture from the ignition system. In extreme difficulty, Holts Cold Start will help to start a car when only a very poor spark occurs.

Chapter 5 Clutch

Contents

Clutch – checking the adjustment	3	Clutch pedal – removal and refitting	5
Clutch – inspection	7	Clutch release bearing and lever – removal and refitting	8
Clutch – refitting	9	Fault diagnosis – clutch	10
Clutch – removal	6	General description	1
Clutch cable – removal and refitting	4	Routine maintenance	2

Specifications

Type .... Single dry plate, diaphragm spring, cable operated. Automatic adjustment

Clutch plate

Diameter .... 216 mm (8.5 ln), 228.6 mm (9.0 in) or 242 mm (9.5 in)

Lining thickness (new) .... 7.25 mm (0.285 in), 8.40 mm (0.330 in), 9.30 mm (0.366 in) or 3.7 mm (0.145 in) *

MT75 transmission only

Torque wrench settings

	Nm	lbf ft
Clutch cover:		
MT75 transmission models	21 to 28	16 to 20
Other transmission types	16 to 20	12 to 15

1 General description

The clutch is of single dry plate type with a diaphragm spring pressure plate. The unit is dowelled and bolted to the rear face of the flywheel.

The clutch plate (or disc) is free to slide along the splined input shaft and is held in position between the flywheel and the pressure plate by the pressure of the pressure plate spring. Friction lining material is riveted to the clutch plate and it has a spring cushioned hub to absorb transmission shocks and to help ensure a smooth take off.

The circular diaphragm spring is mounted on shoulder pins and held in place in the cover by two fulcrum rings. The spring is also held to the pressure plate by three spring steel clips which are riveted in position.

The clutch is actuated by a cable controlled by the clutch pedal. Wear of the friction linings is compensated for by an automatic pawl and quadrant adjuster on the top of the clutch pedal. The clutch release mechanism consists of a ball bearing which slides on a guide sleeve at the front of the gearbox, and a release lever which pivots inside the clutch bellhousing.

Depressing the clutch pedal actuates the clutch release lever by means of the cable. The release lever pushes the release bearing forwards to bear against the release fingers so moving the centre of the diaphragm spring inwards. The spring is sandwiched between two annular rings which acts as fulcrum points. As the centre of the spring is pushed in, the outside of the spring is pushed out, so moving the pressure plate backwards and disengaging the pressure plate from the clutch plate.

When the clutch pedal is released the diaphragm spring forces the pressure plate into contact with the friction linings on the clutch plate

and at the same time pushes a fraction of an inch forwards on its splines so engaging the clutch plate with the flywheel. The clutch plate is now firmly sandwiched between the pressure plate and the flywheel so the drive is taken up.

2 Routine maintenance

1 The only maintenance requirement as far as the clutch is concerned is to periodically lubricate the cable to release lever connection. Working underneath the vehicle, peel back the rubber boot to gain access to the cable end, then smear the cable to lever connection with grease. Refit the rubber boot to complete.

2 Although not normally required during routine service procedures, the clutch adjustment can be checked as described in the following Section.

3 Clutch – checking the adjustment

1 The clutch has an automatic adjustment mechanism located in the pedal assembly and this eliminates the need for periodic adjustment. However, it is possible for the self-adjusting mechanism on the clutch pedal to malfunction, because of incorrect setting previously, so that the clutch cable is over-tensioned. Clutch slip and rapid wear will result. The adjustment may be checked as follows.

Chapter 5 Clutch

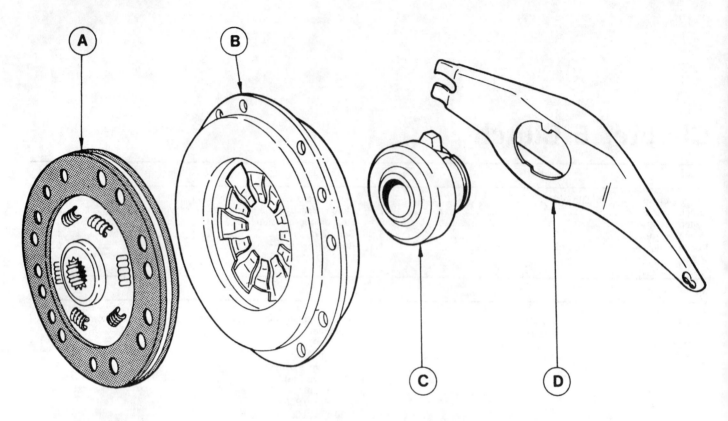

Fig. 5.1 Principal clutch components (Sec 1)

A Clutch plate
B Cover assembly
C Release bearing
D Release lever

2 Have an assistant lift the clutch pedal to the top of its travel. With the pedal held in this position, grasp the outer cable where it meets the bulkhead and check that it can be pulled away from the bulkhead by hand. If so, adjustment is correct.
3 If the cable tension is excessive, try to free the adjuster pawl by jerking the inner cable sharply away from the bulkhead. If this does not work, a new cable will be required.
4 Free the old cable by cutting through the inner at the release lever. **Wear safety glasses during this procedure.** Remove the cable from the release lever and pedal.
5 Fit the new cable with reference to Section 4, and have the assistant hold the clutch pedal up against the stop until the inner has been connected, both to the pedal and to the release arm.
6 Operate the pedal a few times, then check the adjustment as just described.

4 Clutch cable – removal and refitting

1 Raise the front of the vehicle and support it on axle stands at a height which will allow comfortable working space underneath. Check that the handbrake is fully applied and chock the rear wheels.
2 Unhook the inner cable from the toothed segment of the automatic adjustment mechanism on the clutch pedal (photo).
3 From under the vehicle, detach the clutch release lever rubber boot from the side of the clutch housing and slide it up the cable (photo).
4 Slip the cable out of the slot on the release lever then slide the rubber boot off the cable.
5 Withdraw the cable through the bulkhead and remove it from the engine compartment side.
6 To refit the cable, feed it through the bulkhead then slide the rubber

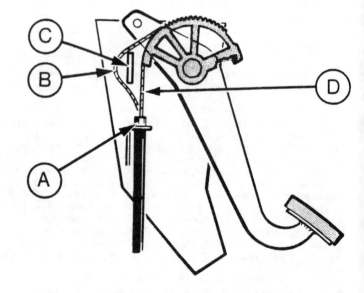

Fig. 5.2 Clutch cable routing (Sec 4)

A Cable abutment
B Incorrect cable route
C Pawl disengagement bracket
D Correct cable route

boot over the cable at the release lever end. Attach the cable to the release lever and refit the rubber boot.
7 Position a suitable wooden block under the clutch pedal to raise it

Chapter 5 Clutch

Fig. 5.4 Correct orientation of clutch pedal pawl spring (C) – pawl not shown for clarity (Sec 5)

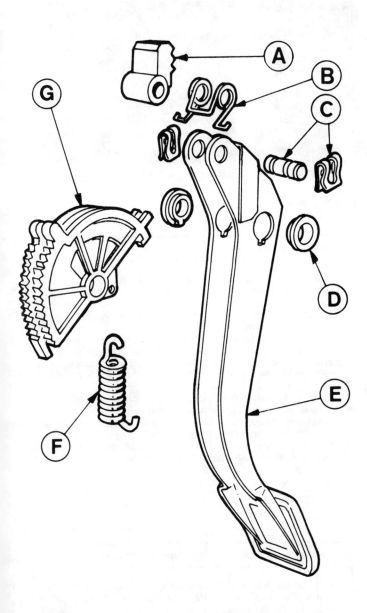

Fig. 5.5 Clutch pedal pawl orientation (Sec 5)

A Correct fitting
B Incorrect fitting
C Pawl

Fig. 5.3 Exploded view of the clutch pedal and automatic adjuster components (Sec 5)

A Pawl
B Spring
C Pawl pin and clip
D Pedal shaft bush
E Clutch pedal
F Toothed segment tension spring
G Toothed segment

fully so that the automatic adjuster pawl is held clear of the toothed segment.
8 Connect the cable to the toothed segment ensuring that it is correctly routed as shown in Fig. 5.2.
9 Remove the wooden block and depress the pedal several times to adjust the cable. Lower the vehicle to the ground.

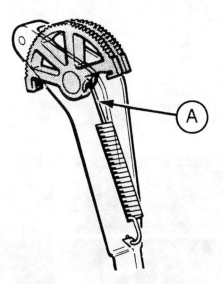

Fig. 5.6 Correct fitting of toothed segment tension spring (A) with long end section toward segment (Sec 5)

5 Clutch pedal – removal and refitting

1 Unhook the clutch inner cable from the toothed segment of the automatic adjustment mechanism on the clutch pedal.
2 Prise free the clutch pedal shaft retainer clip, remove the washer and then withdraw the pedal from the shaft.

3 To dismantle the pedal, extract the two bushes, withdraw the toothed adjuster segment and the tension spring. Extract the clips, withdraw the pin and remove the pawl and spring (but note how they are fitted).

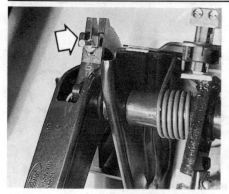
4.2 Clutch cable connection (arrowed) at the pedal end

4.3 Clutch release lever, cable and rubber boot attachments at the gearbox end

6.4 Removing the clutch cover from the flywheel

4 Renew any worn or damaged parts.
5 Fit the pawl, spring, pin and retaining clip. Ensure that the spring and pawl are correctly fitted as shown in Fig. 5.4 and 5.5.
6 Engage the toothed segment tension spring to the segment, and the segment to the pedal. Ensure that the spring is fitted as shown in Fig. 5.6.
7 Locate the new bushes in the pedal.
8 Lift the pawl and pivot the segment so that the pawl teeth rest against the small curved section of the segment, then engage the tension spring with the pedal.
9 Fit the pedal and the automatic adjuster unit into position and secure with the washer and securing clip.
10 Reconnect the cable to the pedal as described in Section 4.

7 Clutch – inspection

1 Examine the surfaces of the pressure plate and flywheel for scoring. If this is only light, the parts may be re-used. But, if scoring is excessive the clutch cover must be renewed and the flywheel friction face reground provided the amount of metal being removed is minimal. If any doubt exists renew the flywheel.
2 Renew the clutch plate if the linings are worn down to or near the rivets. If the linings appear oil stained, the cause of the oil leak must be found and rectified. This is most likely to be a failed gearbox input shaft oil seal or crankshaft rear oil seal. Check the clutch plate hub and centre splines for wear.
3 Examine the clutch cover and diaphragm spring for wear which will be indicated by loose components. If the diaphragm spring has any blue discoloured areas, the clutch has probably been overheated at some time and the clutch cover should be renewed.
4 Spin the release bearing in the clutch housing and check it for roughness. Hold the outer race and attempt to move it laterally against the inner race. If any excessive movement or roughness is evident, renew the release bearing as described in Section 8.

6 Clutch – removal

1 The clutch may be removed by two alternative methods. Either remove the engine (Chapter 1) or remove the gearbox (Chapter 6). Unless the engine requires a major overhaul or the crankshaft rear oil seal requires renewal, it is easier and quicker to remove the gearbox.
2 With a file or scriber, mark the relative positions of the clutch cover and flywheel which will ensure identical positioning on refitting. This is not necessary if a new clutch is to be fitted.
3 Unscrew, in a diagonal and progressive manner, the six bolts and spring washers that secure the clutch cover to the flywheel. This will prevent distortion of the cover and also prevent the cover from suddenly flying off due to binding on the dowels.
4 With all the bolts removed lift the clutch assembly from the locating dowels (photo). Note which way round the clutch plate is fitted and lift it from the clutch cover.

8 Clutch release bearing and lever – removal and refitting

1 With the gearbox and engine separated to provide access to the clutch, attention can be given to the release bearing located in the clutch housing, over the input shaft.
2 If the gearbox is still in the car, remove the rubber boot and disconnect the clutch cable from the release lever with reference to Section 4.
3 Free the release bearing from the lever and withdraw it from the guide sleeve (photo).

8.3 Remove the clutch release bearing ...

8.4 ... and lever

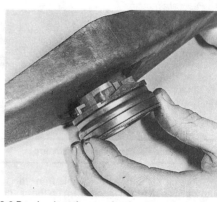
8.6 Bearing location on clutch release lever

Chapter 5 Clutch

9.2 Clutch plate orientation is critical

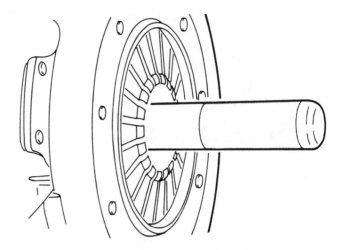

Fig. 5.7 Centralising the clutch plate (Sec 9)

4 Pull the release lever from the fulcrum pin, then withdraw the lever over the input shaft (photo).
5 Check the release bearing as described in Section 7. If there are any signs of grease leakage, renew the bearing.
6 Refitting is a reversal of removal. Ensure that the bearing is correctly located in the release lever (photo) and that the release lever forked end engages with the fulcrum pin.

9 Clutch – refitting

1 It is important that no oil or grease gets on the clutch plate friction linings, or the pressure plate and flywheel faces. It is advisable to refit the clutch with clean hands and to wipe down the pressure plate and flywheel faces with a clean rag before assembly begins.

2 Place the clutch plate against the flywheel, ensuring that it is the correct way round. The projecting torsion spring hub should be furthest from the flywheel, and the 'flywheel side' or 'Schwungradseite' mark towards the flywheel (photo).
3 Fit the clutch cover assembly loosely on the dowels with the previously made marks aligned where applicable. Insert the six bolts and spring washers and tighten them finger tight so that the clutch plate is gripped but can still be moved.
4 The clutch plate must now be centralised so that when the engine and gearbox are mated, the gearbox input shaft splines will pass through the splines in the centre of the clutch plate. Ideally a universal clutch centralising tool should be used or if available an old gearbox input shaft. Alternatively a wooden mandrel can be made.
5 Make sure that the centralising tool is located correctly in the crankshaft spigot bearing then tighten the cover bolts progressively in a diagonal sequence to the specified torque. Remove the tool.
6 Refit the gearbox (Chapter 6) or engine (Chapter 1) as applicable.

10 Fault diagnosis – clutch

Symptom	Reason(s)
Judder when taking up drive	Worn clutch plate friction surfaces or contamination with oil Worn splines on clutch plate or gearbox input shaft Loose engine or gearbox mountings
Clutch drag (failure to disengage)	Clutch plate sticking on input shaft splines Gearbox input shaft seized in crankshaft spigot bearing Faulty automatic cable adjuster
Clutch slip	Worn clutch plate friction surfaces or contamination with oil Weak diaphragm spring due to over-heating Faulty automatic cable adjuster
Noise evident on depressing clutch pedal	Dry or worn release bearing Worn or faulty automatic cable adjuster

Chapter 6
Manual gearbox and automatic transmission

Contents

Part A: Manual gearbox
Extension housing rear oil seal (type F, G and N) – renewal 20
Fault diagnosis – manual gearbox... 23
Gearbox housings (type MT75) – overhaul 17
Geartrains and selectors (type MT75) – overhaul 18
General description.. 1
Routine maintenance 2
Manual gearbox – overhaul requirements 4
Manual gearbox – removal and refitting 3
Manual gearbox (all types) – inspection 6
Manual gearbox input shaft and mainshaft (type N) – dismantling
and reassembly.. 13
Manual gearbox (type F) – dismantling 5
Manual gearbox (type F) – reassembly... 7
Manual gearbox (type G) – dismantling.. 9
Manual gearbox (type G) – reassembly.. 10
Manual gearbox (type MT75) – dismantling 16
Manual gearbox (type MT75) – reassembly... 19
Manual gearbox (type MT75) – special overhaul requirements....... 15
Manual gearbox (type N) – dismantling into major sub- assemblies 12
Manual gearbox (type N) – reassembly 14

Overdrive unit – general 22
Selector housing (type F) – removal, dismantling, reassembly and
refitting.. 8
Selector housing (type G) – removal, dismantling, reassembly and
refitting.. 11
Speedometer driven gear – removal and refitting............................... 21

Part B: Automatic transmission
Automatic transmission – general description 24
Automatic transmission – removal and refitting 35
Automatic transmission fluid level – checking 26
Downshift cable – removal, refitting and adjustment......................... 28
Extension housing rear oil seal – renewal 33
Fault diagnosis – automatic transmission... 36
Inhibitor switch – removal, refitting and adjustment 32
Routine maintenance 25
Selector cable – adjustment 27
Selector mechanism – dismantling and reassembly........................... 30
Selector mechanism – removal and refitting 29
Selector rods – removal, refitting and adjustment.............................. 31
Speedometer driven gear – removal and refitting............................... 34

Specifications

Manual gearbox

General
Type .... Four or five forward gears (depending on model), and one reverse gear. Synchromesh on all forward gears (and reverse on the MT75 type).

Application and designation:
  80 and 100 (1.6 litre models):
    Standard 4-speed .... Type F (M12)
    Heavy duty 4-speed .... Type G (M26)
  80, 100 and 120 (2.0 litre models):
    Heavy duty 4-speed .... Type F (M3)
    5-speed .... Type N (M27)
  100L, 130, 160 and 190 (2.0 litre models):
    Heavy duty 4-speed .... Type G (M26)
    5-speed .... Type N (M27)
  All 1.6 and 2.0 litre models from October 1988:
    5-speed MT75

Chapter 6 Manual gearbox and automatic transmission

Gear ratios (:1)

	Type F[1]	Type F[2]	Type G	Type N	Type MT75[3]	Type MT75[4]
1st	3.96	3.65	4.06	3.90	3.89	4.17
2nd	2.28	1.97	2.16	2.28	2.08	2.24
3rd	1.41	1.37	1.38	1.38	1.34	1.47
4th	1.00	1.00	1.00	1.00	1.00	1.00
5th	–	–	–	0.81	0.82	0.82
Reverse	4.23	3.66	4.29	3.66	3.51	3.76

[1] = *Standard*
[2] = *Heavy duty*
[3] = *Close ratio*
[4] = *Wide ratio*

Overhaul data

Countershaft endfloat:
- Type F .... 0.15 to 0.46 mm (0.006 to 0.018 in)
- Type G .... 0.18 to 0.53 mm (0.007 to 0.020 in)

Countershaft diameter:
- Type F (standard) .... 17.367 to 17.380 mm (0.684 to 0.685 in)
- Type F (h/duty) .... 19.301 to 19.314 mm (0.760 to 0.761 in)
- Type G .... 19.329 to 19.342 mm (0.761 to 0.762 in)
- Type N .... 19.301 to 19.314 mm (0.760 to 0.761 in)

Lubrication

Needle roller bearings and countershaft .... Ford grease type SM1C-115-A
Gear contact and end faces .... Ford grease type SM1C- 4504-C
Input shaft splines .... Ford grease type SM1C-1021- A

Oil type:
- Types F and G .... Gear oil, viscosity SAE 80 EP, to Ford spec SQM-2C-9008-A (Duckhams Hypoid 80)
- Type N .... Gear oil, viscosity SAE 80 EP, to Ford spec ESD-M2C-175-A (Duckhams Hypoid 75W/90S)
- Type MT75 .... Gear oil to Ford spec ESD-M2C - 186-A (Duckhams Hypoid 75W/90S)

Capacity:
- Type F .... 1.45 litres (2.55 pints)
- Type G .... 1.98 litres (3.48 pints)
- Type G (with overdrive) .... 2.50 litres (4.40 pints)
- Type N .... 1.50 litres (2.64 pints)
- Type MT75 .... 1.25 litres (2.20 pints)

Overdrive

Type .... Laycock de Normanville, model J
Application .... Optional equipment on Type G gearbox

Torque wrench settings

Type F, G and N gearboxes

	Nm	lbf ft
Input shaft bearing retainer (guide sleeve) to housing:		
Type F	21 to 25	15 to 18
Type G	17 to 21	13 to 15
Type N	9 to 11	7 to 8
Extension housing to gearbox:		
Type F and N	45 to 49	33 to 36
Type G	54 to 61	40 to 45
Gearbox housing cover:		
Type F	21 to 25	15 to 18
Type G	17 to 21	13 to 15
Type N	9 to 11	7 to 8
Reversing lamp switch	1 to 2	0.7 to 1.5
Selector detent plug:		
Type F	10 to 15	7 to 11
Type G and N	17 to 19	13 to 14
Oil filler plug	23 to 32	17 to 23
Collar nut 5th gear, (type N only)	120 to 150	89 to 110
5th gear locking-plate (type N only)	21 to 26	15 to 19
Clutch housing to gearbox	55 to 65	41 to 48
Mainshaft nut (type G only)	35 to 41	26 to 30
Gearbox to crossmember	70 to 90	52 to 66
Crossmember to floor	40 to 50	30 to 37

Type MT75 gearbox

	Nm	lbf ft
Input shaft guide sleeve	150 to 220	110 to 162
Countershaft bearing retainer:		
Stage 1	20 to 25	14 to 18
Stage 2	Slacken by 60°	Slacken by 60°

90 Chapter 6 Manual gearbox and automatic transmission

Torque wrench settings (continued)

Type MT75 gearbox

	Nm	lbf ft
Bearing retainer locking plate bolt	21 to 29	15 to 21
Reverse idler gear shaft to housing	28 to 36	20 to 26
Reverse gear lock plug	11 to 15	8 to 11
Output shaft flange nut	170 to 230	125 to 170
Drain and filler/level plug	23 to 32	17 to 23
Mainshaft ballbearing retainer plate	20 to 27	15 to 20
Gear lever interlock unit plug	12 to 15	9 to 11
Front to rear gearbox housing bolts	21 to 28	15 to 20
Gearbox to engine bolts	30 to 40	22 to 29
Clutch housing cover plate	21 to 28	15 to 20
Gear lever to lever extension	41 to 58	30 to 42
Reversing lamp switch	30 to 35	22 to 26
Gearbox crossmember to side member	41 to 58	30 to 42
Gearbox mounting insulator to crossmember bolt	70 to 97	51 to 71
Gearbox mounting insulator to gearbox nut	41 to 58	30 to 42
Engine mounting to gearbox brace	51 to 73	37 to 54

Automatic transmission

Type — Ford (Bordeaux) C3 three forward speeds and one reverse, epicyclic gear train with hydraulic control and torque converter

Ratios

1st	2.47:1
2nd	1.47:1
3rd	1.00:1
Reverse	2.11:1

Lubrication

ATF to Ford spec SQM-2C 9010-A (Duckhams D-Matic)

Capacity — 6.3 litres (11.08 pints)

Torque wrench settings

	Nm	lbf ft
Driveplate to converter	36 to 41	27 to 30
Sump to transmission	16 to 23	12 to 17
Downshift cable bracket	16 to 23	12 to 17
Selector lever nut (outer)	10 to 15	7 to 11
Selector lever nut (inner)	41 to 54	30 to 40
Starter inhibitor switch	16 to 20	12 to 14
Converter drain plug	27 to 40	20 to 29
Oil cooler pipe to connector	16 to 20	12 to 14
Oil pipe to connector	9 to 14	6 to 10
Oil pipe connector to transmission	14 to 20	10 to 14
Transmission to engine	30 to 37	22 to 25

PART A: MANUAL GEARBOX

1 General description

A four or five-speed manual gearbox is fitted, depending on model and specification. On all models the unit is mounted in-line and to the rear of the engine. A propeller shaft transfers the drive to the rear axle. All gearbox types are of constant mesh design.

F, G and N type gearboxes

With these types, all forward gear selection is by synchromesh. All forward gears are helical, the reverse gear is straight cut.

The clutch housing, selector housing (F and G types), and the extension housing are all bolted to the main gearbox case.

The input shaft and mainshaft are in line and rotate on ball-bearings. The mainshaft spigot runs in needle roller bearings, as does the counter-shaft (layshaft) gear assembly.

The synchronisers are of baulk ring and blocker bar type and operate in conjunction with tapered cones machined onto the gears. When engaging a gear, the synchroniser sleeve pushes the baulk ring against the tapered gear cone by means of the three spring tensioned blocker bars. The drag of the baulk ring causes the gear to rotate at the same speed as the synchroniser unit, and at this point further movement of the sleeve locks the sleeve, baulk ring and gear dog teeth together.

In the F type gearbox the 1st/2nd gear synchroniser hub is a tight press fit onto the mainshaft and cannot be removed during overhaul. The mainshaft endfloat is controlled by selective circlips.

Reverse gear is obtained by moving the reverse idler gear into mesh with the countershaft gear and the spur teeth of the 1st/2nd synchroniser.

With the F and G types, a three-rail selector mechanism is used, whereby the rails are located in the housing and move in accordance with the gear lever selection position. The selector forks are located in position on the rails and these actuate the gear selection.

On the type F gearbox the selector forks of 1st/2nd and 3rd/4th gears differ in that they are slotted into their respective guides fixed to the selector rails.

With the N type gearbox a single rail selector mechanism is utilized.

MT75 gearbox

This gearbox type was first introduced in the latter part of 1988. The identification designation MT75 stems from manual transmission, and the distance between the shaft centre lines (75 mm).

The unit is a constant mesh type with five forward speeds and one reverse. The input shaft and mainshaft are in-line and rotate on ball bearings in the front and rear gearbox housings. Caged needle roller bearings are used to support the mainshaft spigot, the countershaft gear assembly and the gears on the mainshaft. The synchromesh

Chapter 6 Manual gearbox and automatic transmission

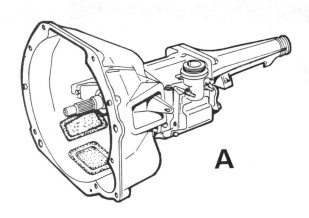

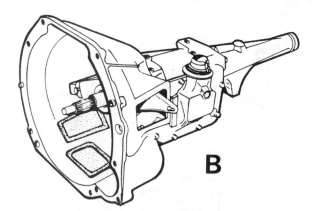

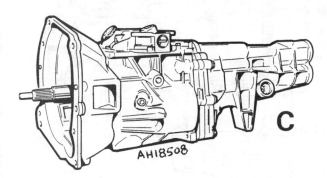

Fig. 6.1 Type F, G and N gearbox units (Sec 1)

A Type F 　　C Type N
B Type G

2.2 Oil level/filler plug (A) and drain plug (B) in the MT75 gearbox

2 Routine maintenance

The following maintenance operations must be carried out at the intervals specified in Routine maintenance *at the beginning of this manual*

1 Carry out a visual inspection of the gearbox for any signs of oil leakage. If any serious leaks are apparent, they must be rectified with reference to the relevant Sections of this Chapter.
2 Check the oil level in the gearbox with the vehicle level. Unscrew the filler plug and check that the lubricant is up to the base of the filler plug orifice. On the type F and G gearboxes the filler plug is on the right-hand side of the casing while on the type N and MT75 units the plug is on the left (photo). If necessary, top up with the specified lubricant as given in the Specifications at the beginning of this Chapter. On five-speed models, allow the level to settle for a few minutes, then recheck and top up further if necessary. Refit the filler plug on completion.
3 Renewal of the gearbox oil is not a service requirement but if it is necessary to drain the lubricant as part of a repair operation this can be done by placing a suitable container beneath the drain plug, located on the same side as the filler plug. Unscrew the plug and allow the oil to drain into the container. Refit the plug on completion. Note that on the type N gearbox, a drain plug is not provided and the oil can only be drained with the unit removed from the vehicle. To do this, remove the top cover, invert the gearbox and allow the oil to drain into a container. After draining, refit the top cover using a new gasket if necessary.

3 Manual gearbox – removal and refitting

Note: *The manual gearbox can be removed as a unit with the engine as described in Chapter 1, then separated from the engine on the bench. However, if work is only necessary on the gearbox or clutch unit, it is better to remove the gearbox on its own from underneath the vehicle. The latter method is described in this Section. A trolley jack will be required, and the aid of an assistant during the actual removal (and refitting) procedures.*

1 Position the vehicle over an inspection pit, or apply the handbrake, raise it at the front and support it on axle stands at a height which will provide sufficient working clearance beneath the vehicle. On type F, G and MT75 units, drain the gearbox oil with reference to Section 2 then refit the drain plug.
2 Disconnect the battery earth lead.
3 Working in the engine compartment, remove the air cleaner unit as described in Chapter 3.
4 Remove the starter motor as described in Chapter 12.
5 On models fitted with the type MT75 gearbox, prise free the gear

operation is by blocker bars and baulk rings similar to that used on the other gearbox types.
　The gear selector lever is mounted on top of the gearbox and engages direct with the selector forks.
　The gearbox case is in two halves, the front half combining the clutch housing, and the rear half which contains the main gear assemblies and the selector units.
　A general view of this gearbox is shown in Fig. 6.2. Before undertaking any major repair or overhaul, reference should first be made to Section 15 for the special tools and procedures required.

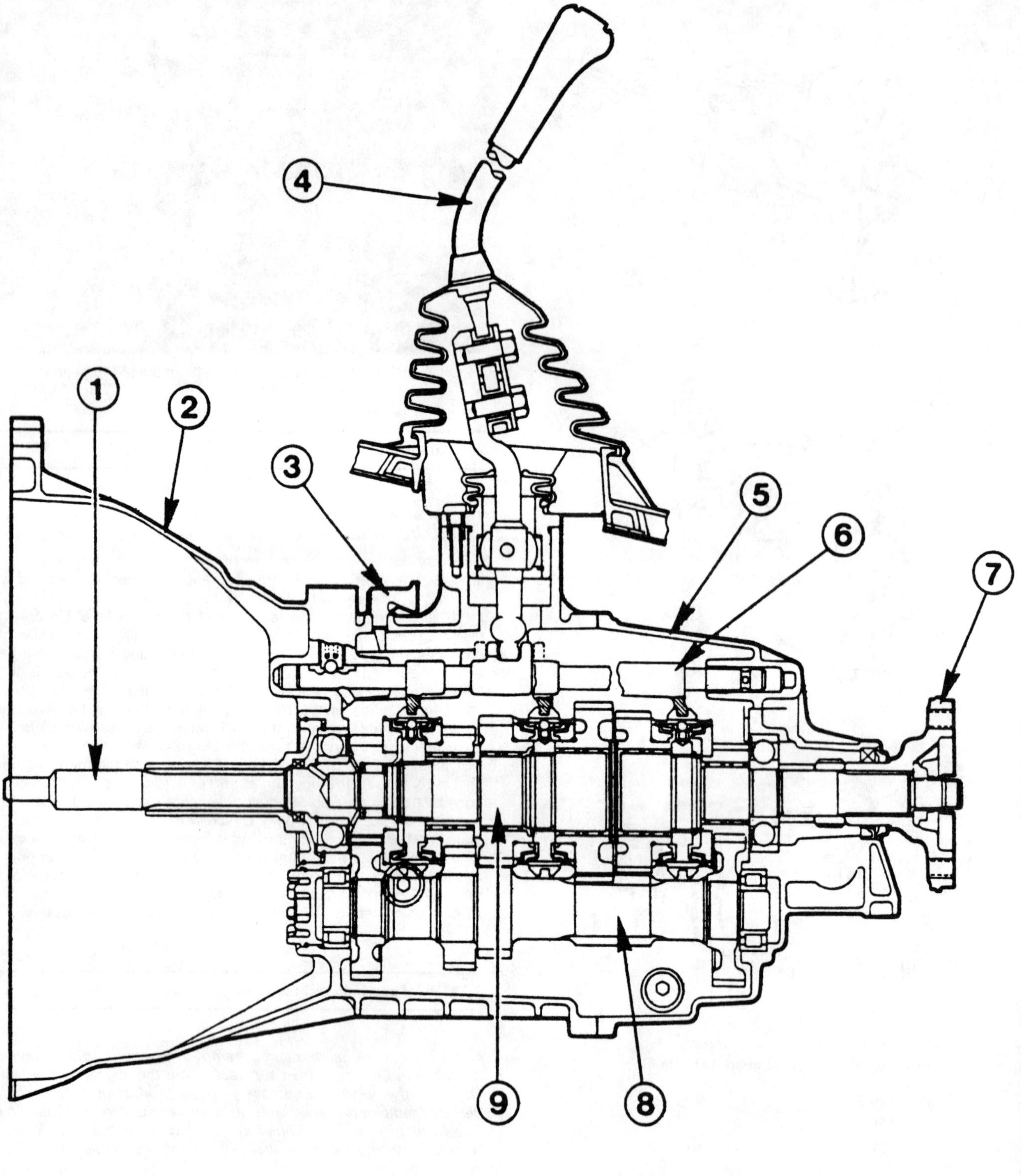

Fig. 6.2 Type MT75 gearbox (Sec 1)

1	Input shaft	4	Gear lever	6	Selector shafts	8 Countershaft cluster gears
2	Front housing	5	Rear housing	7	Drive flange	9 Main shaft
3	Vent					

Chapter 6 Manual gearbox and automatic transmission

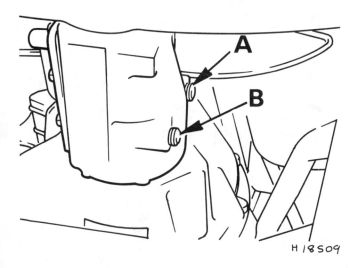

Fig. 6.3 Type F and G gearbox oil level/filler plug (A) and drain plug (B) (Sec 2)

lever gaiter and pull it up the lever to allow access to the lever base fixing. Undo the two retaining bolts and remove the lever extension (photo).
6 On other five-speed gearbox types, remove the gear lever knob, unscrew the lever and then remove it from underneath the vehicle.
7 On four-speed models, unhook and detach the gear lever spring from its bracket (working from underneath), then slide the gaiter up the lever, and unscrew the lever to detach it from the gearbox.
8 Detach the reversing lamp switch lead connector and move the lead out of the way.

9 Remove the propeller shaft as described in Chapter 7.
10 Undo the retaining bolt and remove the speedometer drive cable retaining plate. Withdraw the speedometer cable from the gearbox and position it out of the way (photos).
11 If an electronic tachograph is fitted, detach its lead and position it out of the way.
12 Disconnect the clutch cable from the release lever as described in Chapter 5.
13 Unbolt and detach the lower cover plate(s) from the clutch housing.
14 Position a trolley jack under the gearbox to support it.
15 Undo the gearbox crossmember/mounting unit bolts and remove the crossmember (photos).
16 Unscrew and remove the engine to gearbox retaining bolts, noting the engine earth strap connection location under one of the bolts.
17 Position a second jack or blocks under the engine sump to support it whilst the gearbox is removed.
18 With the gearbox detached and ready for removal, check that all fixings are fully disconnected and positioned out of the way. Enlist the aid of an assistant to help steady the gearbox as it is withdrawn, then pull it rearwards whilst simultaneously supporting its weight and detach it from the engine. At no time during its removal (and subsequent refitting), allow the weight of the gearbox to rest on the input shaft. Where applicable, it may be necessary initially to prise free the clutch housing from the engine location dowels.
19 When the gearbox is fully clear of the engine, lower it, and withdraw it from underneath the vehicle.
20 Refitting is a reversal of the removal procedure. Before lifting the unit into position, check that the clutch release bearing is correctly positioned and lightly grease the input shaft with the specified grease. If the clutch was renewed it should be centralised as described in Chapter 5, so that the input shaft can engage with the clutch plate splines and the spigot bearing in the rear end of the crankshaft.
21 Once the gearbox is fully engaged with the engine, insert a couple of retaining bolts, then refit the crossmember/mounting unit. If working on the MT75 unit, ensure that the crossmember is fitted with the red marker dot towards the front of the vehicle. Tighten all bolts to the specified torque wrench settings.
22 Refer to Chapter 5 for details on reconnecting the clutch cable.

3.5 Gear lever extension retaining bolts (MT75 gearbox)

3.10A Undo the retaining bolt and remove the bracket ...

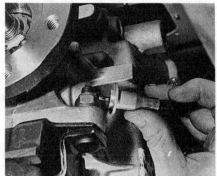

3.10B ... then withdraw the speedometer cable (MT75 gearbox)

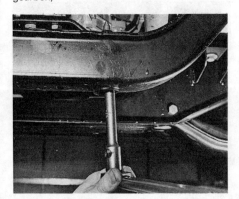

3.15A Undoing the transmission to crossmember-bolts (MT75 gearbox)

3.15B Crossmember to body retaining bolts (MT75 gearbox)

5.20 Input shaft bearing circlip removal

23 Refer to Chapter 7 for details on reconnecting the propeller shaft.
24 On completion, refill/top up the gearbox oil with the specified grade and quantity according to type.

4 Manual gearbox – overhaul requirements

Before deciding to dismantle and overhaul the gearbox, first consider if this course of action is practical. If the unit is known to have covered a high mileage for example, the chances are that the items requiring renewal are numerous and therefore in view of the time and cost of parts probably required, it may well prove more practical to replace the gearbox with a service exchange unit.

If any problems are suspected of being more minor (for example, worn synchro baulk rings), it is then practical to overhaul the gearbox providing the following tools are available:

(a) *Good quality circlip pliers, 2 pairs – 1 expanding and 1 contracting*
(b) *Soft-headed mallet*
(c) *Drifts (steel and brass) 9.525 mm (3.75 ins) diameter*
(d) *Suitable containers for small components, particularly the needle rollers*
(e) *Engineer's vice on a firm workbench*
(f) *Selection of metal tubing (for driving out bearings and seals)*
(g) *A bearing puller*

With the MT75 gearbox, further special tools are required. Refer to Section 15 for details.

5 Manual gearbox (type F) – dismantling

1 Read the whole of this Section before starting work.
2 The internal parts of the gearbox are shown in Fig. 6.5 and the housing components in Fig. 6.6.
3 First drain any remaining oil from the gearbox.
4 Unscrew and remove the four clutch housing retaining bolts and withdraw the housing from the gearbox front face.
5 Disconnect the clutch release lever by pulling it sideways, then remove it from the housing (Fig. 6.7).
6 Unbolt and remove the gear selector housing, which is secured by 7 bolts. If the housing is stuck, tap it lightly with a block of wood to break the seal.
7 Extract the selector forks (1st/2nd and 3rd/4th), noting their location positions (Fig. 6.8).
8 Use a suitable implement and prise out the rear extension oil seal, taking care not to damage the housing. If the extension housing bush is to be removed you will need a suitable extractor, or if available, use the Ford special tool number 16-025 (Fig. 6.9). Should you not possess a

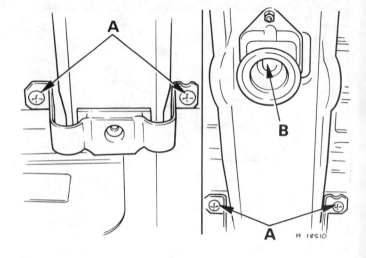

Fig. 6.4 Gearbox and mounting crossmember – Type F, G and N gearboxes (Sec 3)

A – Crossmember to floor mountings
B – Crossmember to gearbox mountings

suitable tool for removing the bush, get your Ford dealer to remove it for you. *It is important to note that if the Ford special tool is being used both the extension seal and bush must be removed and renewed only when the extension housing and mainshaft are in their assembled positions.*
9 Unscrew and remove the four bolts retaining the extension housing to the gearbox. Partially pull the extension from the gearcase seat to the point where the extension can be rotated sufficiently to align the cutaway section in the flange with the countershaft, see Fig. 6.10.
10 Use a suitable diameter drift (see Section 4) and drive the countershaft rearwards out of the gearbox (Fig. 6.11). If possible use an old countershaft for this purpose. Cut it to a length of 177 mm (7 ins). The drift must follow the countershaft through as it is driven out so that the needle rollers are retained in position. Drive the drift through just far enough to locate the countershaft, its bearings and thrustwashers, allowing them to drop in unison to the bottom of the gearbox.
11 Unbolt and withdraw the input shaft bearing retainer from the gearbox front face. It is retained by three bolts.
12 Reach down in the gearbox, move the countershaft to one side and extract the input shaft through the front face of the casing.
13 The rear extension housing complete with mainshaft can now be withdrawn rearwards from the gearcase (Fig. 6.12).
14 Lift out the countershaft gears together with the thrustwashers, but leave the dummy countershaft (drift) in position in the shaft to prevent the bearings from falling out until ready for inspection.

5.22 Circlip removal from bearing outer race

5.26 Circlip removal from the front end of the mainshaft

5.29 Remove the mainshaft rear bearing

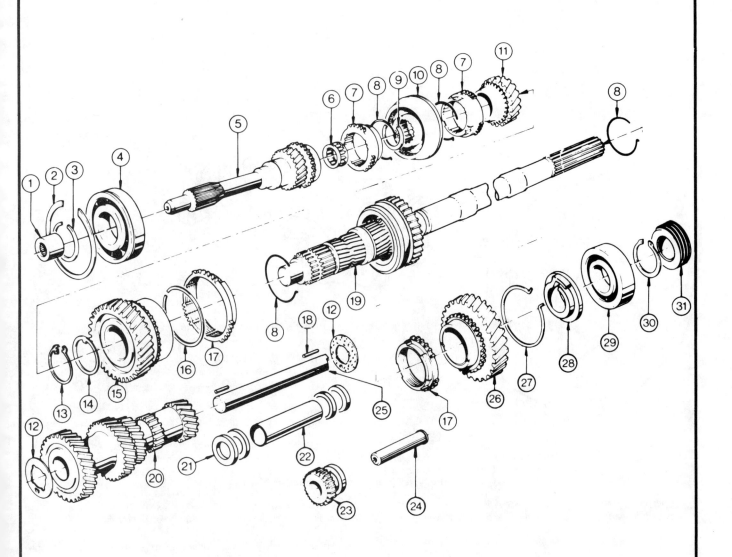

Fig. 6.5 The type F gearbox gear assemblies (Sec 5)

1 Pilot bearing (crankshaft)
2 Circlip
3 Circlip
4 Ball-bearing (grooved)
5 Input shaft
6 Needle bearing
7 3rd/4th synchroniser baulk ring
8 Synchroniser springs
9 Circlip
10 3rd/4th synchroniser hul unit
11 3rd gear
12 Thrustwasher
13 Circlip
14 Thrustwasher
15 2nd gear
16 Circlip
17 1st/2nd gear synchroniser baulk
18 Needle rollers (19 or 21)
19 Mainshaft
20 Countershaft gears
21 Spacer shim
22 Spacer tube (standard unit only)
23 Reverse idler gear
24 Reverse idler gear shaft
25 Countershaft
26 1st gear
27 Circlip
28 Oil scoop ring
29 Ball-bearing
30 Circlip
31 Speedometer worm gear

96   Chapter 6   Manual gearbox and automatic transmission

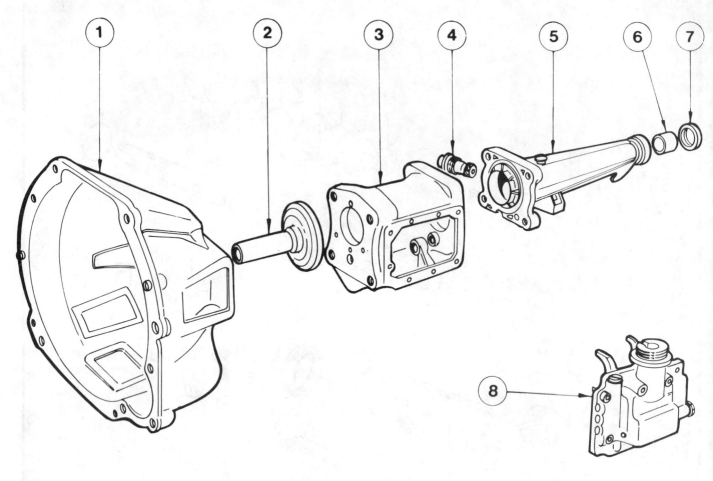

Fig. 6.6 The type F gearbox main assemblies (Sec 5)

1 Clutch housing
2 Input shaft bearing retainer
3 Gearbox casing
4 Speedometer drive pinion
5 Gearbox extension housing
6 Extension housing bush
7 Oil seal
8 Selector housing

Fig. 6.7 Withdraw the clutch release lever by pulling it sideways (Sec 5)

15 To remove the reverse idler gear, use a suitable length soft drift and drive the gear idler shaft out rearwards, then lift out the gear. Note which way round it is fitted (grooved shoulder section to rear).
16 Unscrew and remove the speedometer drive from the extension housing and then remove the bearing from the drivegear.
17 With the gearbox main assemblies now separated, they can be cleaned for inspection and further dismantling as necessary.
18 If it is apparent at this stage that the gear assemblies and their associated components are badly worn and have possibly suffered damage, it will almost certainly be more economical and practical to purchase a new or good secondhand gearbox instead of trying to repair the existing one.
19 The dismantling procedures for the main component assemblies are given in the following sub-sections.

Input shaft dismantling

20 Support the input shaft in a soft-jawed vice taking care not to damage the splines, then prise open the circlip retaining the bearing in position and remove it (photo). If suitable circlip pliers are not available for this purpose, lever the ends of the circlip apart using two screwdrivers.
21 Using suitable tubing, the input shaft bearing should preferably be pressed from the shaft. An alternative method is to rest the bearing outer track on a soft-jawed vice and to drive the shaft through the bearing using a soft-headed mallet.
22 If the bearing is to be renewed, remove the circlip from the bearing outer race and transfer it to the new bearing (photo).

Countershaft gears dismantling

23 Withdraw the dummy countershaft from the gear unit and place the gear on the bench, together with its thrust washers and bearings which should be kept adjacent to their respective fitted positions ready for inspection. According to the gearbox type there will be 19 or 21 needle rollers at each end.

Mainshaft dismantling

24 First extract the mainshaft bearing retaining circlip from its groove

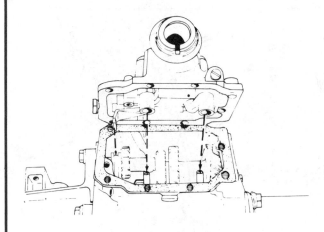

Fig. 6.8 Remove the selector housing and extract the selector forks (Sec 5)

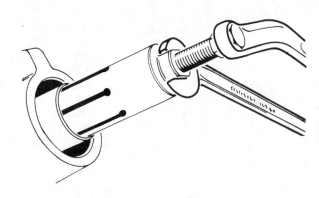

Fig. 6.9 Special tool No 16-025 used for removal of the extension housing bush (Sec 5)

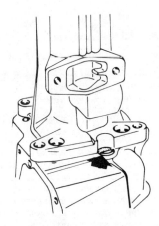

Fig. 6.10 Rotate extension housing sufficiently to enable the countershaft and cutaway to align as shown (Sec 5)

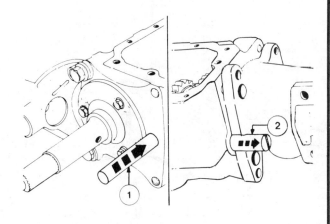

Fig. 6.11 Use a dummy shaft (1) to drive out the countershaft rearwards (2) (Sec 5)

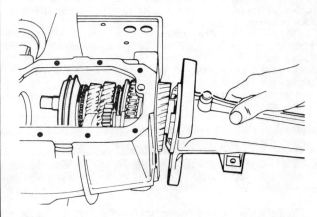

Fig. 6.12 Withdrawing the mainshaft assembly (Sec 5)

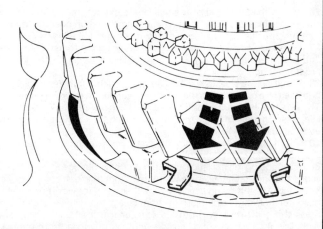

Fig. 6.13 Extract the retaining circlip from the extension housing (Sec 5)

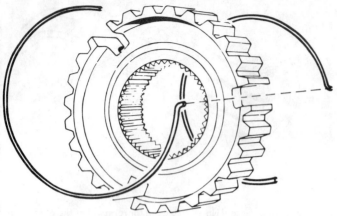

Fig. 6.14 Synchroniser spring location positions (Sec 5)

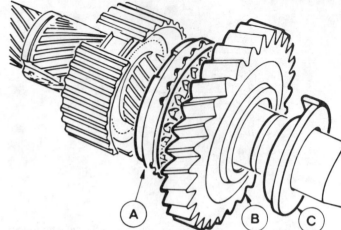

Fig. 6.15 Assemble 1st gear (B) with baulk ring (A) and oil scoop ring (C) (Sec 7)

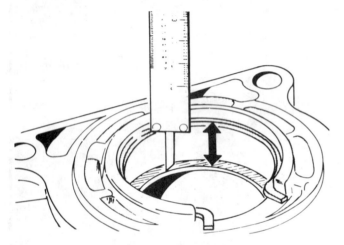

Fig. 6.16 Check depth for circlip requirement (Sec 7)

in the extension housing (Fig. 6.13).
25   Support the extension housing and using a soft-headed mallet, drive the mainshaft from it, taking care not to damage the shaft or housing.
26   From the front end of the mainshaft, extract the small diameter circlip which secures the 3rd/4th gear synchro hub in position (photo). The synchro hub can now be withdrawn as a unit, followed by 3rd gear.
27   Prise apart and remove the circlip retaining the thrustwasher and 2nd gear, then remove the washer and gear from the shaft and lay out in order of fitting.
28   Next on the shaft is the 1st/2nd gear synchro hub assembly. The hub itself is in unit with the mainshaft and cannot therefore be removed. The synchroniser springs and blocker bars can be removed though, but keep them in order of fitting and note how each spring is fitted relative to each other (Fig. 6.14).
29   Working at the rear end of the shaft, remove the circlip retaining the mainshaft bearing (photo).
30   Depending on facilities the speedometer drivegear, bearing oil scoop ring and 1st gear can be withdrawn from the shaft collectively or individually. If removing collectively, either use a tube drift of suitable diameter and length to support the front face of the 1st gear and drive or press the shaft through from the rear, or use a suitable puller and withdraw the gear, bearing and speedometer drivegear from the rear. To remove individually, either use a puller or fabricate suitable U-shaped plates with which to support each component on its front face and with each plate in turn supported in a vice, drive the shaft through the respective components to remove them from the shaft.
31   To dismantle the 3rd/4th gear synchro hub unit, remove the baulk ring, the blocker bars and springs, keeping in order of fitting.
32   For dismantling of the selector housing refer to Section 8.

6   Manual gearbox (all types) – inspection

1   With the gearbox dismantled the respective components can be cleaned, dried and inspected.
2   Examine all gears for excessively worn, chipped or damaged teeth. Any such gears should be renewed. It will usually be found that if a tooth is damaged in the countershaft geartrain, the mating gear teeth on the mainshaft will also be damaged. In any case it is not good policy to mesh new gears with worn gears as they are usually noisy when in operation.
3   Check the synchro-hubs and components for signs of wear or damage. The baulk rings will almost definitely need renewing, the tell-tale signs of excessive wear being if the oil reservoir grooves are worn smooth or uneven. Also, when fitted to their mating cones – as they would be when in operation – there should be no rock. This would signify ovality, or lack of concentricity. One of the most satisfactory ways of checking is by comparing the fit of a new ring on the hub with the old one. If the grooves of the ring are obviously worn or damaged (causing engagement difficulties) the ring should be renewed.
4   The synchro-hubs themselves are also subject to wear and where the fault has been failure of any gear to remain engaged, or actual difficulty in engagement, then the hub is one of the likely suspects.
5   The ends of the splines are machined in such a way as to form a 'keystone' effect on engagement with the corresponding mainshaft gear. Do not confuse this with wear. Check also that the blocker bars (sliding keys) are not sloppy and move freely. If there is any rock or backlash between the inner and outer sections of the hub, the whole assembly must be renewed, particularly if there has been a complaint of jumping out of gear.
6   The thrustwashers at the ends of the countershaft gear train should also be renewed as they will almost certainly have worn if the gearbox is of any age (as will the bearings).
7   The caged bearing between the input shaft and the mainshaft will usually be found in good order, but if in any doubt, renew the bearing.
8   All ball-bearings should be checked for chatter. It is advisable to renew these anyway, even though they may not appear to be too badly worn.
9   Circlips, which in the F type gearbox are all important in locating bearings, gears and hubs, should also be checked to ensure that they are not distorted or damaged. In any case a selection of new circlips of varying thickness should be obtained to compensate for variations in new components fitted, or wear in old ones.
10   Check and if necessary dismantle for overhaul the selector housing as described in Section 8 or 11 as applicable.

Chapter 6 Manual gearbox and automatic transmission

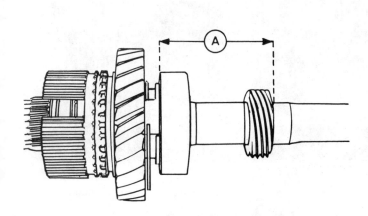

Fig. 6.17 Check bearing to speedometer worm gear clearance (Sec 7)

A = 82.25 mm (3.24 in)

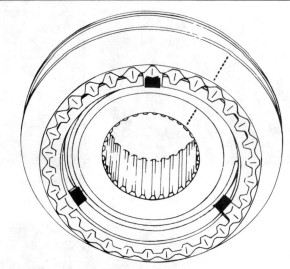

Fig. 6.18 Assembled synchroniser hub with markings aligned (Sec 7)

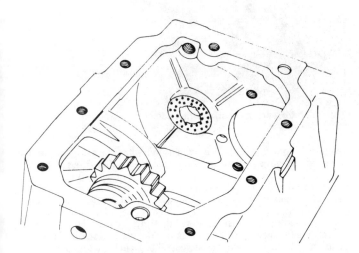

Fig. 6.19 Locate the front thrustwasher for the countershaft gear (Sec 7)

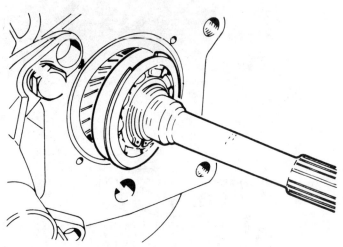

Fig. 6.20 Locate the input shaft (Sec 7)

7 Manual gearbox (type F) – reassembly

1 When reassembling the gearbox, rebuild each sub-assembly in turn, taking care not to damage any parts. Lubricate each part as it is assembled using the specified grease where indicated.

Mainshaft assembly

2 Commence by assembling the 1st gear synchroniser hub, inserting the blocker bars and retaining them with the synchroniser springs. Arrange the springs each side so that their hooked ends are offset relative to each other as shown in Fig. 6.14.
3 Fit the baulk ring and 1st gear to the mainshaft.
4 Slide the oil scoop ring, larger diameter to the rear of the mainshaft, and the bearing onto the mainshaft (Fig. 6.15).
5 Using a piece of suitable diameter tube drift the bearing into position on the mainshaft.
6 If a new mainshaft rear bearing or extension housing are being fitted, the location circlip thickness must be calculated. To do this, fit a circlip into the housing groove and press it outwards so that it butts against the shoulder. Use a depth gauge to measure the distance from the circlip top edge to the bearing stop flange – see Fig. 6.16. Measure the total width of the new bearing and deduct this from the previously noted height measurement to give the necessary thickness of circlip required. Circlips are available in varying thicknesses and it is important to fit one that when in position gives no axial play.

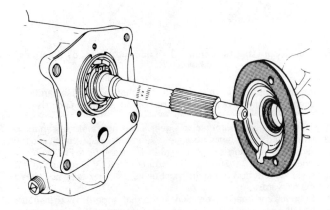

Fig. 6.21 Fit the input shaft bearing retainer with gasket – align groove with oil hole in gearbox front face (Sec 7)

Chapter 6 Manual gearbox and automatic transmission

7.10 Fit 2nd gear baulk ring and 2nd gear

7.12 Fit 3rd gear, baulk ring and synchro-hub

7.15A Insert spacer shim into countershaft, followed ...

7.15B ... by the needle rollers, smeared with grease

7.15C Fit 2nd gear spacer shim when needle rollers are located

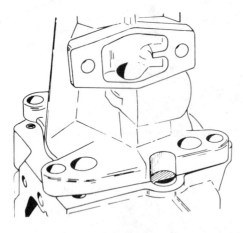

Fig. 6.22 Ensure that the countershaft end section is positioned as shown (Sec 7)

7   Smear the bearing location on the mainshaft with grease and refit the bearing, pressing it into position. The bearing retaining circlip is then fitted.
8   Relocate the speedometer drivegear, pressing it onto the shaft so that when fitted the distance from the speedometer gear rear face to the front face of the bearing is 82.25 mm (3.24 in) as shown in Fig. 6.17.
9   Check that the 1st/2nd synchro hub is correctly assembled. The sliding sleeve and hub markings must align, with the selector groove pointing forwards.
10   Slide the 2nd gear into position on the shaft together with its baulk ring and thrust washer (photo). Make secure by fitting the circlip, ensuring that it is fully located.
11   Reassemble the 3rd/4th gear synchroniser hub, fitting the blocker bars and springs so that when in position the springs are staggered but with their opposing ends located in alignment in the same blocker bar. The sliding sleeve and hub markings must align when fitted (Fig. 6.18).
12   Slide the 3rd gear together with its baulk ring into position on the mainshaft and then locate the 3rd/4th synchro hub assembly. The long side of the hub unit must face the forward end of the shaft. Make secure by fitting the circlip, ensuring that it is fully engaged (photo).
13   The extension housing is now fitted onto the rear of the mainshaft. Warm the housing first by immersing in hot water. Locate the housing and then fit the previously loosely fitted circlip into its extension housing groove.

Countershaft reassembly

14   Insert the dummy countershaft into the countershaft gears and then fill the cavity at each end between the shaft and bore with the specified grease. Do not forget to fit the spacer tube (if applicable).
15   The spacer shims and bearings can now be refitted. Ensure that the longer needle roller bearings are inserted into the rear end of the shaft. With the needle rollers installed, fit the second spacer shims (photos).

Input shaft

16   Smear the ball-bearing location on the input shaft with grease, then locate the new bearing and press it into position using a suitable size tube drift of the same diameter as the bearing inner race. The bearing outer race circlip groove must be offset to the front.
17   With the bearing in position on the shaft, locate the small circlip to retain it in position and locate the large circlip into the bearing outer race groove.

Gearbox general assembly

18   If not already fitted, insert the new oil seal into the input shaft bearing retainer, ensuring that the seal lip faces the gearbox casing. The seal lip must be smeared with a multi-purpose grease prior to assembly.
19   Support the reverse idler gear in line with its shaft location holes in the gearbox and then pass the shaft through, driving it in with a soft-headed mallet until it is 0.2 to 0.8 mm (0.008 to 0.032 in) below the gearbox rear face. When fitted the gear must have its grooved hub section offset to the rear.
20   Smear the faces of the countershaft thrustwashers with the

Chapter 6 Manual gearbox and automatic transmission

specified grease. Locate the front one in position as shown in the inside of the gearbox (Fig. 6.19).

21 Carefully lower the countershaft gear into position in the gearbox so that the thrust washer is not disturbed. Stick the rear thrust washer to the rear of the countershaft gear.

22 The mainshaft together with extension housing can now be fitted from the rear. The flange gasket should be located and stuck to the extension flange face by smearing with grease. This ensures that when the housing is turned after fitting the countershaft, the gasket is not damaged. Initially the extension housing must be located so that the recess for fitment of the countershaft is aligned with the countershaft aperture in the rear gearbox face. This is shown in Fig. 6.22.

23 Lubricate the needle bearing and locate it into the input shaft.

24 Locate the baulk ring onto the cone of the input shaft and then carefully fit the shaft into position through its location aperture in the front face of the gearbox (Fig. 6.20). When in position ensure that the circlip in the bearing outer race is butting against the gearbox front face.

25 Locate the bearing retainer flange gasket, smearing with grease for security of location during fitting. To avoid damaging the oil seal as the housing is fitted, temporarily tape the splines of the input shaft.

26 Fit the bearing retainer into position, ensuring that the oil channel in the gasket and retainer is aligned with the oil hole in the gearbox (Fig. 6.21). Smear the retaining bolts with a sealing compound, insert and tighten to the specified torque. Remove the protective tape from the input shaft splines.

27 Position the gearbox so that the countershaft gear can be engaged

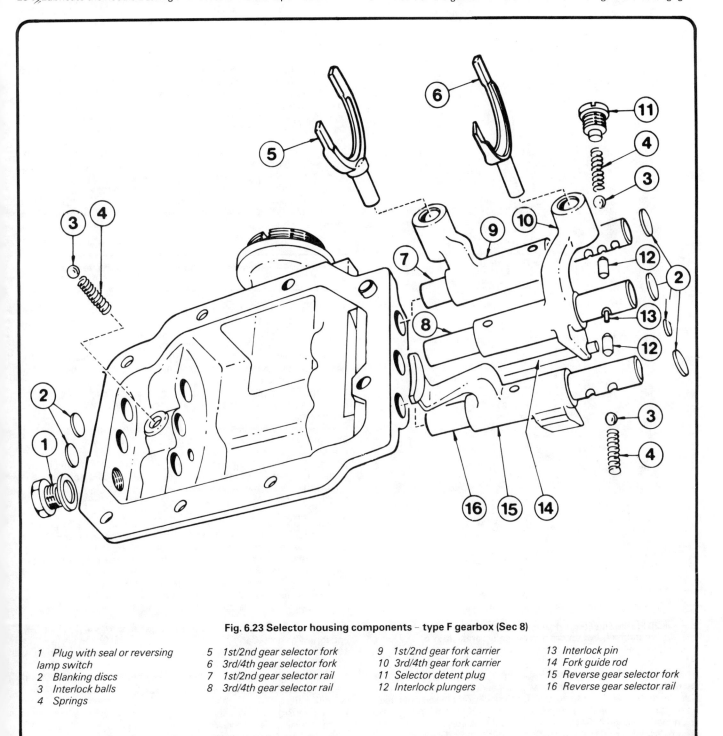

Fig. 6.23 Selector housing components – type F gearbox (Sec 8)

1 Plug with seal or reversing lamp switch
2 Blanking discs
3 Interlock balls
4 Springs
5 1st/2nd gear selector fork
6 3rd/4th gear selector fork
7 1st/2nd gear selector rail
8 3rd/4th gear selector rail
9 1st/2nd gear fork carrier
10 3rd/4th gear fork carrier
11 Selector detent plug
12 Interlock plungers
13 Interlock pin
14 Fork guide rod
15 Reverse gear selector fork
16 Reverse gear selector rail

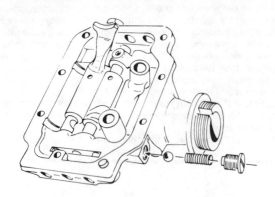

Fig. 6.24 Extract the spring and ball (Sec 8)

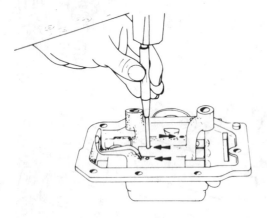

Fig. 6.25 Drive out the selector fork retaining pins (Sec 8)

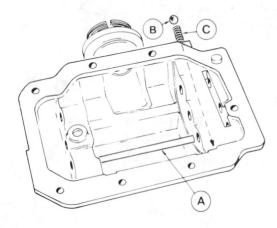

Fig. 6.26 Fit the selector fork guide rod (A). Spring (C) and ball (B) must be fitted before reverse selector rail (Sec 8)

be turned to align its bolt holes with those in the rear gearbox face. Smear the retaining bolts with sealing compound, insert them and tighten to the specified torque.

29 If a new extension housing bush is to be fitted it can now be driven into position using a suitable tube drift, or if available. Ford special tool No 16-015. It is most important that when fitted the oil return groove is located at the bottom of the housing to the rear, with the notch in the bush **not** situated over the oilway.

30 Smear the extension housing oil seal with grease and carefully tap it into position so that it butts against its location shoulder in the housing.

31 Locate the selector housing gasket onto the gearbox, aligning the bolt holes.

32 The 1st/2nd and 3rd/4th gear selector forks are now located into the grooves of their respective hubs. Ensure that each fork is fitted so that the casting number faces forwards.

33 Lower the selector housing into position, engaging the selector fork rods. The reverse gear selector fork must engage with the hub groove of the reverse idler gear (Fig. 6.8). Insert the retaining bolts (smear the threads with sealant) and tighten securely.

34 Reassemble the clutch release lever and bearing and locate them in the clutch housing.

35 Refit the clutch housing to the front face of the gearbox, smear the retaining bolts with sealant, then fit and tighten them to the specified torque.

36 With the gearbox now reassembled, check that all gears can be selected by temporarily fitting the gear lever and rotating the mainshaft with each gear engaged in turn.

37 On refitting, do not forget to refill the gearbox with the correct grade and quantity of oil. If new components have been fitted, the gearbox should be run-in as if it were new for the first few hundred miles.

8 Selector housing (type F) – removal, dismantling, reassembly and refitting

1 The gearbox selector housing can be removed with the gearbox in position. First drain the oil from the gearbox, then raise and support the vehicle so that it is safe to work underneath.

2 The selector housing is located on the side face of the gearbox and is secured with bolts. Before unscrewing these bolts, disconnect the gear lever from its location on top of the housing. Detach the reversing light switch wires (where fitted).

3 Remove the selector housing retaining bolts. Prise or tap free the housing from the gearbox and take it to the bench for cleaning and inspection.

4 Having cleaned the housing and its associated components, check for signs of obvious wear or damage to the selector forks, the fork carriers and selector rails. The selector fork carriers should have a precise movement along the rails with a distinctly 'notchy' feel when aligned for gear engagement. If a sloppy movement is apparent, dismantle the housing components as follows. Refer to Fig. 6.23.

5 Unscrew and remove the detent plug and extract the spring and ball (Fig. 6.24).

6 From the housing end remove the blanking discs and plug (or reversing light switch as applicable), then support the housing and drive out the selector fork carrier retaining pins (Fig. 6.25).

7 The respective rails can now be withdrawn and the carriers lifted out of the housing. Remove 1st/2nd rail, then the 3rd/4th rail, and as they are withdrawn collect the interlock ball and spring.

8 Extract the interlock plungers and then withdraw the reverse gear selector rail and fork. Remove the ball and detent spring, then the guide rod.

9 Clean and check the respective components for wear or damage and renew as necessary.

10 Commence reassembly by inserting the selector fork guide rod as shown in Fig. 6.26.

11 Locate the reverse gear selector rail and fork, having first inserted the spring and ball. Compress the ball and spring to enable the selector rail to be fitted.

12 Fit the 3rd/4th gear selector rail and fork carrier in a similar manner to the reverse rail and fork. Additionally, fit an interlock pin through the hole in the 3rd/4th selector rail. The reverse to 3rd/4th interlock plunger must also be located.

with the corresponding mainshaft and input shaft gears, aligning the countershaft gear with the thrustwashers and countershaft location holes in the gearbox front and rear faces. Carefully refit the countershaft from the rear, with the flat section on the rear end of the shaft positioned as shown in Fig. 6.22. Drive the shaft through, keeping it against the dummy shaft in order that the needle bearings are not dislodged.

28 With the countershaft fully located, the rear extension housing can

Chapter 6 Manual gearbox and automatic transmission

13 Insert the 1st/2nd selector rail and fork carrier. Locate the interlock plunger between 3rd/4th and 1st/2nd gear selector rails before pressing home the rail.
14 Drive the retaining pins into position to locate the respective rails and selector fork carriers.
15 Smear the blanking discs with a sealant and carefully fit them into the end of the housing. Screw the blanking plug (or reversing light switch) into position.
16 Insert the 1st/2nd gear selector rail spring and ball and screw the detent plug in to secure. Smear the plug thread with sealant prior to fitting.
17 Align the selector rails in the neutral position before refitting to the gearbox.
18 Use a new gasket between the gearbox and selector housing joint faces and check that all traces of the old gasket are removed from the flange faces.

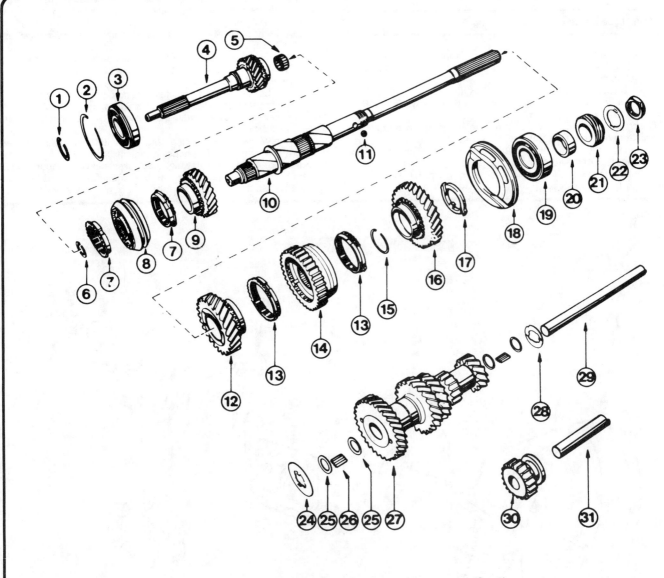

Fig. 6.27 The type G gearbox internal components (Sec 9)

1 Circlip	9 3rd gear	15 Circlip	24 Thrust washer
2 Circlip	10 Mainshaft	16 1st gear	25 Spacer shims
3 Ball-bearing	11 Speedometer worm locking ball	17 Oil scoop ring	26 Needle rollers(22)
4 Input shaft	12 2nd gear	18 Bearing retainer	27 Countershaft gears
5 Needle bearing	13 Synchroniser baulk ring	19 Ball-bearing	28 Thrust washer
6 Circlip	14 1st/2nd gear synchroniser hub unit	20 Spacer sleeve	29 Countershaft
7 Synchroniser hub unit		21 Speedometer worm	30 Reverse idler gear
8 3rd/4th gear synchroniser		22 Lockwasher	31 Idler shaft
		23 Mainshaft nut	

104 Chapter 6 Manual gearbox and automatic transmission

19 Refit the housing, taking care to align and locate the selectors with their respective forks. The forks must be fitted with their casting marks facing forwards.
20 Smear the housing retaining bolt threads with sealant, then insert the bolts and tighten evenly to the specified torque.
21 Reconnect the gear lever to the selector housing.
22 Refill the gearbox with the specified quantity and grade of oil, then test drive the vehicle to check gear selection.

9 Manual gearbox (type G) – dismantling

1 Refer to Section 4 for general details and tool requirements, then read through the overhaul instructions to assess the procedures involved before starting.
2 Drain any remaining oil from the gearbox housing.
3 The gearbox components and their layout are shown in Fig. 6.27 and 6.28.
4 Unscrew and remove the clutch housing bolts and withdraw the housing from the front face of the gearbox.
5 Disconnect the clutch release bearing lever by pulling it sideways then remove it from the housing.
6 Unbolt and remove the gear selector housing which is secured by eight bolts. Detach the gasket.

7 Before removing the rear extension housing, consideration must be given as to whether to renew the oil seal and housing bush. *If the Ford special tool is to be used, they must only be removed and refitted with the housing in position.*
8 The oil seal should always be renewed and this can be prised from the housing, taking care not to damage the housing.
9 The extension housing bush is best removed using Ford Special tool number 16-011. This is similar to the tool shown in Fig. 6.9 (the bush removal method is also the same as that for the F type gearbox). Remove the four extension housing retaining bolts and remove the housing. Fit the new bush and seal when the extension housing is relocated on assembly.
10 Use a drift of suitable diameter and length and remove the countershaft rearwards from the gearbox. The drift or dummy shaft should be the same length as the countergear, so that on removal it can remain in the countershaft gear cluster to retain the needle roller bearings in position at each end (photo).
11 Allow the countershaft gear cluster to drop down in the gearbox so that it is disengaged from the input and mainshaft gears.
12 The rear extension and mainshaft can now be withdrawn from the rear of the gearbox.
13 Unscrew the three retaining bolts and withdraw the input shaft bearing retainer housing. This housing contains the front oil seal so if it is intended to re-use the seal (unwise), tape the splines of the input shaft to prevent damaging the seal during removal and refitting of the housing. Pull or drive out the input shaft to remove it from the gearbox (photo).

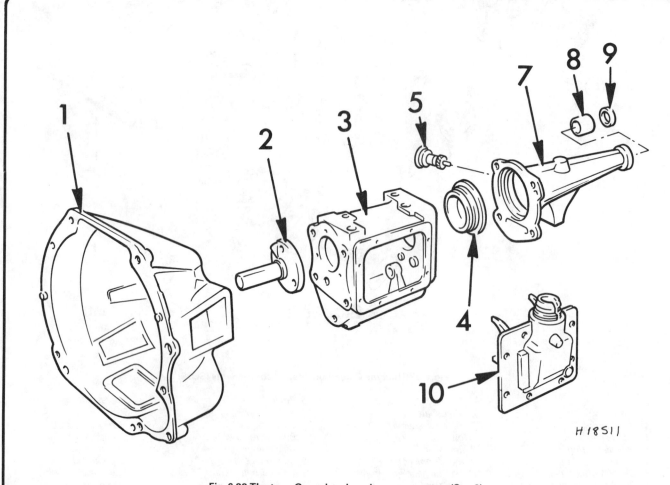

Fig. 6.28 The type G gearbox housing components (Sec 9)

1 Clutch housing	4 Bearing retainer	7 Extension housing	9 Oil seal
2 Input shaft bearing retainer	5 Speedometer drive pinion	8 Extension housing bush	10 Selector housing
3 Gearbox casing			

Chapter 6 Manual gearbox and automatic transmission

9.10 Countershaft removal using a drift

9.13 Prise bearing outer track from main casing with a screwdriver

10.16 Insert the spacer shim, needle rollers and outer spacer shim to the countershaft gear at each end

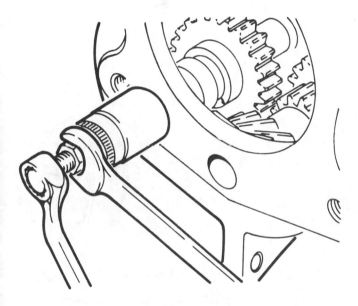

Fig. 6.29 Reverse gear idler shaft removal method (Sec 9)

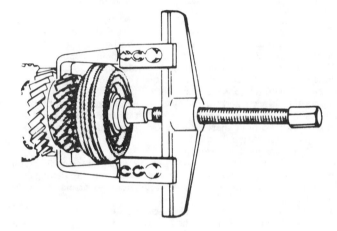

Fig. 6.30 Use a puller to withdraw the 3rd/4th gear synchro hub and 3rd gear (Sec 9)

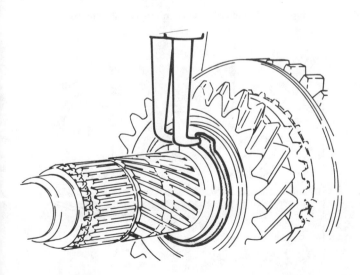

Fig. 6.31 Remove the circlip retaining 1st/2nd gear synchro hub (Sec 9)

14 To remove the reverse idler gear and shaft, draw it from the gearbox by screwing an M8 x 60 mm bolt into the end of the shaft. The bolt must be fitted with a nut and a suitable socket to act as a spacer – see Fig. 6.29. Hold the bolt still and tighten the nut to draw out the shaft. Remove the gear, noting its direction of fitting.
15 With the main assemblies removed from the gearbox they can be cleaned and dried for inspection and further dismantling as necessary.
16 Should it be apparent at this stage that the gear assemblies and their associated components are badly worn and possibly damaged, it will almost certainly be more economical and practical to purchase a new or good secondhand gearbox instead of trying to repair the existing one.
17 The dismantling procedures for the main assemblies are given in the following sub-sections.

Input shaft dismantling
18 Refer to Section 5, paragraphs 20 and 21. The dismantling instructions are identical.

Countershaft gear dismantling
19 Withdraw the dummy countershaft from the gear cluster and lay the gear on the bench together with the thrust washers, spacer shims and needle roller bearings. There are 22 needle roller bearings at each end. Keep them separate and safe if they are likely to be refitted.

Mainshaft dismantling
20 Start at the front end of the shaft. Prise open the circlip retaining the 3rd/4th synchroniser hub and remove the clip.
21 To remove the 3rd/4th synchroniser hub and 3rd gear you will need a suitable puller which must be located as shown in Fig. 6.30. Draw the hub and 3rd gear from the shaft.

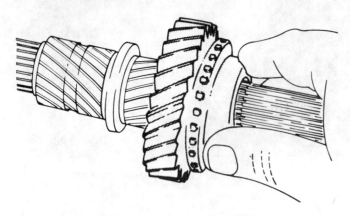

Fig. 6.32 Locate 2nd gear as shown (Sec 10)

22 Support the mainshaft in a soft-jawed vice and remove the mainshaft nut at the rear, having released its lockwasher tab. The nut will be fairly tight so ensure that you have the shaft securely supported.
23 Depending on facilities, the speedometer drivegear can be withdrawn from the shaft using a suitable puller, or by fabricating a U-shaped plate which will slot in front of the gear. Whilst the plate is supported in a vice, and the mainshaft assembly supported by hand, use a soft-headed mallet and drive the shaft forwards through the gear to release it from its shaft location. On removing the speedometer drivegear, retain its interlock ball.
24 Withdraw the spacer sleeve.
25 The bearing and bearing retainer can now be pressed or driven off rearwards using a suitable tube drift, but take care not to damage the retainer or 1st gear. Remove the ball-bearing from the retainer when it is withdrawn.
26 Withdraw the oil scoop and 1st gear rearwards.
27 Prise open and remove the circlip retaining the 1st/2nd gear synchroniser hub in position. The synchroniser hub can then be removed from the shaft together with 2nd gear (Fig. 6.31).
28 To dismantle the synchroniser hubs first note the alignment markings on the hub and sleeve, then remove the sleeves and blocker bars, keeping them in their relative positions.
29 With the gearbox fully dismantled, the respective components can be cleaned and inspected as detailed in Section 6.
30 To dismantle the selector housing refer to Section 11.

10 Manual gearbox (type G) – reassembly

1 When reassembling the gearbox, rebuild each sub-assembly in turn, taking care not to damage any parts. Lubricate each part as it is assembled with the specified grease where indicated.

Mainshaft assembly

2 Commence by reassembling the synchroniser hubs. Locate the blocker bars and retain with the springs, which must be fitted so that their gaps are offset to each other but with the opposing hooked ends aligned as shown in Fig. 6.14. The check marks on the sleeve and hub must be in alignment when fitted.
3 Slide 2nd gear into position on the shaft together with its baulk ring (Fig. 6.32).
4 Locate the 1st/2nd gear synchroniser hub unit and make secure by fitting the retaining circlip (ensuring that it is fully located). The peripheral groove of the sliding sleeve must face rearwards when fitted.
5 Slide 1st gear into position together with its synchroniser hub, followed by the oil scoop ring. The scoop ring is fitted with its oil groove facing the gear as shown in Fig. 6.33.
6 Smear the mainshaft bearing location in its retainer with grease. Similarly lubricate the ball-bearing, then locate the bearing into its retainer. The bearing and retainer are then pressed or drifted into position using a suitable piece of tubing.
7 Check that the bearing and spacer are fully located and then slide the spacer into position on the shaft.

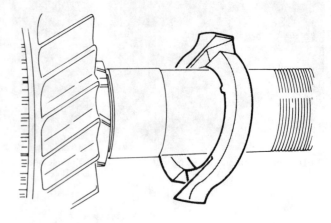

Fig. 6.33 Fit the oil scoop ring facing as shown (Sec 10)

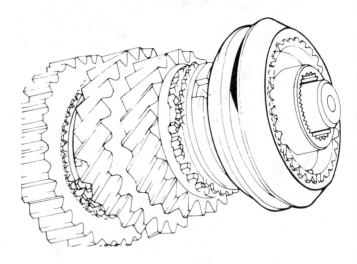

Fig. 6.34 Fit the 3rd/4th synchroniser hub (Sec 10)

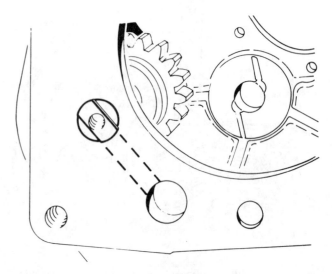

Fig. 6.35 Reverse idler shaft to be positioned as shown (Sec 10)

Chapter 6 Manual gearbox and automatic transmission

Fig. 6.36 Locate the countershaft thrustwashers in the gearbox (Sec 10)

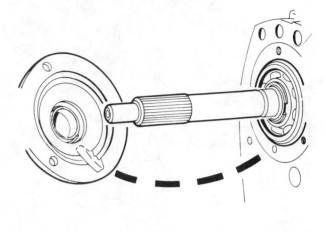

Fig. 6.37 Align the oil return ports as indicated (Sec 10)

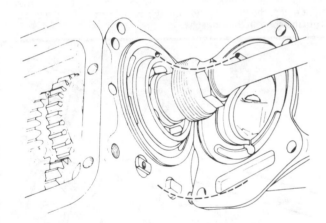

Fig. 6.38 Align the bearing retainer hole with the guide pin (Sec 10)

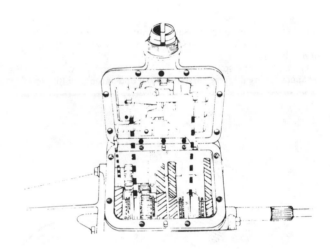

Fig. 6.39 Locate the selector forks and housing (Sec 10)

8 Grease and locate the interlock ball and then slide the speedometer gear into position on the shaft and over the interlock ball.
9 Slide the lockwasher into position, engaging its tab into the recess in the speedometer drivegear. Fit the mainshaft nut and tighten it to the specified torque whilst supporting the mainshaft in a soft-jawed vice. Bend over the lockwasher tabs to secure the nut. (If a suitable torque wrench adaptor is not available, the correct torque for the mainshaft nut may have to be estimated.)
10 Working at the front end of the shaft, fit the 3rd gear together with its baulk ring.
11 The 3rd/4th synchroniser hub is fitted next by pressing into position on the shaft using a length of tube of the correct diameter, ie, to press on the hub inner shoulder. The synchroniser hub is fitted with the selector fork groove offset to the rear (Fig. 6.34).

Input shaft assembly

12 If the bearing has been removed for renewal, locate and press the new bearing into position on the front of the shaft. Smear the shaft with grease to assist fitment of the bearing.
13 With the bearing in position, fit the retaining circlip, ensuring that it is secure in its groove.
14 The larger circlip can also be fitted at this stage into the groove in the periphery of the outer bearing race.

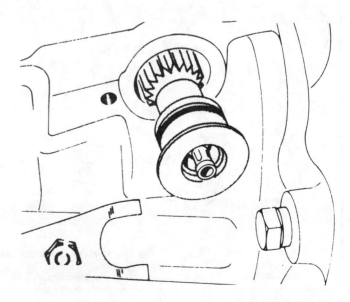

Fig. 6.40 Insert the speedometer drivegear fitted with a new O-ring seal (Sec 10)

Chapter 6 Manual gearbox and automatic transmission

10.22 Drift input shaft into position in front of gearbox

Countershaft assembly

15 Insert the dummy countershaft into the countergear and then fill the cavity at each end with the specified grease (between the dummy shaft and bore).
16 Locate the inner spacer shim, the 22 needle roller bearings and the outer spacer shim, then repeat at the other end of the gear (photo).

Gearbox general assembly

17 If not already fitted, insert a new oil seal into the input shaft bearing retainer housing so that when fitted the seal lip faces the gearbox side of the housing. Smear the seal lip with grease.
18 Before refitting any of the gear assemblies to the gearbox, check that it is thoroughly clean.
19 Hold the reverse idler gear in position in the gearbox with its selector groove in the hub section facing rearwards. Insert the idler gear shaft (lubricated with grease), and drive it into position from the rear. When fitted the shaft must be located so that its rear flat end aligns with the countershaft bore – see Fig. 6.35.
20 Smear the faces of the countershaft thrustwashers with grease and locate them into their recesses in the gearbox. The large washer fits at the front end of the case and the tabs of both washers must point towards the casing when fitted (Fig. 6.36).
21 The countershaft gear and dummy shaft can now be carefully lowered into the gearbox. Take care not to dislodge the thrustwashers.
22 Locate the input shaft into the front face of the gearbox so that the bearing outer circlip butts against the gearbox (photo).
23 The input shaft bearing retainer housing can now be fitted, but first locate its flange gasket. Smear the gasket with grease to enable it to stick in position on the flange. It is most important to ensure that when fitted the gasket and bearing housing are correctly located, with the oil hole in the front of the gearbox aligned with the grooved section of the gasket and retainer housing (Fig. 6.37). Temporarily tape the input shaft splines to avoid damaging the retainer oil seal.
24 The retainer housing bolts must be smeared with a sealing compound prior to fitting. Tighten them to the specified torque.
25 Lubricate the needle roller bearing and fit it onto its location on the front end of the mainshaft (or into the input shaft). Slide the baulk ring onto the cone on the input shaft.

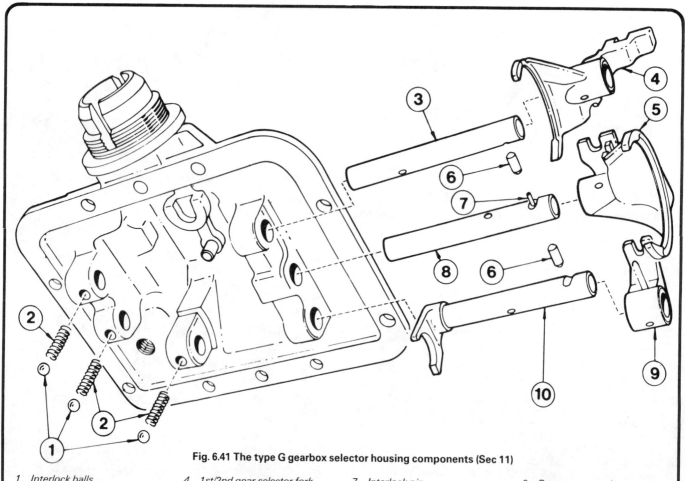

Fig. 6.41 The type G gearbox selector housing components (Sec 11)

1 Interlock balls	4 1st/2nd gear selector fork	7 Interlock pin
2 Springs	5 3rd/4th gear selector fork	8 3rd/4th gear selector rail
3 1st/2nd gear selector rail	6 Interlock plunger	9 Reverse gear selector dog
		10 Reverse gear selector rail with fork

Chapter 6 Manual gearbox and automatic transmission

26 Smear the extension housing gasket with grease and locate it in position on the rear gearbox face.
27 The mainshaft can now be carefully fitted into the gearbox and the front bearing guided into the input shaft.
28 Turn the gearbox onto its side and align the countergear with the shaft location holes in the gearbox end faces, with the respective gears engaged with those of the input and mainshaft assemblies.
29 Check that the end thrustwashers are in position and then push the countershaft through from the rear to locate the countergear. As the shaft is pushed through, keep it against the dummy shaft to ensure that the needle rollers are retained in position.
30 When in position the countershaft must be located with the shaft flat end section pointing along the recess for the reverse idler shaft in the extension housing.
31 Align the bearing retainer hole with the guide pin in the extension housing, locate the housing (Fig. 6.38) and make secure with the four retaining bolts. The retaining bolts must be coated with a suitable sealing compound before fitting.
32 With the extension housing fitted, a new bush can be driven into position using a suitable piece of tubing. When in position ensure that the bush oil return groove is at the bottom of the extension housing to the rear and that the notch in the bush is **not** situated over the oilway.
33 Tap a new oil seal carefully into position in the end of the extension housing, lubricating its seal lip before fitting.
34 Refit the gear selector housing. Smear the housing gasket with grease and with the gears and selector forks in neutral, lower the housing onto the gearbox. The selector forks must fit into the synchroniser hub grooves as shown in Fig. 6.39. Tighten the bolts securely.
35 Refit the speedometer drive gear unit, with a new O-ring seal (Fig. 6.40).
36 Reassemble the clutch release lever and bearing and locate in the clutch housing.
37 Refit the clutch housing to the front face of the gearbox, smear the threads of the retaining bolts with sealant, then insert and tighten them to the specified torque.
38 With the gearbox now reassembled, check that all gears can be selected by temporarily fitting the gear lever and turning the mainshaft with each gear engaged in turn.
39 On refitting the gearbox, do not forget to refill it with the correct grade and quantity of oil. If new components have been fitted, the gearbox should be run-in gently for the first few hundred miles.

11 Selector housing (type G) – removal, dismantling, reassembly and refitting

1 To remove the housing and initially inspect it, refer to Section 8, paragraphs 1 to 4 inclusive. If the assembly is to be dismantled, refer to Fig. 6.41 and proceed as follows.
2 Extract the 1st/2nd gear selector fork to rail lockpin, then withdraw the rail and detach the selector fork, catching the ball and spring as they are removed.
3 Repeat the above operation for the 3rd/4th gear selector fork and rail.
4 Remove the reverse gear selector rail and fork in a similar manner to that given in paragraph 2.
5 Extract the interlock plungers by turning the housing onto its side.
6 Reassembly is a reversal of the removal procedure, but care must be taken to fit the selector forks correctly. Use new retaining pins for their location on the rails. Don't forget to insert the interlock fitting the respective rails and forks.
7 Refer to Section 8, paragraphs 17 to 22 inclusive, for details of housing fitment, but take note that the selector forks differ and instead of aligning the forks with the selectors, the selector forks must align with the grooves in the synchroniser hubs.

12 Manual gearbox (type N) – dismantling into major sub-assemblies

1 Clean the exterior of the gearbox with paraffin and wipe dry.
2 Remove the clutch release bearing and lever with reference to Chapter 5.
3 Unscrew and remove the reversing light switch (photo).
4 Unbolt the clutch bellhousing from the front of the gearbox. Remove the gasket (photos).

12.3 Removing the reversing light switch

12.4A Clutch bellhousing bolts (arrowed)

12.4B Removing the clutch bellhousing

12.5A Removing the input shaft bearing retainer ...

12.5B ... and gasket

12.6A Unscrew the retaining bolts ...

Chapter 6 Manual gearbox and automatic transmission

5   Unscrew the bolts and withdraw the input shaft bearing retainer and gasket from the front of the gearbox (photos).
6   Unscrew the bolts and remove the top cover and gasket (photos). As the cover is removed, take care not to change the angle of the oil deflector plate.
7   Invert the gearbox and allow the oil to drain, then turn it upright again.
8   Unscrew the bolts and lift the 5th gear locking plate from the extension housing (photo).
9   Extract the 5th gear locking spring and pin from the extension housing (photos). Use a screw to remove the pin.
10   Working through the gear lever aperture, use a screwdriver or small drift to tap out the extension housing rear cover (photo).
11   Select reverse gear and pull the selector shaft fully to the rear. Support the shaft with a piece of wood then drive out the roll pin and withdraw the connector from the rear of the selector rod (photos).
12   Unbolt and remove the extension housing from the rear of the gearbox. If necessary tap the housing with a soft-faced mallet to release it from the dowels. Remove the gasket (photos).
13   Prise the cover from the extension housing and withdraw the speedometer driven gear (photo).
14   Select neutral then using an Allen key, unscrew the selector locking mechanism plug from the side of the main casing then extract the spring and locking pin if necessary using a pen magnet (photos).
15   Drive the roll pin from the selector boss and selector shaft.
16   If necessary the selector shaft centralising spring and 5th gear locking control may be removed. Using a small screwdriver push out the pin and plug and slide the control from the selector shaft (photos).
17   Note the location of the selector components then withdraw the selector shaft from the rear of the gearbox and remove the selector boss and locking plate, 1st/2nd and 3rd/4th selector forks, and 5th gear selector fork and sleeve. Note that the roll pin hole in the selector boss is towards the front (photos).
18   Extract the circlip and pull the 5th gear synchroniser unit from the main casing leaving it loose on the mainshaft (photos).
19   Slide the 5th driven gear from the synchroniser unit hub (photo).
20   Select 3rd gear and either 1st or 2nd gear by pushing the respective synchroniser sleeves – this will lock the mainshaft and countershaft gear cluster.
21   Unscrew and remove the 5th driving gear retaining nut while an assistant holds the gearbox stationary (photo). The nut is tightened to a

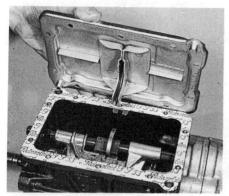

12.6B ... remove the top cover ...

12.6C ... and gasket

12.8 Removing the 5th gear locking plate

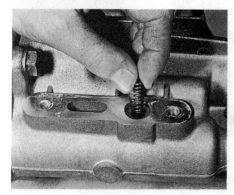

12.9A Extracting the 5th gear lock spring ...

12.9B ... and pin

12.10 Removing the extension housing rear cover

12.11A Drive out the roll pin ...

12.11B ... and remove the selector rod connector

12.11C Selector rod connector

12.12A Removing the extension housing ...

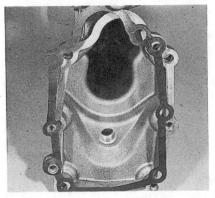

12.12B ... and gasket

12.13 Removing the speedometer driven gear

12.14A Unscrew the plug ...

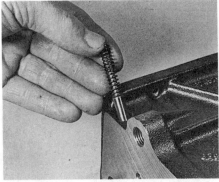

12.14B ... to remove the selector lock spring and pin

12.16A Insert a small screwdriver ...

12.16B ... and push out the plug ...

12.16C ... and pin from the selector shaft centralising spring and 5th gear locking control

12.17A Removing the selector boss and locking plate ...

12.17B ... 1st/2nd selector fork ...

12.17C ... 3rd/4th selector fork ...

12.17D ... 5th gear interlock sleeve ...

12.17E ... and 5th gear selector fork

12.18A Extract the circlip ...

12.18B ... and remove the 5th gear synchroniser dog hub ...

12.18C ... and 5th gear synchroniser unit

12.19 Showing 5th driven gear

12.21 Removing the 5th driving gear retaining nut

12.22A Removing the washer from the 5th driving gear

12.22B Pull the 5th driving gear from the splines with a puller ...

12.22C ... and remove the 5th driving gear from the countershaft gear cluster

12.22D Removing the spacer ring

12.23 Removing the countershaft gear cluster bearing retaining circlip

12.24 Using a screwdriver to remove the countershaft gear cluster bearing from the intermediate housing

Chapter 6 Manual gearbox and automatic transmission

12.26 Removing the input shaft

12.27A Removing the 4th gear baulk ring

12.27B Removing the input shaft needle roller bearing

12.28A Removing the mainshaft and intermediate housing ...

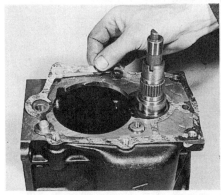
12.28B ... and gasket

12.29A Removing the countershaft and gear cluster

12.29B View of the countershaft and gear cluster

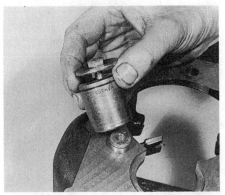
12.30A Method of removing the reverse gear idler shaft

12.30B Removing the reverse idler gear

high torque setting and an additional extension bar may be required.
22   Remove the washer and pull the 5th driving gear from the countershaft gear cluster using a two-legged puller and socket in contact with the cluster. Remove the spacer ring (photos). Select neutral.
23   Extract the circlip retaining the countershaft gear cluster bearing in the intermediate housing (photo).
24   Using a soft-faced mallet tap the intermediate housing free of the main casing and pull the intermediate housing rearwards as far as possible. Using a screwdriver inserted between the intermediate housing and main casing prise the bearing from the shoulder on the countershaft gear cluster and remove it from the intermediate housing (photo).
25   Using a soft metal drift from the front of the main casing, drive the countershaft rearwards sufficient to allow the gear cluster to be lowered to the bottom of the casing.
26   Ease the input shaft from the front of the casing, if necessary using a small drift inside the gearbox to move the bearing slightly forwards, then using levers beneath the bearing circlip (photo).
27   Remove the 4th gear baulk ring. Remove the input shaft needle roller bearing from the end of the mainshaft or from the centre of the input shaft (photos).
28   Remove the mainshaft and intermediate housing from the main casing. Remove the gasket (photos).
29   Withdraw the countershaft and gear cluster from the main casing (photos).
30   Insert a suitable bolt into the reverse gear idler shaft, and using a nut, washer and socket pull out the idler shaft. Note the fitted position of the reverse idler gear then remove it (photos).
31   Remove the guide from the reverse relay lever then extract the circlip and remove the relay lever from the pivot (photos).
32   Remove the magnetic disc from the bottom of the main casing. Also remove any needle rollers which may have been displaced from the countershaft gear cluster (photo).
33   With the gearbox dismantled into major sub-assemblies, each assembly can be inspected for excessive wear and damage as described in Section 6. Note however that there are five synchro baulk rings, no countershaft gear cluster thrustwashers, two ball bearings and one roller bearing.
34   Renew the housing oil seals shown in the accompanying photos.

Chapter 6 Manual gearbox and automatic transmission

12.31A Showing location for reverse idler gear guide in relay lever

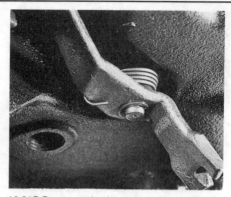
12.31B Reverse relay lever and pivot

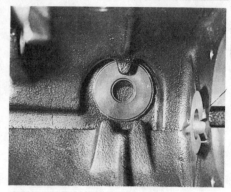
12.32 Magnetic disc location in the bottom of the casing

12.34A Prising out the oil seal from the input shaft bearing retainer

12.34B Fitting a new oil seal to the bearing retainer

12.34C Location of the speedometer driven gear oil seal in the extension housing

12.34D Rear view of the extension housing showing the mainshaft oil seal and bush

13.1 Extract the circlip from the input shaft ...

13.3 ... and remove the bearing

13 Manual gearbox input shaft and mainshaft (type N) – dismantling and reassembly

Input shaft
1 Extract the small circlip from the input shaft (photo).
2 Locate the bearing outer track on top of an open vice, then using a soft-faced mallet, drive the input shaft down through the bearing.
3 Remove the bearing from the input shaft noting that the groove in the outer track is towards the front splined end of the shaft (photo).
4 Place the input shaft on a block of wood and lightly grease the bearing location shoulder.
5 Locate the new bearing on the input shaft with the circlip groove facing the correct way. Then using a metal tube on the inner track drive the bearing fully home (photo).
6 Refit the small circlip.

Mainshaft
7 Extract the circlip and slide the 3rd/4th synchroniser unit together with the 3rd gear from the front of the mainshaft, using a two-legged puller where necessary. Separate the gear and unit, then remove the 3rd gear baulk ring (photos).
8 Remove the outer ring from the 2nd gear then extract the thrust washer halves (photos).
9 Slide the 2nd gear from the front of the mainshaft and remove the 2nd gear baulk ring (photos).
10 Mark the 1st/2nd synchroniser unit hub and sleeve in relation to each other and note the location of the selector fork groove, then slide the sleeve forward from the hub and remove the blocker bars and springs. Note that the synchroniser hub cannot be removed from the mainshaft (photos).
11 Using a suitable puller pull the speedometer drivegear off the rear of the mainshaft (photo).
12 Extract the circlip then remove the 5th gear synchroniser unit and 5th driven gear from the mainshaft.
13 Extract the circlip retaining the mainshaft bearing then support the intermediate housing on blocks of wood and drive the mainshaft through the bearing with a soft-faced mallet (photos).
14 Remove the oil scoop ring, 1st gear, and 1st gear baulk ring (photos).

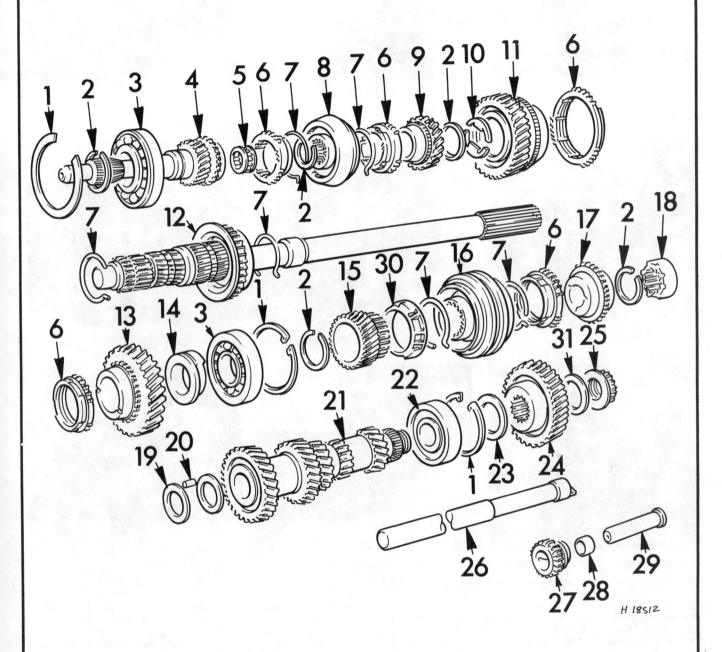

Fig. 6.42 Exploded view of the type N gearbox internal components (Sec 13)

1 Circlip	9 3rd gear	17 5th gear synchroniser hub	25 12 sided nut
2 Circlip	10 Thrust half washer	18 Speedometer drivegear	26 Countershaft
3 Ball bearing	11 2nd gear	19 Spacer washers	27 Reverse idler gear
4 Input shaft	12 Mainshaft with	20 Needle rollers	28 Bush
5 Needle roller bearing	synchroniser unit	21 Countershaft gear cluster	29 Idler shaft
6 Synchroniser baulk ring	13 1st gear	22 Roller bearing	30 Blocker bar retainer
7 Retaining spring	14 Oil scoop ring	23 Thrust washer	31 Washer
8 3rd/4th gear synchroniser unit	15 5th driven gear	24 5th driving gear	
	16 5th gear synchroniser unit		

13.5 Using a metal tube to fit the bearing to the input shaft

13.7A Extract the circlip ...

13.7B ... and remove the 3rd/4th synchroniser and baulk ring

13.7C ... and 3rd gear

13.8A 2nd gear thrustwashers and retaining ring location

13.8B Removing the outer retaining ring

13.8C Removing the thrustwasher halves

13.9A Removing 2nd gear ...

13.9B ... and baulk ring

15 If necessary extract the circlip and drive the ball bearing from the intermediate housing using a metal tube (photo). Also the synchroniser units may be dismantled, but first mark the hub and sleeve in relation to each other. Slide the sleeve from the hub and remove the blocker bars and springs.
16 Clean all the components in paraffin, wipe dry and examine them for wear and damage. Obtain new components as necessary. During reassembly lubricate the components with gearbox oil and where new parts are being fitted lightly grease contact surfaces.
17 Commence reassembly by assembling the synchroniser units. Slide the sleeves on the hubs in their previously noted positions, then insert the blocker bars and fit the springs as shown in Fig. 6.14.
18 Support the intermediate housing then, using a metal tube on the outer track, drive in the new bearing and fit the circlip (photo).
19 Fit the blocker bar spring to the rear of the 1st/2nd synchroniser hub followed by the 1st gear baulk ring (photo).
20 Slide the 1st gear and oil scoop ring (with the oil groove towards 1st gear) onto the mainshaft.
21 Using a metal tube on the mainshaft bearing inner track, drive the intermediate housing onto the mainshaft and fit the circlip (photo). Make sure that the large circlip is towards the rear of the mainshaft.
22 Locate the 5th driven gear and 5th gear synchroniser with circlip, loose on the mainshaft. Tap the speedometer drivegear lightly onto its shoulder – its final position will be determined later (photo).
23 Fit the 1st/2nd synchroniser sleeve to the hub in its previously noted position with the selector groove facing forward then insert the blocker bars and fit the springs as shown in Fig. 6.14.
24 Fit the 2nd gear baulk ring to the 1st/2nd synchroniser unit with the blocker bars located in the slots.
25 Slide the 2nd gear onto the front of the mainshaft and retain with the thrustwasher halves and outer ring (photo).
26 Slide the 3rd gear onto the front of the mainshaft then locate the baulk ring on the gear cone.
27 Locate the 3rd/4th synchroniser unit on the mainshaft splines with the long side of the hub facing the front (photo). Tap the unit fully home using a metal tube then fit the circlip. Make sure that the slots in the 3rd gear baulk ring are aligned with the blocker bars as the synchroniser unit is being fitted.

Chapter 6 Manual gearbox and automatic transmission

13.10A Removing 1st/2nd gear synchroniser sleeve ...

13.10B ... and blocker bars

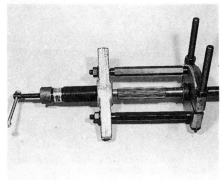

13.11 Removing the speedometer drivegear with a puller

13.13A Extracting the mainshaft bearing circlip

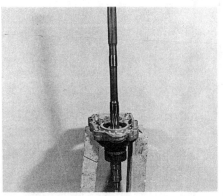

13.13B Method of driving the mainshaft through the intermediate housing and bearing

13.14A Removing the oil scoop ring, 1st gear and baulk ring

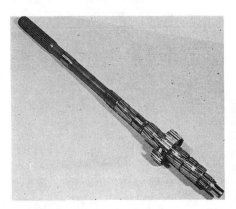

13.14B The dismantled mainshaft

13.15 Extracting the bearing retaining circlip from the intermediate housing

13.18 Intermediate housing and mainshaft bearing

14 Manual gearbox (type N) – reassembly

1   Locate the magnetic disc in the bottom of the main casing.
2   Fit the reverse relay lever onto the pivot and retain with the circlip. Fit the guide to the lever.
3   Position the reverse idler gear in the main casing with the long shoulder facing the rear and engaged with the relay lever. Slide in the idler shaft and tap fully home with a soft-faced mallet.
4   Smear the specified grease inside the end of the countershaft gear cluster then fit the spacers and needle roller bearings – there are 21 needle rollers. Make sure that there is sufficient grease to hold the needle rollers in position during the subsequent operation (photos).
5   Insert the countershaft in the gear cluster until the front end is flush with the front gear on the cluster (photo).
6   Locate the countershaft and gear cluster in the bottom of the main casing.
7   Position a new gasket on the main casing then fit the mainshaft and intermediate housing, and temporarily secure with two bolts.
8   Fit the input shaft needle roller bearing to the end of the mainshaft or in the centre of the input shaft (photo).
9   Fit the 4th gear baulk ring to the 3rd/4th synchroniser unit with the cut-outs over the blocker bars, then fit the input shaft assembly and tap the bearing fully into the casing up to the retaining circlip (photo).
10   Invert the gearbox so that the countershaft gear cluster meshes with the input shaft and mainshaft gears.
11   Using a soft metal drift drive the countershaft into the main casing until flush at the front face – the flat on the rear end of the countershaft must be horizontal (photo).
12   Using a metal tube tap the countershaft gear cluster bearing into

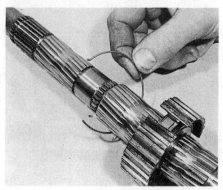
13.19 Fitting the blocker bar spring to the rear of the 1st/2nd synchroniser hub

13.21 Fitting the intermediate housing and bearing

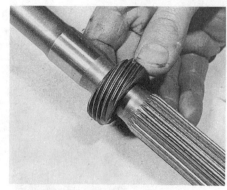
13.22 Fitting the speedometer drivegear

13.25 Showing location hole for 2nd gear thrustwasher halves

13.27 Fitting the 3rd/4th synchroniser unit

14.4A Inserting the spacers in the countershaft gear cluster ...

14.4B ... followed by the needle rollers ...

14.4C ... and outer spacers

14.5 Showing the countershaft inserted in the gear cluster

14.8 Fitting the input shaft needle roller bearing on the mainshaft

14.9 Fitting the input shaft

14.11 Correct position of countershaft before driving into the main casing

Chapter 6 Manual gearbox and automatic transmission

14.12 Fitting the countershaft gear cluster bearing

14.14A Tightening the 5th driving gear nut

14.14B Using a chisel to peen the nut collar

14.14C 5th driving gear nut locked to the countershaft gear cluster

14.16A Fitting the spacer to the 5th gear synchroniser unit

14.16B Fitting the 5th gear synchroniser unit to 5th driven gear

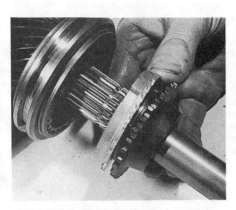

14.16C Fitting the 5th gear baulk ring and dog hub to the mainshaft

14.16D Fitting the circlip to the dog hub

14.17 Checking the circlip to speedometer drivegear distance

the intermediate housing and secure with the circlip (photo).
13 Fit the spacer ring then, using a metal tube, tap the 5th driving gear onto the splines of the countershaft gear cluster.
14 Fit the thrustwasher and retaining nut. Select 3rd gear and either 1st or 2nd gear by pushing the respective synchroniser sleeves. While an assistant holds the gearbox stationary tighten the nut to the specified torque, then lock it by peening the collar on the nut into the slot in the gear cluster (photos).
15 Select neutral then slide the 5th driven gear into mesh with the driving gear.
16 Slide the 5th gear synchroniser unit complete with spacer onto the 5th driven gear. Then using a metal tube, drive the dog hub and 5th baulk ring onto the mainshaft splines while guiding the baulk ring onto the blocker bars. Fit the circlip (photos).

17 Heat up the speedometer drivegear using a hot air blower or by immersing in boiling water, then slide it into its correct position on the mainshaft – the distance between the gear and the 5th gear dog hub circlip should be 121.5 to 122.5 mm (4.787 to 4.826 in) (photo).
18 Locate the 5th gear selector fork in its synchroniser sleeve and locate the interlock sleeve in the groove (short shoulder to front), then insert the selector shaft through the sleeve and selector fork into the main casing (photo).
19 Locate the 1st/2nd and 3rd/4th selector forks in their respective synchroniser sleeves, position the selector boss and locking plate, and insert the selector shaft through the components into the front of the main casing. The roll pin hole in the selector boss must be towards the front.
20 If removed refit the selector shaft centralising spring and 5th gear

Chapter 6 Manual gearbox and automatic transmission

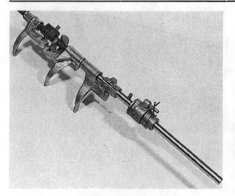

14.18 Selector shaft and components assembled on the bench

14.21 Driving the roll pin into the selector boss

14.23 Fitting the speedometer driven gear cover

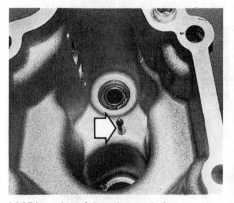

14.25 Location of the selector shaft centralising spring pin (arrowed)

14.26 Tightening the extension housing bolts

14.30 Applying sealer to 5th gear locking plate location

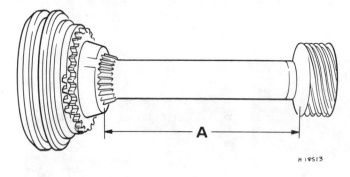

Fig. 6.43 Speedometer drivegear fitting distance (A) (Sec 14)

For distance A, see text

locking control by inserting the pin and plug.
21   Align the holes then drive the roll pin into the selector boss and selector shaft (photo).
22   Insert the selector locking pin and spring, apply sealer to the plug threads, then insert and tighten the plug using an Allen key.
23   Fit the speedometer driven gear to the rear extension housing. Apply a little sealer to the cover then press it into the housing (photo).
24   Remove the temporarily fitted bolts from the intermediate housing then select 4th gear.
25   Stick a new gasket to the extension housing with grease, and fit the housing to the intermediate housing. Take care not to damage the rear oil seal, and make sure that the selector shaft centralising spring locates on the pin (photo).
26   Insert the bolts and tighten them to the specified torque in diagonal sequence (photo). Before inserting the three bolts which go right through the main casing, apply sealer to their threads.
27   Select reverse gear and locate the connector on the rear of the selector rod. Support the rod with a piece of wood then drive in the roll pin. Select neutral.
28   Press the rear cover into the extension housing.
29   Check that the 5th gear interlock sleeve is correctly aligned, then insert the 5th gear locking pin and spring.
30   Apply some sealer to the 5th gear locking plate, locate it on the extension housing, and insert and tighten the bolts to the specified torque (photo). Ensure that sealant is not allowed to enter and block the drillings.
31   Fit the gearbox top cover together with a new gasket and tighten the bolts to the specified torque in diagonal sequence.
32   Fit the input shaft bearing retainer (oil slot downwards) together with a new gasket and tighten the bolts to the specified torque in diagonal sequence. Where necessary apply sealer to the bolt threads.
33   Fit the clutch bellhousing to the front of the gearbox together with a new gasket. Apply sealer to the bolt threads, then insert the bolts and tighten them to the specified torque in diagonal sequence.
34   Insert and tighten the reversing light switch in the extension housing.
35   Fit the clutch release bearing and lever with reference to Chapter 5.

15 Manual gearbox (type MT75) – special overhaul requirements

1   This gearbox cannot easily be dismantled, overhauled and reassembled without the use of the following special tools (photo):

Ford tool No	Description
16-040	Input shaft guide sleeve removal/installation tool
16-041	Input shaft removal and installation tool
16-041-01	Adaptor (input shaft removal)
16-042A	Adaptor (mainshaft/coupling installer)
16-042A-01	Mainshaft and coupling installer
16-047	Dummy selector shaft

These tools are available from the following company:
V.L. Churchill Ltd, P.O. Box No.3, London Road, Daventry, Northants, NN11 4NF.

Chapter 6 Manual gearbox and automatic transmission

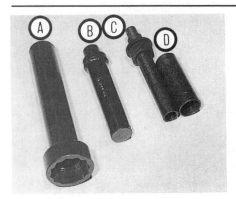

15.1 Some of the Ford special tools used to overhaul the MT75 gearbox
A 16.040
B 16.044 (Guide sleeve oil seal installer)
C 16.042A.01
D 16.042A

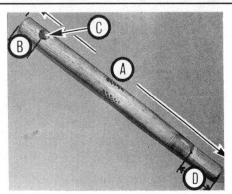

15.4 Dummy selector shaft 16-047 can be fabricated from a 15 mm dia dowel rod as shown to the following dimensions
A Total length = 217.7 mm
B End of rod to hole centre = 25.4 mm
C Hole diameter = 6.3 mm
D Length of rebate = 31.7 mm

16.12 Withdrawing the rear drive flange with a puller

16.13 Extracting the old rear oil seal

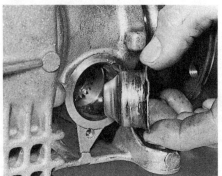

16.14A Extract the blanking plug ...

16.14B ... and remove the speedometer drive pinion

2 Unfortunately the fabrication of such tools is considered beyond the scope of the average DIY mechanic, and unless substitute tools are produced to a high and exacting standard, the gearbox can easily be damaged beyond repair using sub-standard copies.
3 The gearbox will also need to be suitably supported, and if available, use special tool No 16-045 for this purpose. Failing this a suitable metal bracket plate can easily be fabricated to suit, but ensure that both the support plate and the mounting vice are strong and well supported as the gearbox is quite heavy.
4 The dummy selector shaft (16-047) can easily be substituted by fabricating a copy using a 15 mm diameter length of dowel rod, cut and shaped to suit (photo).
5 Unless the aforementioned special tools are available, it is recommended that the overhaul of this gearbox be entrusted to a Ford dealer or a transmission repair specialist. If the gearbox is known to have covered a high mileage, a service exchange unit should be seriously considered as a more practical solution in the long run. The overhaul instructions in the following Sections assume that the special tools mentioned are available.

16 Manual gearbox (type MT75) - dismantling

Note: *During the dismantling procedures keep the various components in order of removal to avoid possible confusion during reassembly. Snap-rings and circlips in particular must be retained in order of fitting as they are of selective thickness and must be renewed during reassembly with equivalents of equal thickness.*

1 If not already drained, remove the drain plug and drain the oil from the gearbox into a suitable container for disposal. Refit the drain plug.

2 Remove the clutch release bearing and lever as described in Chapter 5.
3 Relieve the lock staking of the output shaft flange nut, then with an assistant supporting the gearbox, unscrew and remove the retaining nut. Prevent the flange from turning by bolting a bar to the flange as shown (Fig. 6.44). The nut must be renewed during reassembly.
4 Working at the clutch housing end, mark the relative positions of the input shaft guide sleeve and the front face of the gearbox (as a general guide when reassembling), then unscrew and remove the guide sleeve using tool No 16-040 (Fig. 6.45). Remove the thrustwasher.
5 Extract the input shaft circlip.
6 Mark the relative position of the countershaft bearing retainer to the front face of the housing, then unbolt and remove the retainer locking plate. The countershaft bearing retainer can now be unscrewed using a 17 mm Allen key.
7 Unscrew the three selector plugs from the three interlock mechanisms (Fig. 6.46). Extract the springs, pins, sleeves and balls, but keep them together and marked for position of fitting.
8 Unscrew and remove the reversing lamp switch.
9 Refer to Fig. 6.47 to identify the location of the two reverse idler gear shaft bolts (colour coded blue), then loosen them both off but only remove the front bolt at this stage.
10 Unscrew and remove the ten front to rear housing retaining bolts, then carefully lever the rear housing from the front, but only lever against the reinforced ribs (not between the mating surfaces). When the seal is broken, the two housings may well be held still by the two dowels on the mating face. It should be possible, however, with a little additional leverage and by gently tapping back the input shaft with a soft faced mallet, to fully separate the two housings. Alternatively, withdraw the front housing from the rear housing by screwing special tool No 16-041-01 into the threaded end of the guide sleeve location aperture, (having first ensured that the threads are clean). Tighten the tool to withdraw the front housing from the rear. Excessive force must not be applied as the 4th gear baulk ring could easily be damaged. Note that the

Chapter 6 Manual gearbox and automatic transmission

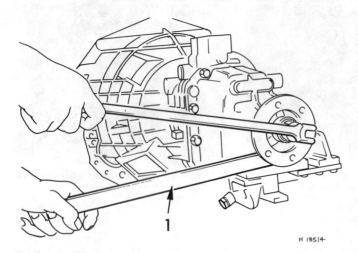

Fig. 6.44 Using a suitable bar (1) to hold the output flange while undoing the retaining nut (Sec 16)

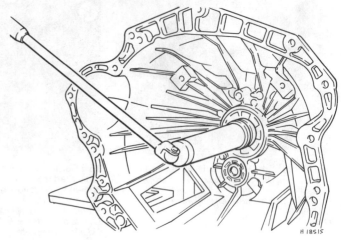

Fig. 6.45 Input shaft guide sleeve removal using special tool 16-040 (Sec 16)

dowels must not be removed.

11 If available, bolt the support plate bracket (tool No 16-045 or equivalent) to the rear housing and support the unit firmly in a vice.
12 Withdraw the rear drive flange from the mainshaft using a suitable puller as shown (photo).
13 Withdraw the rear oil seal from the rear housing using suitable screwdrivers as levers (photos).
14 Carefully prise out the speedometer drive pinion blanking plug and withdraw the speedometer drive pinion. If damaged or distorted during removal, the plug must be renewed when reassembling (photos).
15 Withdraw the speedometer drive worm and spacer sleeve from the rear end of the mainshaft. Note that the blue mark on the worm faces the rear flange (photos).
16 Locate and secure the rear housing assembly so that its rear end is down.
17 Using a suitable drift, drive out the roll pins from the central selector shaft, (but support the shaft as the pins are knocked out). Note the orientation and fitting positions of the selector fork and driver on the central shaft, then withdraw them from the unit. As the shaft is removed take care not to loose the lock plunger (photo).
18 Insert dummy selector shaft (tool No 16-047 or fabricated equivalent, see Section 15 for details) into the vacated centre selector location in the housing (photo).
19 Unscrew the remaining reverse idler gear shaft retaining bolt from the housing, then attach a puller as shown to remove the mainshaft from the bearing in the housing (photo).
20 Before removing the mainshaft from the bearing, check that the remaining two selector shafts are set in the neutral position and the dummy shaft is also correctly set. This is necessary to ensure that the selector shafts do not get jammed up by the selector plungers during removal of the combined gear assemblies.
21 Strap together the mainshaft and countershaft assemblies using plastic cable ties as shown at the front and rear ends (photo). Do not pass these ties around the selector shafts.
22 Locate a third cable around the gear assemblies and the selector shafts, but do not tighten this cable too tight or the selector shafts will be tilted making withdrawal difficult.
23 The combined assemblies can now be withdrawn as a unit, comprising the mainshaft, countershaft, reverse idler unit and the two remaining selector shafts. As they are withdrawn from the rear housing the dummy selector shaft is left in position in the housing. If during removal the selector baulk rings start to move to the rear on their hubs, it is indicative that the selector shafts have jammed, and although the mainshaft is free, it is prevented from further movement by the restraining action of the selector forks. **Do not** try to force the shafts out if this happens, but push the assembly back into the housing to free the selector forks in their rings.
24 As the combined shafts/selector forks are withdrawn, the reverse idler gear will need to be collected as it falls free.
25 To separate the gear assemblies, and selector shafts and forks for further dismantling and overhaul, cut the cable ties.
26 Remove the interlock plungers from the rear housing by extracting them using a pen magnet via the selector shaft bores or by removing a side access plug.

16.15A Remove the speedometer drive worm ...

16.15B ... and spacer sleeve

16.17 Drive out the roll pin from the central selector shaft/fork

Chapter 6 Manual gearbox and automatic transmission

16.18 Insert the dummy selector shaft

16.19 Attach the puller as shown to push out the mainshaft

16.21 Strap the gear assemblies together as shown using cable ties for convenience

17 Gearbox housings (type MT75) – overhaul

Rear housing

1 The countershaft rear roller bearing unit can be withdrawn from the housing using a conventional internal puller/extractor tool.
2 The mainshaft rear bearing is secured in position by a retainer plate. Undo the three retaining bolts to remove the plate then drive out the bearing (from the outside inwards) using a suitable tube drift (photo).
3 The selector shaft bushes can be removed from the rear housing using a suitable withdrawal tool, preferably Ford tool No 21-036A. Ensure the tool is supported with the deeper ports by using a spacer bolt (Fig. 6.48).
4 Drive the new bush into position using a suitable drift so that when fitted the bush is flush with the end of the housing (Fig. 6.49).
5 Drive the new mainshaft bearing into position using a suitable drift and/or puller. If available use Ford tools 15-035 (spindle only), 15-064 and 15-068. With the bearing fully fitted, relocate the retainer plate and fit the securing bolts, tightening them to the specified torque wrench setting.
6 Position the new countershaft bearing in the housing and drive it into place, as far as the stop using a suitable tubular drift.

Front housing

7 If required, renew the selector shaft front end bushes as described previously for the rear end bushes (Fig. 6.50).
8 The right and left-hand gear lever plungers can be removed by first undoing the lock screw using a T40 Torx socket, then withdraw the spring and ball. Now unscrew the plug from each side of the housing using an 8 mm hexagon socket and remove the springs and plungers. Note that the left and right-hand plungers differ (photos).
9 Remove the input shaft bearing from the housing, by driving it out using a suitable drift, towards the clutch housing. Remove the circlip from the bearing as this must be renewed during reassembly.

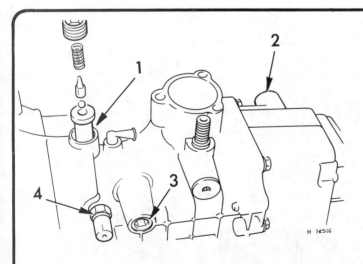

Fig. 6.46 Selector shaft interlock assemblies (Sec 16)

1 Central shaft (3rd/4th)
2 Right-hand shaft (1st/2nd)
3 Left-hand shaft (5th/reverse gear)
4 Reversing lamp switch

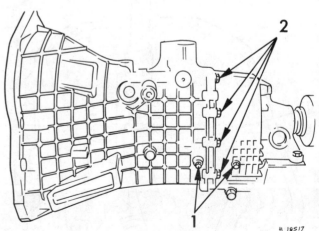

Fig. 6.47 View showing reverse idle shaft bolts (1) and the gearbox housing bolts (2) (Sec 16)

124   Chapter 6   Manual gearbox and automatic transmission

17.2 Rear housing showing the countershaft bearing (bottom) and mainshaft bearing with retainer plate (top)

17.8A Undo the lock screw (arrowed) ...

17.8B ... and remove the gear lever plunger unit

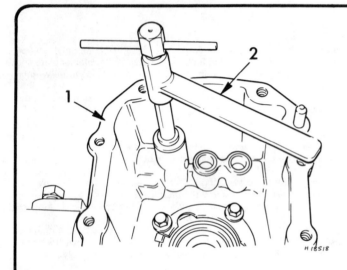

Fig. 6.48 Selector shaft bush removal from rear housing (1) using special tool 21-036A (2) (Sec 17)

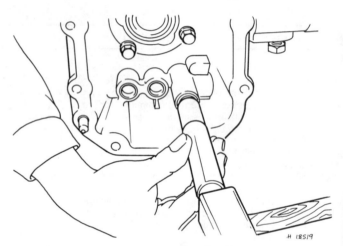

Fig. 6.49 Fitting a new selector shaft bush using special tool 21-044A (Sec 17)

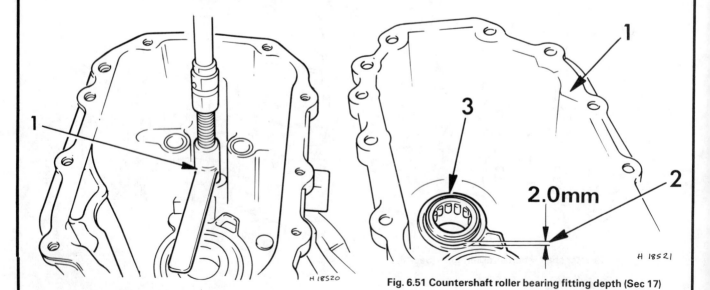

Fig. 6.50 Selector shaft bush removal from front housing using special tool 21-036A (1) (Sec 17)

Fig. 6.51 Countershaft roller bearing fitting depth (Sec 17)

1   Front housing   3   Countershaft roller bearing
2   Protrusion

Chapter 6 Manual gearbox and automatic transmission

17.8C Removing the gear lever right-hand plunger ...

17.8D ... and coil spring

17.8E Removing the gearlever left-hand plunger plug and coil spring ...

17.8F ... and the plunger

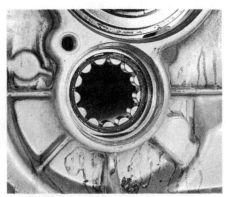
17.10 Countershaft bearing in the front housing

17.14 Input shaft ball bearing

10 Working from the clutch housing side, drive out the countershaft bearing using a suitable drift but take care not to damage the housing thread (photo).
11 If the housing is damaged it must be renewed.
12 Lubricate the gear lever plungers with the specified lubricant, then insert them with their springs into their respective housings. Smear the plug threads with sealant, then screw them into position and tighten them to the specified torque wrench setting.
13 Lubricate the countershaft roller bearing and insert it into position in the front housing, from the geartrain side, so that it protrudes by 2 mm (0.08 in). If the original bearing is being used, press it back about 2 mm (0.08 in) using the countershaft bearing retainer (Fig. 6.51).
14 Locate a new circlip into the groove in the periphery of the input shaft ball bearing, then drive or press the bearing into the housing from the clutch side (photo).

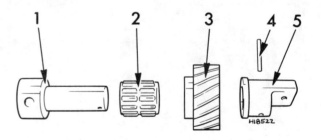

Fig. 6.52 Reverse idler gear components (Sec 18)

1 Shaft
2 Needle roller bearing
3 Idler gear
4 Roll pin
5 Support block

18 Geartrains and selectors (type MT75) – overhaul

Countershaft
1 Withdraw the bearing inner race from the countershaft using a conventional puller, or if available, Ford tool No 15-050 and 16-050.
2 The countershaft cannot be dismantled further. If renewing the countershaft bearing, renew the inner and outer race as they are matched during manufacture.
3 To fit the bearing inner race to the countershaft, heat it up to a temperature of 100°C (212°F) using a hot air blower, or by immersing it in boiling water. (Do not use a gas flame and do not overheat the bearing.) Fit the race onto the shaft (photos).

Input shaft and guide sleeve
4 If the input shaft is damaged or excessively worn it must be renewed. Unless known to be fairly new, always renew the input shaft/mainshaft spigot bearing.
5 Check and if necessary renew the pilot bearing in the rear end of the crankshaft.
6 The oil seal in the input shaft guide sleeve should always be renewed as a matter of course. Prise out the old seal using a suitable screwdriver as a lever, but take care not to damage the sleeve. As it is

Chapter 6 Manual gearbox and automatic transmission

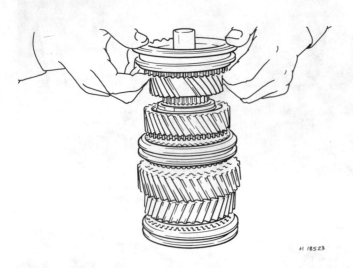

Fig. 6.53 Removing 3rd/4th synchro unit together with 3rd gear (Sec 18)

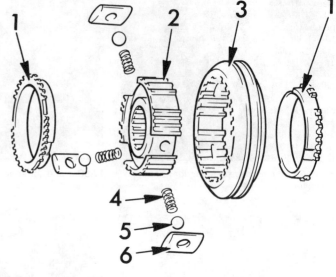

Fig. 6.54 Exploded view of a synchro unit (Sec 18)

1 Synchro baulk rings
2 Synchroniser hub
3 Synchro sleeve
4 Spring
5 Ball
6 Blocker bar

withdrawn, note the direction in which the seal is fitted. Drive the new seal into position (correctly orientated) using a suitable tube drift (photo).

Reverse idler gear unit

7 Drive out the roll pin and withdraw the support block, idler gear and needle roller bearing from the shaft. Renew any defective or excessively worn items. Always renew the roll pin (photo and Fig. 6.52).

8 Lubricate the respective components and refit them in the reverse order of removal, but ensure that the angle of the support block is correct and align the threaded holes (photo).

Mainshaft

9 Before dismantling the mainshaft, it should be noted that the respective baulk rings are all identical and therefore if they are likely to be re-used, they must be marked for identification as they are separated

18.3A Countershaft bearing inner race

18.3B General view of the countershaft

18.6 Input shaft guide sleeve showing oil seal location (and O-ring)

18.7 Driving out the roll pin from the reverse idler gear unit

18.8 Reverse idler gear unit

18.13A Align the grooves with the recesses in the bearing inner race ...

Chapter 6 Manual gearbox and automatic transmission

and kept with their synchro units. Also mark the synchro-hubs and sleeves for the same reason.
10 Withdraw the 5th gear together with its baulk ring and roller bearing from the rear end of the mainshaft.
11 Invert the shaft and support its rear end in a vice fitted with soft jaws to avoid damaging it.
12 Using suitable circlip pliers, extract the snap-ring then remove the 3rd/4th gear synchro unit together with 3rd gear and its needle roller bearing. Take care not to allow the synchro unit to become dismantled during removal as the springs, blocker bars and balls will fly out if released suddenly. Pull on the gear side to withdraw – not on the synchro sleeve (see Fig. 6.53).
13 To remove the third gear bearing inner race it must first be heated up to a temperature of 100°C (212°F) using a hot air blower or failing this, by immersing in hot water. Do not use a gas flame or other similar method to heat up the bearing and take care that it is not overheated. Depending on the method of heat, this can be a time consuming operation due to heat transference from the bearing to the shaft. Once the bearing is sufficiently heated, turn the 2nd gear to align its oil grooves with the recesses in the bearing inner race, then prise the race clear using two screwdrivers as levers (photo). Keep the bearing with 3rd gear once removed.
14 Withdraw the 2nd gear together with its needle roller bearing and baulk ring. Keep the bearing and gear together.
15 Remove the snap-ring and remove the 1st/2nd gear synchro unit (with 1st gear) from the shaft, again keeping them together as a unit, and taking care not to accidentally dismantle the synchro unit as they are withdrawn. When removed, keep the 1st/2nd gear unit away from the 5th/reverse unit as they are identical. Mark them for identification to avoid possible confusion on reassembly.

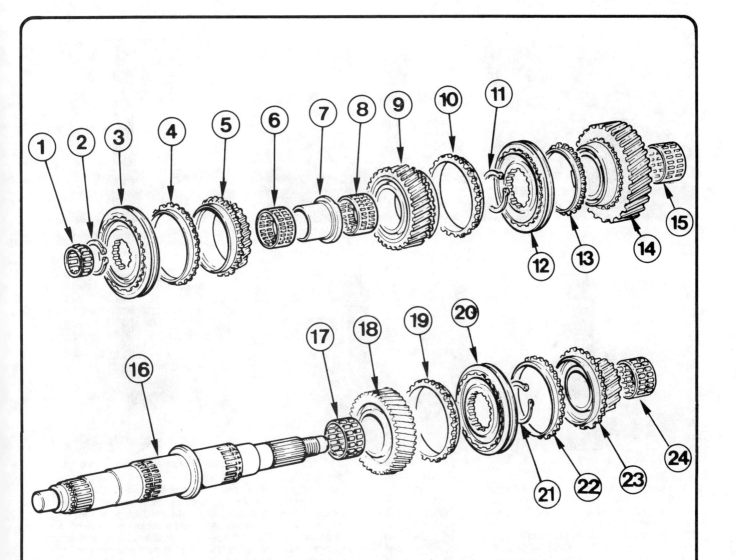

Fig. 6.55 Exploded view of the mainshaft and associated components (Sec 18)

1 Needle bearing (input shaft)	7 Inner race (needle bearing 3rd gear)	13 Synchroniser baulk ring (1st gear)	19 Synchro baulk ring (5th gear)
2 Circlip	8 Needle bearing (2nd gear)	14 1st gear	20 Synchro unit (5th/reverse gear)
3 Synchro unit 3rd/4th gear	9 2nd gear	15 Needle bearing (1st gear)	21 Circlip
4 Syncho baulk ring 3rd gear	10 Synchro baulk ring (2nd gear)	16 Main shaft	22 Synchro baulk ring (5th gear)
5 3rd gear	11 Circlip	17 Needle bearing (reverse gear)	23 5th gear
6 Needle bearing (3rd gear)	12 Synchro unit (1st/2nd gear)	18 Reverse gear	24 Needle bearing (5th gear)

128 Chapter 6 Manual gearbox and automatic transmission

18.13B ... then lever free the 3rd gear bearing inner race

18.17 Inspect the synchro baulk rings for excessive wear

18.18 Keep the synchro-hub springs, balls and blocker bars in their respective fitting positions

18.20A Compress the springs and blocker bars to fit the sleeve

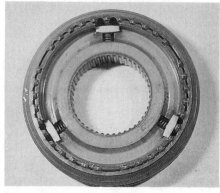

18.20B Reassemble the synchro unit

18.21 Check the snap-rings and circlips for fit in their grooves

18.23A Fit the needle roller bearing ...

18.23B ... the reverse gear ...

18.23C ... the reverse gear synchro baulk ring ...

16   Invert the mainshaft in the vice, then remove the snap-ring and withdraw the 5th/reverse gear synchro unit from the shaft, (again taking care not to accidentally dismantle the synchro unit). Keep the needle bearing with this unit.
17   With the mainshaft dismantled into sub-assemblies, the synchro units can be inspected. An exploded view of a synchro unit is shown in Fig. 6.54. The most likely item to require renewal is the synchro baulk rings. Inspect their contact (friction) faces for signs of excessive wear (photo).
18   Before separating the synchro-hubs from their sleeves, mark them for identification and orientation. To avoid loosing the springs and balls as the units are separated, wrap the unit in a clean cloth to catch the components in as the units are dismantled. Slide the sleeve slowly from the hub and allow the blocker bars, springs and balls to eject into the cloth. Try to keep them in order of fitting for inspection (photo).
19   Any items which on inspection are found to be damaged or excessively worn must be renewed.
20   To reassemble, refit the balls, springs and blocker bars into the hub, compress them and carefully slide the sleeve in position over them, ensuring that it is correctly orientated as marked during removal. Lubricate the assemblies as they are reconstructed. When reassembled, ensure that the sleeve slides positively over the hub in each direction, but take care not to accidentally dismantle the unit (photos).
21   Renew the mainshaft and/or any associate components which are excessively worn or damaged. Renew all snap-rings and circlips with replacements of the same thickness. If new adjacent parts are fitted, check that the snap-rings and circlips are a snug fit in their location grooves. Both snap-rings and circlips are available in varying selective thicknesses to suit. Finally check that all parts are clean and well lubricated as they are refitted (photo).
22   Secure the mainshaft in the vice with the input shaft end down.
23   Fit the needle roller bearing, reverse gear, the 5th/reverse gear synchro unit and the synchro baulk ring onto the output end of the shaft. Drift or press them into position and then secure them with the selective snap-ring (photos).
24   Release the mainshaft from the vice, invert it and refit it into the vice with the input end up.
25   Fit the 1st gear needle roller bearing, 1st gear, 1st/2nd synchro unit (with synchro baulk ring), and secure in position with a selective snap-ring (photos).

18.23D ... and the 5th/reverse synchroniser unit

18.23E ... locate the selective snap-ring ...

18.23F ... into its groove to secure the unit

18.25A Locate the needle roller bearing ...

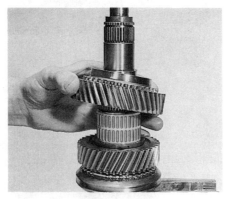

18.25B ... 1st gear ...

18.25C ... the baulk ring ...

18.25D ... the 1st/2nd synchro hub unit ...

18.25E ... and the selective snap-ring

18.26A Locate 2nd gear synchro baulk ring ...

18.26B ... followed by the 2nd gear needle roller bearing ...

18.26C ... and 2nd gear

18.27A Heat up the 3rd gear bearing inner race in a can filled with hot water ...

Chapter 6 Manual gearbox and automatic transmission

18.27B ... then slide it onto the mainshaft

18.28A Locate the 3rd gear needle roller bearing ...

18.28B ... and 3rd gear

18.29A Fit the 3rd gear synchro baulk ring ...

18.29B ... and the 3rd/4th synchro-hub unit

18.29C Fit the selective snap-ring

18.30A Fitting the 5th gear roller bearing

18.30B ... the synchro baulk ring ...

18.30C ... and 5th gear

26   Locate the 2nd gear synchro baulk ring, then the needle roller bearing and 2nd gear (photos).
27   The 3rd gear bearing inner race must be heated up to a temperature of 100°C (212°F) before it can be fitted into position on the mainshaft. Heat it using a hot air blower or immerse it in boiling water for a few minutes to bring it up to this temperature (photo). Do not heat it using a gas flame. When sufficiently heated, the bearing inner race will slide easily onto the mainshaft, but ensure that it abuts the shoulder (photo). Allow the inner race to cool off before proceeding further.
28   Assemble the needle roller bearing onto the inner race, then fit 3rd gear (photos).
29   Fit the synchro baulk ring against 3rd gear (photo), then locate the 3rd/4th gear synchro unit, ensuring that the smaller hub boss is towards the input shaft end (photo). Secure with a selective snap ring (photo).
30   Remove the mainshaft unit from the vice, then locate the 5th gear needle roller bearing, 5th gear baulk ring and 5th gear onto the output end of the shaft (photos).

Selector shafts and forks
31   To dismantle the selector shafts, first note the position and orientation of each selector fork, then support the shafts, drive out the roll pins and slide the forks from their shafts.
32   Refit in the reverse order of removal, using new roll pins to secure the forks. When reassembled, the selector shafts and forks must be as shown in Fig. 6.56.

Gear lever unit
33   To dismantle the gear lever, secure its lower part in a vice, then press down on the ball cup and simultaneously extract the snap-ring from the top end. Withdraw the lever, spring and ball cup from the housing.
34   If the ball locating pin is worn, drive it out and renew it. Renew the coil spring if it is weak or broken.
35   Reassemble in the reverse order of dismantling. Smear the lever and housing with grease prior to assembly.

Chapter 6 Manual gearbox and automatic transmission

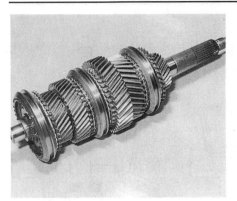
18.30D The assembled mainshaft unit

19.1 Insert the input shaft needle roller bearing

19.2 Fit the input shaft to the mainshaft

19.4 Strap the gear assemblies and selectors together ready for assembly

19.5 Plate bolted to rear housing to support it in the required position for reassembly

19.6 Dummy shaft inverted with hole towards front (up)

19.8A View showing correct orientation of reverse idler gear in rear housing (with bolt holes aligning) when fitted

19.8B Engage reverse idler gear as shown and align bolt holes accordingly

19.10 Special Ford tool used to draw the mainshaft into position in the rear housing

19 Manual gearbox (type MT75) – reassembly

1 Lubricate the input shaft needle roller bearing and insert it into position in the shaft (photo).
2 Locate the 4th gear synchro baulk ring to the hub or friction cone on the input shaft, then assemble the input shaft to the mainshaft (photo).
3 Align the mainshaft and countershaft assemblies so that their corresponding gears are in mesh, then secure them together for refitting using cable ties.
4 Locate the 1st/2nd and 5th/reverse gear selector shaft/fork assemblies into position on the gear assemblies and secure them with a cable tie (but not too tightly) (photo).
5 Position the gearbox rear housing in an upright position with its rear face down. Attach a support plate and locate the housing securely in a vice, but failing this get an assistant to support it in this position (photo).
6 Insert the dummy shaft into one of the outer selector shaft bores in the housing to assist refitting the interlock pins. If the side access plugs are still in position, and a suitable pen magnet is available, the interlock pins can be inserted via the central selector shaft bore. The two pins are located between the outer and centre selector shaft bores in the transverse port (Fig. 6.57). If a pen magnet is not available, one of the

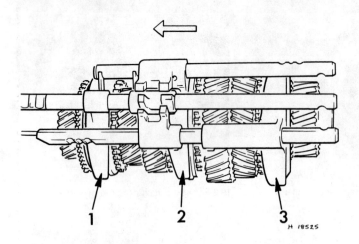

Fig 6.56 Selector shaft and fork assemblies showing arrangement when assembled to the geartrain (Sec 18)

1 3rd/4th gear
2 1st/2nd gear
(Arrow indicates front)
3 5th/reverse gear

Fig. 6.57 Sectional views showing interlock pin fitting using dummy shaft (Sec 19)

A Insert lock plungers through dummy shaft hole
B Dummy shaft located with chamfered end down

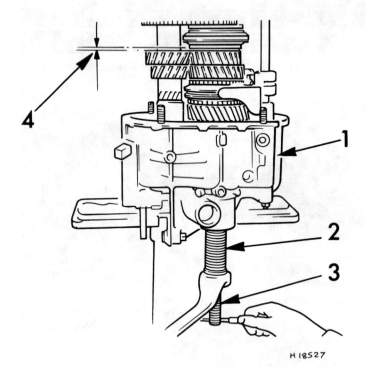

Fig. 6.58 Mainshaft assembly to the rear housing (Sec 19)

1 Rear housing
2 Tool No 16-042A
3 Tool No 16-042A-01
4 Distance X (a clearance must exist)

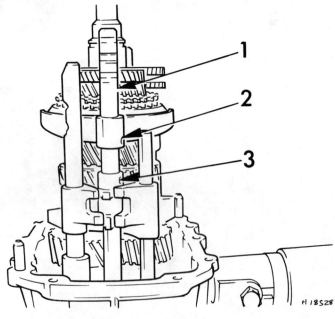

Fig. 6.59 Locate the central selector shaft (1), fork (2) and driver (3) (Sec 19)

side access plugs will need removing to allow the pins to be inserted from the side. In this instance the hole in the dummy shaft must be aligned with the interlock ports to allow the pins to be passed through and into position. Insert the first pin through the dummy shaft to the location port on that side, then relocate the dummy shaft into the centre bore, and then push the pin further through the shaft and into its interlock port. Refit the dummy shaft into the original outer bore and insert the second pin into its interlock port. With this achieved, withdraw the dummy shaft from the outer bore, then invert it so that the hole is to the front and then reinsert it into the centre bore and align the indents with the interlock pins (photo).

7 With the interlock pins and dummy centre shaft fitted, apply a small amount of grease to the outboard sides of the interlock pins using a suitable screwdriver as a spatula, and simultaneously press the pins against the dummy shaft indents. This will ensure that the outer ports are clear to allow refitting of the selector shafts, the grease assisting in holding the pins in position. If removed, do not refit the access plug to the interlock port yet.

Chapter 6 Manual gearbox and automatic transmission

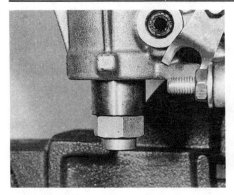

19.11 Alternative method of drawing the mainshaft into the rear housing using spacers and nut

19.14 Relocate the magnetic disc into the slotted section in the housing

19.19 Front housing supported on spacers (arrowed) to ensure that it is parallel

19.27A Insert the interlock assemblies (3rd/4th shown) ...

19.27B ... and their retaining plugs

19.28 Refit the reversing light switch

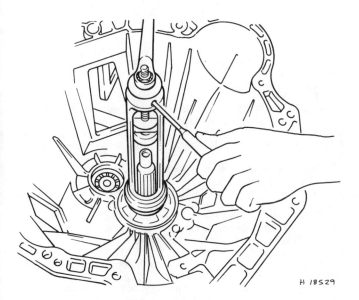

Fig. 6.60 Drawing the front housing in position using special tools 16-041 and 16-042 (Sec 19)

8   Check the orientation of the reverse idler gear by temporarily positioning it in the housing so that the side with the flat face is uppermost and the holes are aligned with those in the housing (photo). Maintain this orientation and secure the reverse idler gear to the semi-assembled gear clusters using a cable tie (photo). Support the gear cluster assemblies and carefully fit them into position in the rear housing. As they are lowered into position, get an assistant to help in guiding the selector shafts into their bores in the rear housing. Having provisionally located the gear assemblies into the rear housing, there are two ways to complete fitting depending on the refitting tools available, but whichever method is employed, it is essential that no excessive force is used to draw the mainshaft and associate assemblies into position in the rear housing.

9   First fit the spacer sleeve and speedometer drive gear onto the rear end of the mainshaft.

10   If available, locate Ford tool Nos 16-042A and 16-042A-01 (photo) onto the rear of the mainshaft and carefully draw the mainshaft into the rear housing. As they are drawn into position, ensure that the countershaft correctly enters its roller bearing and the selector shafts their bores. When fitted check that a clearance exists at point 'X' in Fig. 6.58.

11   If the Ford special tool mentioned above is not available, the mainshaft and associate assemblies can be drawn into position using suitable bushes, a large flat washer and the old retaining nut. The bushes will need to be of suitable diameter so that they bear against the speedo gear without damaging it but allowing the mainshaft to be drawn through it. The washer fits over the end of the shaft and locates between the bush and the nut. As the nut is tightened, the shaft is drawn into position. When the nut reaches the end of its thread it will need to be removed and a shorter length of bush fitted (or extra washers) to allow the shaft to be drawn fully into position (photo).

12   Fit, but hand tighten only at this stage, the reverse idler gear shaft retaining bolt.

13   Remove the special tools used to fit the mainshaft and cut free the cable ties. Remove the speedometer drive worm and spacer.

14   Insert the magnetic disc into its location (photo).

15   Carefully withdraw the dummy shaft from the central selector bore.

16   Ensure that the interlock pins are still correctly located, insert the interlock pin into the central selector shaft, fit the shaft, fork and driver into position. Support the shaft and drive new roll pins into position to secure the fork and driver on the shaft (Fig. 6.59).

17   Check that the mating surfaces of the front and rear housings are clean. Ensure that the selector shafts and gears are set in neutral.

18   The front housing is now ready for fitting, but the method employed is dependent on the refitting tools available. If available, draw the housing into position using Ford tool No 16-041 and 16-042 (less the spindle), and follow the procedure described in Method 1 below. If this tool is not available, follow the alternative fitting method described in Method 2 (paragraph 21).

Chapter 6 Manual gearbox and automatic transmission

19.30A Fitting spacer sleeve and speedometer drive worm

19.30B Gear selectors and lever plungers viewed through lever turret

19.31 Locate the new rear oil seal ...

19.32A ... fit the drive flange and ...

19.32B ... tighten the retaining nut (note tool fitted to flange to prevent it from turning)

19.33 Fit the new input shaft circlip

Method 1

19 Lift and lower the front housing into position, initially engaging it over the input shaft, then as it is drawn down, engage the countershaft and selector shafts. The selector shafts will need to be guided into their bores in the housing so engage the help of an assistant for this. Three temporary spacer bushes must be fitted between the two housings to keep them about 25 mm (1.0 in) apart prior to finally drawing them together (we used three identical ⅜th drive sockets for this purpose). The spacers ensure that the housings are parallel to each other and also prevent the possibility of the 4th gear synchro baulk ring becoming jammed and stretched (photo).

20 Draw the two housings together as far as the spacers using the Ford special tools (16-041/16-042), then remove the three spacers and apply an even coating of sealant to the mating surface. Draw the front housing down into position, engaging it with the dowel pins and ensuring that the countershaft front bearing is not dislodged. Now proceed from paragraph 25.

Method 2

21 If the Ford special tool is not available, the housing can be fitted in the following manner. First remove the input shaft bearing from the housing and then remove the snap-ring from the groove in the bearing periphery.

22 Withdraw the input shaft from the front of the mainshaft but leave the 4th gear synchro baulk ring in engagement with the hub. Heat up the bearing using a hot air blower then fit it into position on the shaft.

23 Relocate the input shaft into position on the front of the mainshaft, ensuring that the needle roller bearing is in position (also the synchro baulk ring).

24 Allow the bearing to fully cool down on the input shaft, then heat up the bearing aperture in the front housing, (using a hot air blower), then when suitably heated, the housing can be carefully lowered into position over the input shaft bearing. Refit the snap-ring into its groove in the bearing periphery. The front housing refitting procedure is otherwise similar to that described previously in Method 1.

Warning note: *If an attempt is made to drift the front housing down into engagement with the rear housing, the 4th gear synchro baulk ring will jam in position on its cone. This will necessitate dismantling to rectify, apart from which the ring could well be distorted.*

25 With the front and rear housings fitted together, insert the retaining bolts and tighten them in a progressive sequence to the specified torque wrench setting.

26 Fit the second reverse idler gear shaft retaining bolt into the front housing, then tighten the two bolts to the specified torque.

27 Insert the selector interlock mechanism sleeves, balls, pins, springs and plugs. Smear the plug threads with sealant as they are fitted, then tighten them to the specified torque setting (photos and Fig. 6.46).

28 Refit the reversing lamp switch and tighten it to the specified torque setting (photo).

29 If removed, refit the selector shaft interlock pin side access plug. Smear it with sealant, then carefully drive it into position in the housing.

30 Relocate the spacer sleeve and the speedometer drive worm, with the blue mark towards the flange, onto the rear end of the mainshaft (photo), then temporarily fit the rear flange, drawing it into position using Ford tool No 16-042A and 16-042A-01. Fit and tighten the securing nut to the specified torque wrench setting, then check that the mainshaft and input shaft rotate freely when in neutral. Select the respective gears in turn via the top turret aperture using a suitable screwdriver as a lever to move the selector rods into the various gear selection positions. Turn the input shaft by hand to check for satisfactory engagement (photo).

31 Having checked that the gears engage in a satisfactory manner, remove the old retaining nut and the rear flange from the mainshaft, then carefully fit the new oil seal into position in the rear end of the gearbox housing, using a suitable tube drift to drive it home (photo).

32 Refit the drive flange (as previously described), smear the threads with sealant and then fit the new retaining nut and tighten it to the specified torque setting (photos).

33 Select and fit a new input shaft circlip into the grooves in front of the bearing (photo).

34 Insert a new O-ring seal into the guide sleeve, locate the thrustwasher into the flange face of the sleeve, then fit the guide sleeve over the input shaft and screw it into the front housing. Tighten the sleeve to

Chapter 6 Manual gearbox and automatic transmission

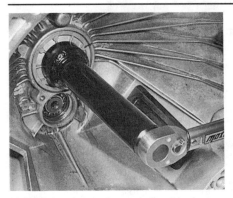
19.34 Tightening the input shaft guide sleeve using special tool 16-040

19.35A Fit a new O-ring seal onto the countershaft bearing retainer ...

19.35B ... screw the retainer into position using an adaptor ...

19.35C ... and tighten to the specified torque

19.37 Countershaft bearing retainer and locking plate/bolt

the specified torque setting using Ford tool No 16-040 (photo).
35 Locate the new O-ring seal onto the countershaft bearing retainer then screw the special retainer into position in the front of the gearbox housing. Tighten it to the specified torque. Note that there are two stages to follow (photos).
36 Using a suitable brass drift, strike the lug each side of the retainer twice to seat the bearing. Check that the countershaft bearing and retainer are in contact. It should not be possible to further screw the bearing retainer in by hand. If it is, repeat the procedures described in paragraphs 35 and 36 until satisfactory.
37 Locate the bearing retainer locking plate and bolt. Tighten the bolt to the specified torque setting (photo).
38 Refit the clutch release lever and bearing (Chapter 5).
39 Insert the speedometer drive gear into its housing, apply sealant to the retaining plug and then carefully drive it into position.
40 Refit the drain plug and tighten it to the specified torque setting.

20 Extension housing rear oil seal (type F, G and N) – renewal

1 Remove the propeller shaft as described in Chapter 7.
2 Carefully prise free the oil seal from the extension housing using a suitable screwdriver as a lever. Note the orientation of the seal in the housing.
3 Press or drive the new seal into the housing using a tubular drift of suitable diameter.
4 Lubricate the seal lips then refit the propeller shaft as described in Chapter 7.
5 Check and if necessary top up the gearbox oil.

21 Speedometer driven gear – removal and refitting

1 Raise the front of the vehicle and support it on axle stands.

2 On the type F, G and N gearboxes, detach the speedometer cable at the gearbox end by undoing the retaining plate bolt and pulling the cable clear. The driven gear and bush assembly can now be withdrawn from the gearbox.
3 On the MT75 gearbox, prise free the plug from the housing on the opposite side of the cable aperture, then withdraw the driven gear from the location port in the side of the gearbox.
4 Allow for a certain amount of oil spillage from the gearbox when the driven gear is withdrawn.
5 If the gear itself is badly worn or damaged, it must be renewed. Where fitted, also renew the O-ring seal on the bush assembly if it is in any way damaged.
6 Refit in the reverse order of removal, having lubricated the gear and cleaned the area around the location aperture in the housing.
7 Check the gearbox oil level and top up on completion.

22 Overdrive unit – general

Certain Transit manual gearbox models are equipped with an overdrive unit as a factory fitted option. The unit is attached to the rear of the gearbox and takes the form of a hydraulically operated epicyclic gear. Overdrive operates on third and fourth gears to provide fast cruising at lower engine revolutions. The overdrive is engaged or disengaged by a driver operated switch which controls an electric solenoid mounted on the overdrive unit. A further switch (inhibitor switch) is included in the electrical circuit to prevent accidental engagement of overdrive in reverse, first or second gears.
Satisfactory fault diagnosis, repair and/or overhaul of the overdrive unit requires specialist knowledge, factory tools and environmentally clean working conditions. For these reasons, it is recommended that the advice of a Ford dealer is sought in the event of any unsatisfactory performance or suspected fault on the unit.

Chapter 6 Manual gearbox and automatic transmission

23 Fault diagnosis – manual gearbox

Symptom	Reason(s)
Weak or ineffective synchromesh	Baulk rings or blocker bars worn or damaged
Jumps out of gear	Selector fork shaft detent springs weak or broken Selector forks or synchroniser and reverse gear location grooves worn Synchroniser and gear dog teeth worn Selector fork loose on shaft
Excessive noise	Incorrect grade oil in gearbox or level too low Worn ball-bearings or countershaft needle rollers Countershaft gear thrust washers worn Gear teeth excessively worn or damaged
Excessive difficulty in engaging gears	Clutch fault Worn gearchange lever or selector fork components

Note: *It is sometimes difficult to decide whether it is worthwhile removing and dismantling the gearbox for a fault which may be nothing more than a minor irritant. Gearboxes which howl, or where the synchromesh can be 'beaten' by a quick gearchange, may continue to perform for a long time in this state. A worn gearbox usually needs a complete rebuild to eliminate noise because the various gears, if re-aligned on new bearings, will continue to howl when different wearing surfaces are presented to each other. The decision to overhaul, therefore, must be considered with regard to time and money available, relative to the degree of noise or malfunction that the driver has to suffer.*

PART B: AUTOMATIC TRANSMISSION

24 Automatic transmission – general description

A three-speed automatic transmission is available as an optional fitting on certain models. The selector lever is centrally located and incorporates a button in the side of the T-handle. The button must be pressed before the selector can be moved from 'N' and between those positions indicated by arrows in Fig. 6.61.

Forward movement is obtained by selecting 'D' (fully automatic) or '1' (low gear) followed by '2' (2nd gear). With the selector lever in 'P' (Park), an internal pawl locks the transmission. Reverse gear is engaged by selecting 'R'.

The system includes a three-element hydrokinetic torque converter which transmits the power from the engine to the transmission unit; the torque converter is capable of variable torque multiplication.

The hydraulically-operated epicyclic gearbox responds to both road speed and throttle pedal demand by means of an internal governor and valve control, and the correct gear for the current conditions is therefore automatically selected.

Should it be necessary to tow the vehicle, the selector lever must be moved to the 'N' (neutral) position but the vehicle must not be towed at speeds in excess of 30 mph or for distances in excess of 30 miles (50 km). The propeller shaft should be disconnected if these limits are to be exceeded.

The transmission fluid is cooled by means of a fin-type oil cooler mounted to the right of the radiator.

An inhibitor switch is fitted to the transmission to prevent inadvertent starting of the engine whilst the selector lever is in any position other than 'N' or 'P'.

Due to the complexity of the automatic transmission unit, if performance is not up to standard, or overhaul is necessary, it is imperative that this be left to a main agent who will have the special equipment and knowledge for fault diagnosis and rectification. The contents of the following Sections are therefore confined to supplying general information and any service information and instructions that can be used by the owner.

A cutaway view of the automatic transmission is shown in Fig. 6.62.

It is most important that the handbrake be fully applied and 'P' selected whenever the vehicle is being worked on, particularly when making under-bonnet adjustments. This is necessary since if left in gear with the engine running, an increase above the normal speed will provoke vehicle 'creep'.

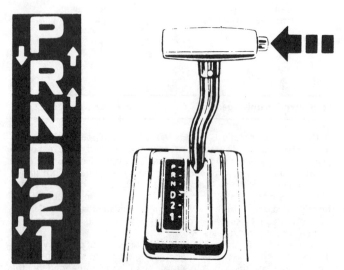

Fig. 6.61 T-handle selector button must be pressed to move selector to respective positions indicated (Sec 24)

25 Routine maintenance

At the intervals specified in Routine maintenance *at the beginning of this manual, carry out the following maintenance operations*
1 Check the automatic transmission fluid level as described in Section 26.
2 Lubricate the selector and downshift cables and linkages with engine oil.
3 Check the fluid cooler hoses, transmission casing joints and oil seal locations for any sign of fluid leakage, damage or deterioration.
4 Have the brake band adjustment checked by a Ford dealer.

Chapter 6 Manual gearbox and automatic transmission

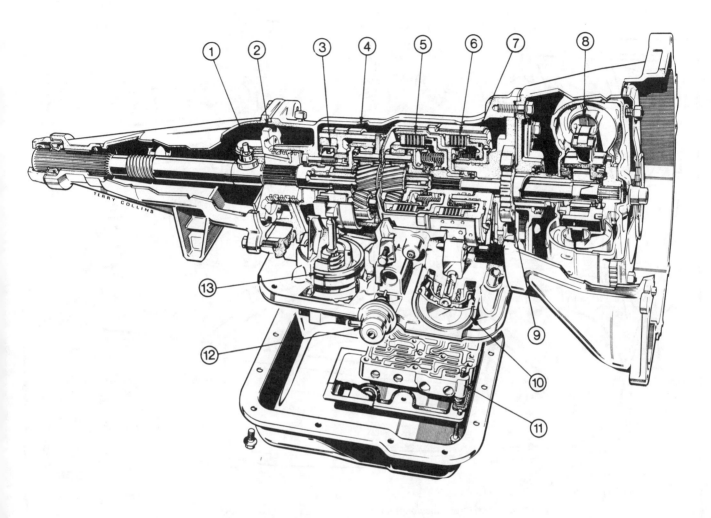

Fig. 6.62 Cutaway view of the automatic transmission (Sec 24)

1 Centrifugal governor
2 Parking gear
3 One way clutch
4 Rear brake band
5 Forward clutch
6 Reverse and top gear clutch
7 Front brake band
8 Torque converter
9 Hydraulic pump
10 Front servo
11 Valve body
12 Vacuum diaphragm
13 Rear servo

26 Automatic transmission fluid level – checking

1 Fluid level should be checked with the transmission at operating temperature (after a run) and with the vehicle parked on level ground.
2 With the engine idling and the handbrake and footbrake applied, move the selector lever through all positions three times, finishing up in position 'P'.
3 Wait one minute then with the engine still idling, withdraw the transmission dipstick. Wipe the dipstick with a clean lint-free cloth, re-insert it fully and withdraw it again. Check that the fluid level lies within the detent on the dipstick.
4 If topping up is necessary, do so via the dipstick tube using clean transmission fluid of the specified type. Do not overfill.
5 Refit the dipstick and switch off the engine.

27 Selector cable – adjustment

1 Remove the selector indicator housing and engage the selector lever in 'D' position.
2 Prise the grommet from the side face of the selector housing and insert a feeler gauge between the selector pawl stub and notch (see Fig. 6.64). This clearance (A) must be 0.004 to 0.008 in (0.1 to 0.2 mm) with the lever set in the 'D' position.
3 Where adjustment is necessary, loosen the cable locknut and adjust the cable accordingly by means of the screw head (Fig. 6.65). Tighten the locknut on completion.
4 Refit the grommet and selector housing indicator unit.
5 Removal and refitting of the cable is covered in Section 30.

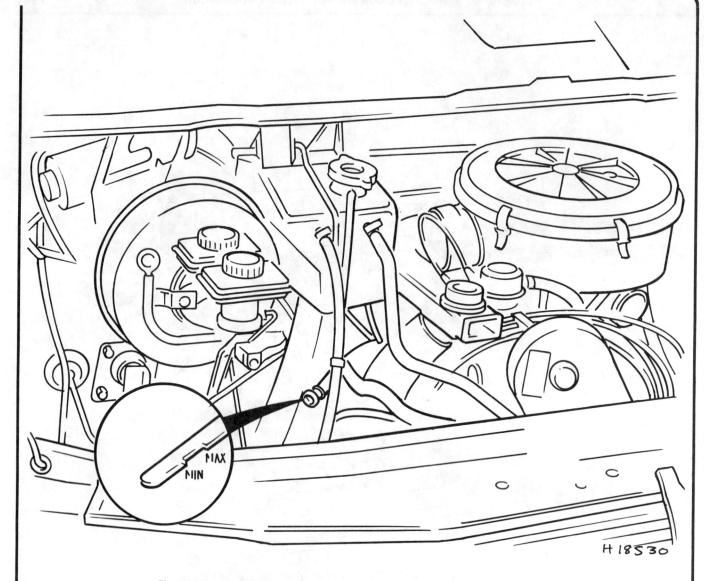

Fig. 6.63 Automatic transmission fluid level dipstick and level markings (Sec 26)

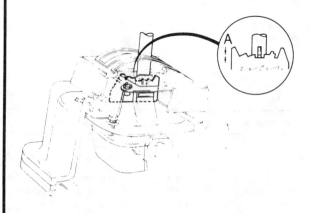

Fig. 6.64 Selector pawl stub to notch clearance (A) (Sec 27)

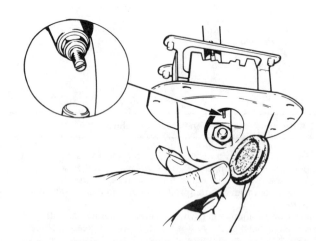

Fig. 6.65 Remove grommet for access to cable adjuster and locknut (Sec 27)

Chapter 6 Manual gearbox and automatic transmission

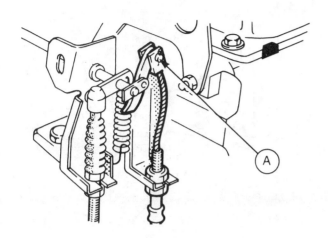

Fig. 6.66 Carburettor downshift cable connection (A) (Sec 28)

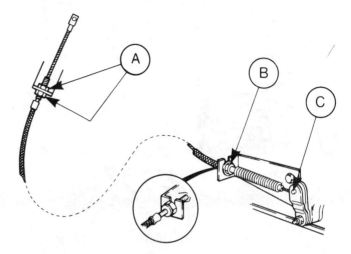

Fig. 6.67 Downshift cable connections (Sec 28)

A Adjuster nuts (upper end)
B Retaining nuts (lower end)
C Downshift lever connection

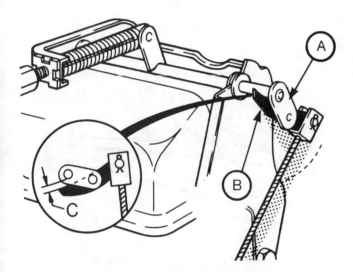

Fig. 6.68 Operating shaft lever (A) and cable connection lever (B). Clearance (C) to be as specified (Sec 28)

28 Downshift cable – removal, refitting and adjustment

1 Open and support the bonnet.
2 Referring to Fig. 6.66, disconnect the downshift cable from the carburettor linkage. The clevis pin is secured by a split pin which when withdrawn enables the pin to be extracted and the cable detached from the linkage.
3 Refer to Fig. 6.67 and disconnect the cable from the slotted bracket by unscrewing the upper nut, screwing the lower nut back fully and then pulling the cable down to unhook it from its support bracket.
4 Fully engage the handbrake and then raise and support the front of the vehicle with axle stands.
5 The cable can now be detached from its transmission location bracket by loosening the locknuts and releasing the cable from its slot in the bracket.
6 Unhook the cable from the downshift lever and withdraw the cable.
7 Refit in the reverse order to removal. Adjust the cable on completion before fully tightening the upper adjustment/retaining nuts.
8 To check the cable adjustment, press the cable connection lever away from the operating shaft lever using a screwdriver and then check the clearance between the two using feeler gauges (Fig. 6.68). The correct clearance is 0.008 to 0.04 in (0.2 to 1.0 mm). If necessary lengthen or shorten the cable accordingly by loosening or tightening the adjuster nut of the cable upper thread.

29 Selector mechanism – removal and refitting

1 Lift clear the indicator housing from the selector lever bracket and then unscrew and remove the three screws securing the sound insulator to the selector lever bracket (Fig. 6.69). Remove the sound insulator.
2 Slide the lever indicator lamp shroud upwards and remove the bulb assembly (Fig. 6.70).
3 Unscrew and remove the four selector lever bracket retaining bolts with washers. Remove the bracket and gasket.
4 Tie a suitable length of cord or wire to the upper selector rod to prevent it from dropping downwards when disconnected. Lift the selector mechanism and remove the retaining clip at the shift rod end.
5 Unhook and withdraw the selector lever unit.
6 Refit in the reverse order to removal. Adjust the selector rods as given in Section 31 if necessary. Tighten the selector bracket retaining bolts to the specified torque.

30 Selector mechanism – dismantling and reassembly

1 With the selector mechanism removed from the vehicle (as given in the previous Section), first prise out the rubber grommet and then remove the nut retaining the lever arm. Press the lever arm from the housing.
2 Unscrew the operating cable locknut and then remove the selector pawl and spring from the selector lever bottom end.
3 From the selector lever handle remove the socket-headed cap screw and then remove the handle.
4 Detach the cable nipple from the roller unit at the lever top end and remove the pushbutton.
5 Use a suitable punch to drive out the operating cable guide bush-to-top linkage retaining pin (Fig. 6.71). Remove the cable guide bush and top linkage from the lever arm bottom end.
6 Inspect all parts for wear and renew as necessary.
7 Reassemble in the reverse order to removal, but note the following special points:

   (a) When refitting the pushbutton into the handle, the long keyway must be fitted first
   (b) Hold the pushbutton against its spring, fit the handle to the lever and retain in position with the cap screw
   (c) Smear some medium grease onto the selector pawl before refitting, with spring and guide bush, into the lever lower end

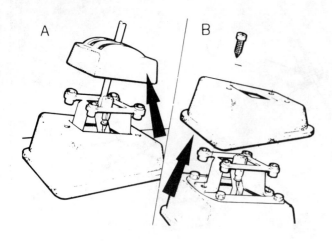

Fig. 6.69 Remove selective indicator housing (A) and sound insulator (B) (Sec 29)

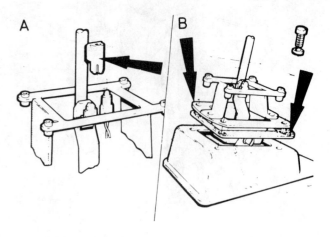

Fig. 6.70 Detach the lamp shroud (A) and selector lever bracket (B) (Sec 29)

Fig. 6.71 Selector lever cable top linkage (Sec 30)

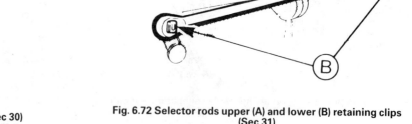

Fig. 6.72 Selector rods upper (A) and lower (B) retaining clips (Sec 31)

(d) Apply a general purpose grease into the selector lever housing slots before fitting the lever arm
(e) Adjust the selector cable as necessary as given in Section 27 on completion

31 Selector rods – removal, refitting and adjustment

1 Both the upper and lower selector rods are secured by means of clips at each end and these can be prised free to release the rods (Fig. 6.72).
2 Renew the rod(s) and bushes if worn or damaged.
3 To refit and adjust the selector rods, first locate the transmission control lever (attached to the forward end of the lower selector rod) in the 'D' position. To ensure that the lever is in the 'D' position, move it forwards to the '1' position, then move it back by two notches when it will be in 'D' (Fig. 6.73). Keep the lever in this position when adjusting the two rods.
4 Fit the fixed end of the upper rod to the forward pin on the relay lever and secure with its clip. The adjustable end of the upper rod should be allowed to rest against the transmission.
5 Fit the lower rod the rear pin on the relay lever and retain with clips.

Check at this point that the selector lever and the transmission control lever are still engaged in 'D'.
6 Fit the upper rod to the location pin of the selector lever arm, if necessary adjusting the end of the rod so that it is in exact alignment with the location pin. Loosen the locknut, rotate the end of the rod accordingly and retighten the locknut. When refitted onto the location pin, refit the retaining clip.
7 Check that full engagement of all gear selector positions is possible.

32 Inhibitor switch – removal, refitting and adjustment

1 Disconnect the wires to the switch, then unscrew and remove the switch from the transmission.
2 Remove the O-ring seal from the switch. This must be renewed.
3 To refit the switch, locate the new O-ring seal, then screw the switch into the housing. Correct adjustment is made automatically as the switch is installed. Reconnect the leads to the switch.
4 Check that the engine only starts when the selector is set in the 'P' or 'N' position, and that the reversing lamp only operates when the selector lever is in the 'R' position.

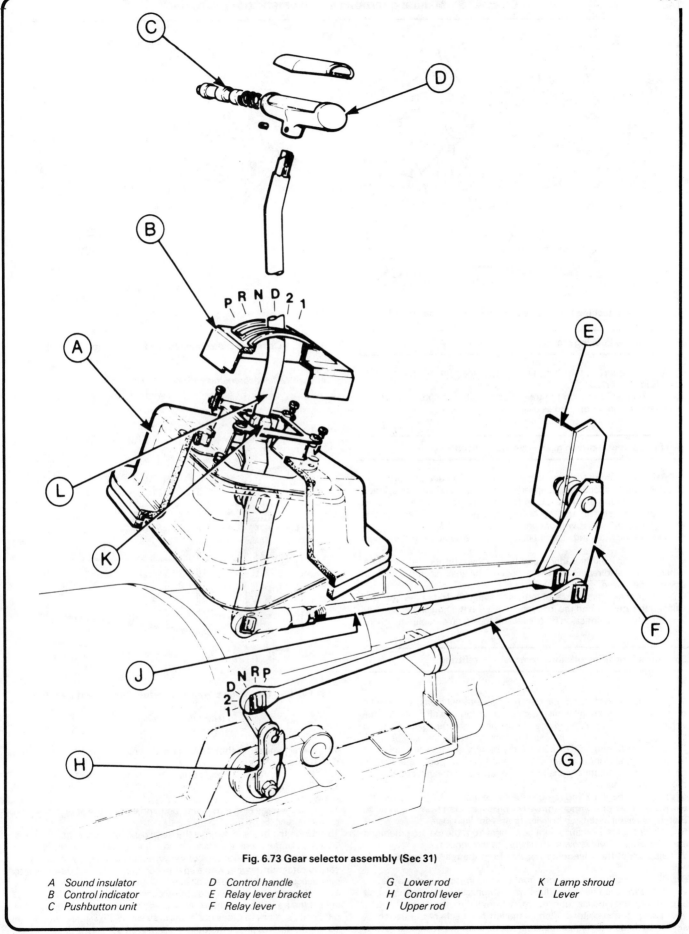

Fig. 6.73 Gear selector assembly (Sec 31)

A Sound insulator	D Control handle	G Lower rod	K Lamp shroud
B Control indicator	E Relay lever bracket	H Control lever	L Lever
C Pushbutton unit	F Relay lever	I Upper rod	

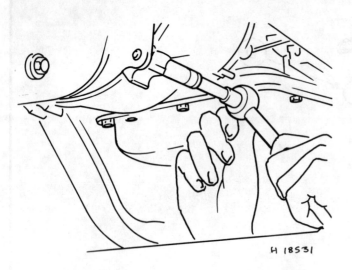

Fig. 6.74 Detaching the engine to transmission brace (Sec 35)

33 Extension housing rear oil seal – renewal

1 The procedure is the same as that described for manual gearbox models in Section 20.
2 On completion top up the transmission with the specified type of automatic transmission fluid as described in Section 26.

34 Speedometer driven gear – removal and refitting

1 Raise the front of the car and support it on axle stands.
2 Undo the bolt securing the speedometer cable retaining plate, lift off the plate and withdraw the cable and driven gear from the transmission.
3 Allow for a certain amount of fluid spillage from the transmission when the driven gear is withdrawn.
4 Extract the retaining clip and disconnect the cable from the driven gear.
5 If the gear itself is badly worn or damaged, it must be renewed. Where fitted, also renew the O-ring seal on the gear assembly if it is in any way damaged.
6 Refit in the reverse order of removal having lubricated the gear and cleaned the area around the location aperture in the housing.
7 Check the transmission fluid level and top up on completion.

35 Automatic transmission – removal and refitting

1 Remember that automatic transmission internal faults can only be diagnosed successfully whilst the transmission is still fitted to the vehicle, and therefore the advice of a suitably equipped garage should be sought before removing the unit, unless renewal is the only object.
2 Open the bonnet and disconnect the battery negative terminal.
3 Jack up the front and rear of the vehicle and support it adequately on stands, or alternatively position it over an inspection pit; ensure that the vehicle is unladen.
4 Refer to Chapter 3 and remove the air cleaner.
5 Refer to Chapter 7 and remove the propeller shaft from the vehicle. Suitably seal the extension housing to prevent fluid loss.
6 Disconnect the exhaust front downpipe from the exhaust manifold and silencer, and withdraw it from beneath the vehicle.
7 Disconnect the speedometer cable from the rear of the transmission.
8 Disconnect the selector rods from the transmission and relay lever.
9 Withdraw the fluid level dipstick, then unscrew the retaining bolt and remove the dipstick tube assembly from the transmission.
10 Locate the inhibitor switch on the left-hand side of the transmission, then detach the wiring block connector from the switch.

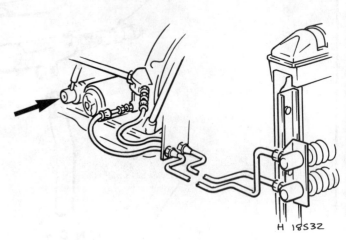

Fig. 6.75 Oil cooler pipe connections to automatic transmission (Sec 35)

Note vacuum unit and hose connection arrowed

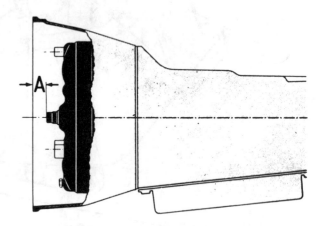

Fig. 6.76 Check the torque converter engagement (Sec 35)

A = 10.0 mm (0.4 in) minimum

11 Detach the downshift cable at the downshift lever and bracket and tie it back out of the way.
12 Refer to Chapter 12 and remove the starter motor.
13 Detach the vacuum pipe from its diaphragm and tie it back out of the way.
14 Unscrew the single retaining bolt and detach the engine to transmission brace (Fig. 6.74).
15 Undo the bolt and remove the earth cable at the transmission.
16 Working through the starter motor aperture, unscrew and remove the four driveplate-to-torque converter retaining nuts; it will be necessary to rotate the engine with a spanner on the crankshaft pulley bolt to gain access to each of these bolts.
17 Wipe the area around the oil cooler connections to the right-hand side of the transmission, then unscrew and remove the unions, plug the pipe ends and tie the pipes out of the way (Fig. 6.75).
18 Take the weight of the transmission with a trolley jack positioned

Chapter 6   Manual gearbox and automatic transmission

143

beneath the transmission sump; to prevent damage to the sump interpose a block of wood between the jack and sump, but make sure that there is no chance of the transmission slipping in subsequent operations.

19 Unscrew and remove the five bolts and remove the transmission support crossmember.

20 Using a further jack and block of wood, support the rear end of the engine beneath the sump.

21 Unscrew and remove the bellhousing retaining bolts and carefully withdraw the transmission rearwards to separate the torque converter spigot from the crankshaft adaptor. It would be wise to enlist the help of an assistant during this operation in order to steady the transmission on the trolley jack and hold the torque converter in the transmission.

22 Carefully withdraw the transmission from the engine and lower it, for removal from underneath the vehicle. As it is withdrawn, retain the torque converter firmly against the transmission to prevent it from falling out. If this should happen, allow for considerable oil spillage as the converter separates from the transmission oil pump. To retain the converter in position once the transmission is removed, locate a suitable retainer bar across the front flange face of the transmission.

23 Refitting the automatic transmission is a reversal of the removal procedure, but the following points should be noted:

(a) When refitting the converter, align the oil drain plug with the driveplate opening. When the converter is fully fitted and engaged with the oil pump drivegear, ensure that the distance between the converter flange end face and the converter housing flange is as shown in Fig. 6.76

(b) When the converter case and engine flanges are flush, check that the converter rotates freely, then insert the flange bolts and tighten to the specified torque

(c) Adjust the selector rods and inhibitor switch as described in Section 31 and 32 then refill the transmission with the correct grade of fluid as given in Section 26

(d) Refit the propeller shaft with reference to Chapter 7

(e) Adjust the downshift cable as described in Section 28 and lower the vehicle to the ground

(f) On completion run the engine and operate the transmission to ensure that it is satisfactory. Check the oil cooler pipes and connections for any signs of leakage

36 Fault diagnosis – automatic transmission

Symptom	Reason(s)

Faults in these units are nearly always the result of low fluid level or incorrect adjustment of the selector linkage or downshift cable. Internal faults should be diagnosed by your main Ford dealer who has the necessary equipment to carry out the work.

Symptom	Reason(s)
Engine will not start with selector in any position	Disconnected or defective inhibitor switch
Engine starts in all selector positions	Defective or short-circuited inhibitor switch
Gearchange speeds incorrect	Selector linkage adjustment incorrect Downshift cable broken or maladjusted
Parking pawl inoperative	Incorrect linkage adjustment Internal linkage fault
Selector lever action stiff	Defective lever, selector linkage or inhibitor switch
Fluid loss	Defective vacuum diaphragm or plug socket seal Rear extension housing to transmission housing seal leak Rear expansion housing oil seal defective Speedometer drive gear ring seal defective Torque converter leak Transmission oil pump seal defective Torque converter/transmission housing seal defective

Chapter 7 Propeller shaft

Contents

Fault diagnosis – propeller shaft	6
General description	1
Propeller shaft – removal and refitting	3
Propeller shaft centre bearing – renewal	4
Propeller shaft rubber 'Guibo' joint – removal and refitting	5
Universal joints and centre bearing – testing for wear	2

Specifications

Type .... Two-piece with centre bearing, centre and rear universal joints, front joint either standard universal or rubber coupling according to model

Torque wrench settings

	Nm	lbf ft
Universal joint bolt	35 to 40	26 to 29
Driveshaft to rear axle	60 to 70	44 to 52
Centre mounting to floor mounting:		
Panel van, Bus and 'Convertible'	26 to 32	19 to 24
Other models	51 to 64	38 to 47
Centre bearing to mounting	26 to 32	19 to 24

1 General description

Drive is transmitted from the manual gearbox or automatic transmission to the rear axle by a finely balanced tubular propeller shaft, split into two halves and supported at the centre by a rubber mounted bearing.

Fitted at the front, centre and rear of the propeller shaft assembly are universal joints, which cater for movement of the rear axle with suspension travel, and slight movement of the power unit on its mountings. On certain models, a rubber 'Guibo' joint is used at the front in place of the universal joint (photo) and, depending on model and transmission type, a constant velocity joint may be fitted in place of the centre universal joint. The universal joints are of the sealed type and cannot be serviced, however it is possible to renew the centre bearing and rubber joint.

Maintenance is limited to checking for wear in the universal joints, and centre bearing at the intervals specified in *Routine Maintenance* at the beginning of this manual. Also, where applicable, the rubber joint should be inspected for wear or deterioration and the sliding joint in the propeller shaft rear section should be lubricated at the grease nipple provided.

2 Universal joints and centre bearing – testing for wear

1 Wear in the universal joints is characterised by vibration in the transmission, or a clicking or knocking noise when taking up drive.
2 To test a universal joint, jack up the vehicle and support it on axle stands. Attempt to turn the propeller shaft, either side of the joint being checked, in opposite directions. Also attempt to lift each side of the joint. Any movement within the universal joint is indicative of considerable wear and if evident the complete propeller shaft must be renewed.
3 Wear in the centre bearing is characterised by a rumbling or grating noise in the transmission.
4 The centre bearing is a little more difficult to test for wear. If bearing movement (as distinct from universal joint or rubber insulator movement) can be felt when lifting the propeller shaft front section next to the mounting bracket, the bearing should be removed as described in Section 4 and checked for roughness while spinning the outer race by hand. If excessive wear is evident, the bearing must be renewed. Also check the rubber insulator itself for any signs of deterioration and renew if necessary.
5 On models so equipped, the rubber 'Guibo' joint at the front of the propeller shaft should be inspected for signs of cracking, splits, or general deterioration of the rubber. This will be most evident around the metal sleeves through which the retaining bolts pass. Should wear be detected, the joint should be renewed as described in Section 5.

1.1 Rubber joint fitted to the front of the propeller shaft on some models

Chapter 7 Propeller shaft

3.3 Propeller shaft rear universal joint to axle flange joint

3.4 Propeller shaft front universal joint and transmission coupling flange (MT75 transmission shown)

3.5 Propeller shaft centre bearing and universal joint

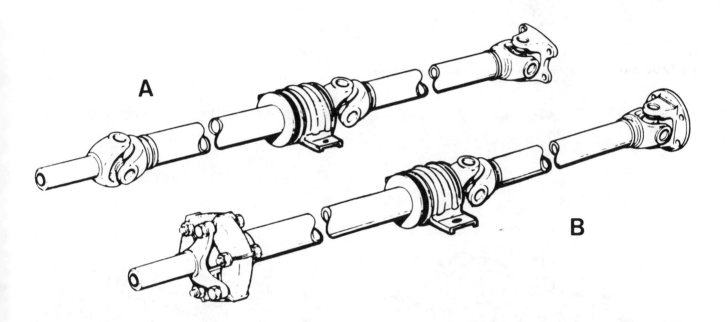

Fig. 7.1 Propeller shaft types used (typical) (Sec 1)

A With standard front universal joint
B With front rubber Guibo joint

3 Propeller shaft – removal and refitting

1 Jack up the rear of the vehicle and support it on axle stands. Chock both front wheels.
2 Mark the rear universal joint and final drive flanges in relation to each other.
3 Unscrew and remove the bolts securing the propeller shaft to the final drive unit while holding the shaft stationary with a long screwdriver inserted between the joint spider (photo). If necessary apply the handbrake as an additional means of holding the shaft stationary.
4 Where applicable, undo the retaining bolts and remove them from the front universal joint to gearbox transmission coupling flange (photo).
5 Support the weight of the combined front and rear propeller shaft sections and then unscrew and remove the two centre bearing retaining bolts from the vehicle underbody (photo).
6 Lower the propeller shaft from the coupling flanges and the centre bearing mounting and remove it from the vehicle. If a sliding front joint is used, pull it rearwards to disengage it from the gearbox/transmission output shaft, then to prevent any loss of oil/fluid from the gearbox a chamfered plastic cap can be inserted into the oil seal (Fig. 7.2). Alternatively a plastic bag can be positioned on the gearbox and retained with an elastic band.
7 Refitting is a reversal of the removal procedure but note the following points:

(a) Where a sliding type front joint is used, lubricate the rear seal by smearing it with grease prior to inserting the propeller shaft. Engage top gear to prevent the output shaft from turning

(b) Align any coupling flange marks made during removal, then reconnect the flange couplings. Tighten the flange bolts in a diagonal progressive sequence to the specified torque setting

(c) Do not fully tighten the centre mounting bracket bolts to the specified torque wrench setting until after the vehicle has been lowered and is free standing (unladen)

(d) Check, and if necessary, top up the gearbox/transmission oil/fluid levels as described in Chapter 6

(e) Where applicable, lubricate the propeller shaft sliding joint(s) via the grease nipple(s)

Chapter 7 Propeller shaft

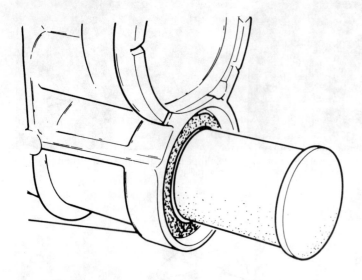

Fig. 7.2 Using a chamfered plastic cap to prevent loss of oil from the gearbox (Sec 3)

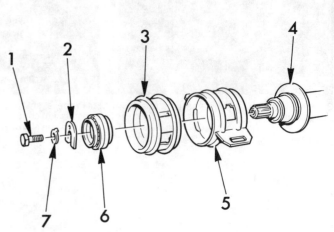

Fig. 7.3 Exploded view of the propeller shaft centre bearing unit (Sec 4)

1 Bolt
2 U-washer
3 Rubber insulator
4 Driveshaft
5 Housing
6 Ball-bearing
7 Lock-plate

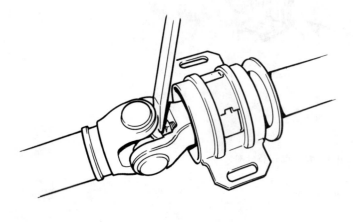

Fig. 7.4 Levering the driveshaft sections apart (Sec 4)

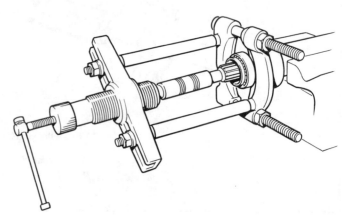

Fig. 7.5 Type of puller required to remove the ball bearing (Sec 4)

4 Propeller shaft centre bearing – renewal

1 Remove the propeller shaft as described in Section 3.
2 Prise up the lock tab from the universal joint retaining bolt, then loosen off the bolt using a suitable flat ring spanner.
3 Extract the U-shaped washer from the bolt head, then insert a suitable screwdriver or metal bar between the bolt head and the universal joint, and lever the front and rear driveshaft sections apart (Fig. 7.4).
4 Remove the housing and rubber insulator from the bearing unit.
5 The bearing can be removed using a puller similar to that shown in Fig. 7.5.
6 To renew the rubber insulator in the housing, bend open the six metal retaining tongues and then remove the old insulator. Insert the new insulator together with the ball bearing into the housing. Ensure that the insulator engages with the channel in the housing.
7 When the insulator and bearing are fitted, bend back the metal retaining tabs using suitable grips.
8 Stand the driveshaft on end and support it on a tube of suitable diameter, then drive the centre bearing home using a tube sleeve as shown in Fig. 7.6 as far as the stop.
9 Refit the bolt together with a new locking plate.

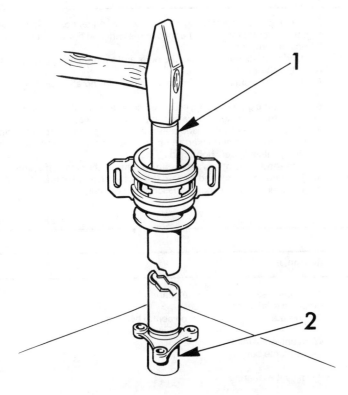

Fig. 7.6 Centre bearing installation. Support on tube (2) and drive bearing home with sleeve (1) (Sec 4)

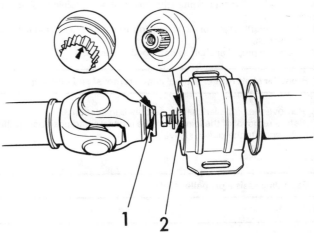

Fig. 7.7 Driveshaft to universal joint alignment (Sec 4)

1 Double width groove mark
2 Master spline

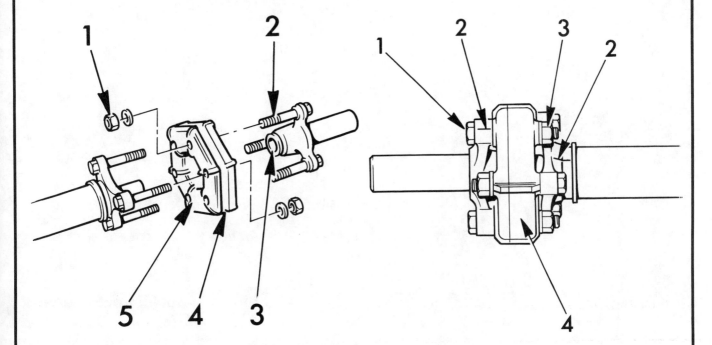

Fig. 7.8 Exploded view of the Guibo type joint unit (Sec 5)

1 Self-locking nut
2 Bolt
3 Guide bearing
4 Steel band
5 Guibo joint

Fig. 7.9 The Guibo joint showing the securing bolts (1 and 3), the alignment marks (2) and the joint (4) (Sec 5)

Chapter 7 Propeller shaft

10   Reassemble the universal unit to the shaft, aligning the master spline with the double width groove, and press them together (Fig. 7.7).
11   Slide the U-shaped washer into position with the pegged side towards the bearing, then tighten the universal joint retaining bolt to the specific torque.
12   Use a length of rod or a suitable drift to tap up the locking tab and secure the bolt.
13   Refit the propeller shaft as described in Section 3.

5 Propeller shaft rubber 'Guibo' joint – removal and refitting

1   Remove the propeller shaft as described in Section 3.
2   Before removal of the joint, mark the relative fitted positions of the propeller shaft to the sliding sleeve.

3   Fit a clamp, comprising of two worm drive hose clips joined together, around the circumference of the joint and tighten it until it just begins to compress the rubber.
4   Undo the six nuts, remove the bolts and separate the joint from the propeller shaft and sliding sleeve.
5   Carefully inspect the joint for signs of deterioration and renew if necessary.
6   Smear the guide bearing in the sliding sleeve and the corresponding journal in the propeller shaft with multi-purpose grease.
7   Align the previously made marks on the propeller shaft and sliding sleeve, fit the joint and insert the bolts. Secure the assembly with the retaining nuts securely tightened.
8   If the original joint has been refitted, remove the clamp. If a new joint has been fitted, cut off and discard the metal retaining band.
9   Refit the propeller shaft as described in Section 3.

6 Fault diagnosis – propeller shaft

Symptom	Reason(s)
Vibration	Worn universal joints or centre bearing
	Propeller shaft out of balance
	Deteriorated rubber insulator on centre bearing
Knock or 'clunk' when taking up drive	Worn universal joints
	Loose flange bolts
Excessive 'rumble' increasing with road speed	Worn centre bearing

Chapter 8 Rear axle

Contents

Differential carrier (H type axle) – removal and refitting	8
Differential unit (G and F type axles) – repair and overhaul	9
Drive pinion oil seal (all types) – renewal	7
Fault diagnosis – rear axle	10
General description	1
Rear axle – removal and refitting	3
Rear axleshaft (halfshaft) – removal and refitting	4
Rear hub (F and H type axles) – removal, overhaul and refitting	5
Rear hub (G type axle) – removal, overhaul and refitting	6
Routine maintenance	2

Specifications

Axle type .... Fully floating or three-quarter floating, hypoid.

Application
1.6 litre	Ford type 32 (F) type axle
1.6 and 2.0 litre	Ford type 34 (H) type axle
2.0 litre	Ford type 51A and 53 (G) type axle

Ratios
32 (F) type axle	4.63 : 1
34 (H) type axle	3.9 : 1, 4.11 : 1, 4.56 : 1 or 5.14 : 1
51A (G) type axle	4.63 : 1 or 5.14 : 1
(G) type axle	4.63 : 1, 5.14 : 1 or 5.83 : 1

Track
32 (F) and 34 (H) type axles	1590 mm (62.6 in)
51A (G) type axle	1700 mm (67.0 in)
53 (G) type axle	1511 mm (59.5 in)

General
Crownwheel/pinion backlash:

32 (F), 51A and 53 (G) type axles	0.12 to 0.22 mm (0.0047 to 0.0086 in)
34 (H) type axle	0.10 to 0.20 mm (0.004 to 0.008 in)
Rear wheel bearing play – G type axle	0.05 to 0.20 mm (0.002 to 0.008 in)

Lubrication
Lubricant type .... Hypoid gear oil, viscosity SAE 90 EP to Ford spec SQM-2C 9002-AA (Duckhams Hypoid 90S)

Capacity
F type axle	1.4 litres (2.46 pints)
H type axle	2.7 litres (4.75 pints)
G type axle	1.7 litres (3.0 pints)

Torque wrench settings

	Nm	lbf ft
Bearing cap to axle case:		
F type axle	60 to 70	44 to 52
G type axle	99 to 118	73 to 87
H type axle	95 to 109	70 to 80
Differential carrier to case (H type axle)	55 to 68	41 to 50
Hub nut (F and H type axle)	200 to 240	148 to 177
Hub locknut (G type axle)	70 to 80	52 to 59
Rear axle housing cover (F and G type axle)	35 to 40	26 to 29
Spring U-bolt nuts:		
F and H type axle	88 to 100	65 to 73
G type axle	120 to 130	89 to 96

Chapter 8 Rear axle

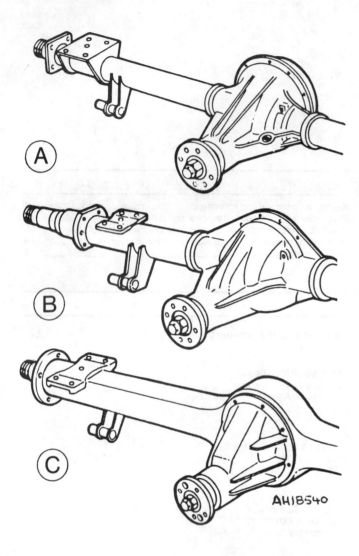

Fig. 8.1 The identifying features of the three rear axle types (Sec 1)

A Type 32 – F axle
B Type 51A and 53 – G axle
C Type 34 – H axle

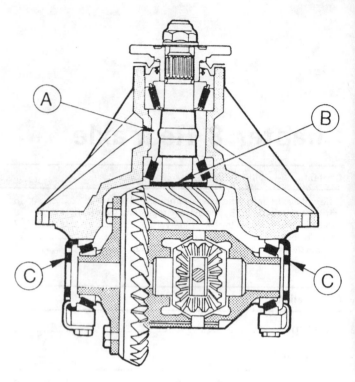

Fig. 8.2 Sectional view of the F type axle unit (Sec 1)

A Collapsible spacer sleeve
B Pinion shim
C Housing shims

torted and the pinion to crownwheel setting be disturbed.

The differential unit runs on taper roller bearings, the adjustment of which is dependent on type. On the F axle the adjustment is by shims located between the differential case and the bearings. On the C type the adjustment is made by adjuster nuts in the axle housing. On the H type adjustment is made by adjuster nuts in the differential carrier.

The outer wheel hubs of the H and F type axles are similar, having ball-bearings as shown in Figs. 8.5 and 8.6. The outer hub of the G type axle is fitted with two taper roller bearings, the free play of which is set by an adjuster nut which is secured by a lockwasher and locknuts. This hub assembly is shown in Fig. 8.7.

On all three axle types, the outer hub bearings are lubricated by rear axle oil.

1 General description

On all models the rear axle is suspended on single semi-elliptic leaf springs. A two-piece propeller shaft transmits the drive from the transmission to the rear axle.

One of three axle types will be fitted according to model, these being identified as the F, G or H type axle. An identifying profile of each type is shown in Fig. 8.1.

On the F and G type axles, the differential unit is a fully floating hypoid (Salisbury) type and is mounted direct into the axle casing, access being through the rear inspection cover (Figs. 8.2 and 8.4).

The H type (Timken) differential is of the three-quarter floating type and is bolted to its own carrier, which is attached to the front face of the axle case (Fig. 8.3).

On all models the pinion bearing spacer is of the collapsible sleeve type. Special care must be taken when renewing the pinion seal not to overtighten the flange retaining nut, or the sleeve could become dis-

2 Routine maintenance

1 Check the oil level in the rear axle casing. Make this check at the intervals specified in the *Routine maintenance* Section at the start of this manual. The vehicle must be standing level for this check, preferably over an inspection pit for convenience.
2 Using a suitable square head key, unscrew and remove the filler/level plug from the axle housing. Allow for a certain amount of spillage as the plug is withdrawn, particularly if the oil is still hot (Fig. 8.8 and photo).
3 The oil level must be topped up to the base of the filler hole. Use a piece of bent wire to check the level if necessary but take care not to lose it inside the housing! Top up the oil level if necessary with the correct grade of oil then refit the filler/level plug.
4 If the level was found to be well below the minimum level, it is advisable to make a visual inspection of the axle casing or cover for any

Chapter 8 Rear axle

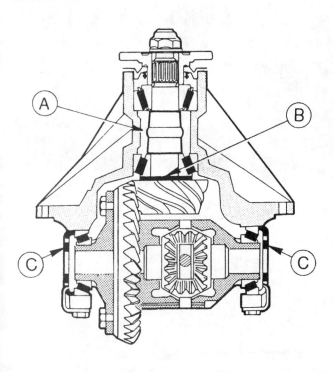

Fig. 8.3 Sectional view of the H type axle unit (Sec 1)

A Collapsible spacer sleeve
B Pinion shim
C Bearing adjusting nuts

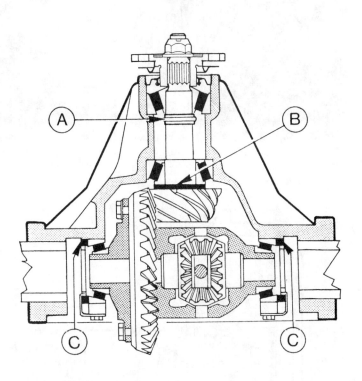

Fig. 8.4 Sectional view of the G type axle unit (Sec 1)

A Collapsible spacer sleeve
B Pinion shim
C Bearing adjusting nuts

signs of leakage past the gasket. If there is no sign of leakage from the joint, check at the hub ends. If an oil seal at the hub is defective it must be renewed as soon as possible as oil could be reaching the brake linings.

5 If for any reason the oil must be drained, then on the G and F types the axle casing cover must be unbolted and lifted away from the casing, as a drain plug is not fitted. Always renew the gasket when refitting the cover and coat the bolt threads with sealant. On the H type the oil can only be drained if the complete differential carrier is removed (Sec 8).

3 Rear axle – removal and refitting

The rear axle removal/refitting details described below are for the removal of the unit on its own. If required, it can be removed together with the roadwheels and rear leaf springs as a combined unit, although this method requires the vehicle to be raised and supported at a greater height (to allow the roadwheels to clear the body during withdrawal of the unit). If the latter method is used, follow the instructions given but ignore the references to removal of the roadwheels and detaching the springs from the axle. Refer to Chapter 10 for details on detaching the springs from the underbody.

1 Ensure that the vehicle is unloaded, then chock the front wheels, jack up the rear of the vehicle and support it adequately beneath the underframe side members in front of the rear springs.

2 Support the weight of the rear axle with a trolley jack positioned beneath the differential housing.

3 Apply the handbrake firmly, unscrew the rear wheel nuts and remove the rear wheels. (Remember that six stud wheels have left-hand thread wheel nuts on the left-hand side).

4 Mark the pinion and propeller shaft drive flanges so that they can be refitted to their original position, then unscrew and remove the four self-locking nuts and bolts and detach the propeller shaft, supporting it on a stand.

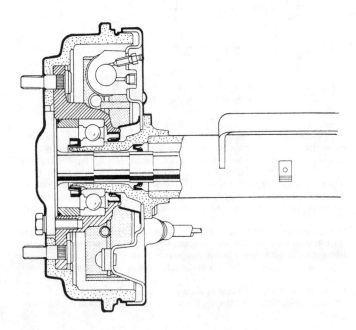

Fig. 8.5 Wheel hub on the F type axle (Sec 1)

Chapter 8 Rear axle

2.2 Oil filler/level plug in the H type axle

3.9A Axle U-bolt and retaining nuts viewed from underneath

3.9B Axle U-bolts, upper centre plate and interleaf viewed from above (H type axle shown)

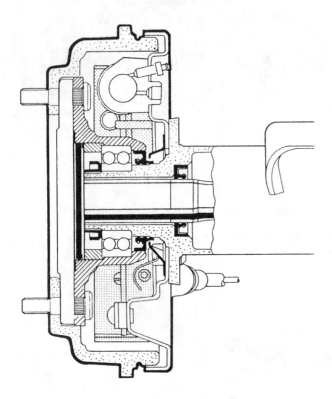

Fig. 8.6 Wheel hub on the H type axle (Sec 1)

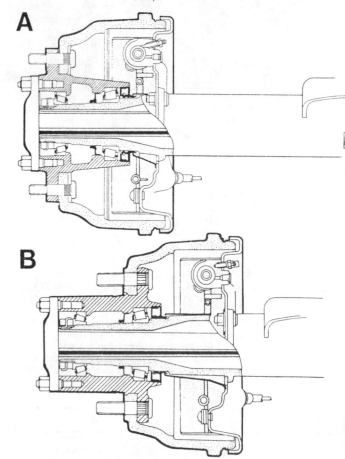

Fig. 8.7 Wheel hub on the G type axle (Sec 1)

A Type 51A   B Type 53

5 Disconnect the flexible hydraulic brake hose from the rigid line connection to each rear brake where they are attached to the axle, (referring to Chapter 9 for details). When disconnected, plug the lines to prevent the ingress of dirt and excessive fluid leakage. Note that where a load apportioning valve (LAV) is fitted, the regulating spring must be detached at its top end by removing the clevis pin. Do not detach it at the lower end or the LAV will require readjustment on reassembly (see Chapter 9) (Figs. 8.9 and 8.10).
6 Disconnect the handbrake cable at the equaliser unit, (see Chapter 9 for details). Also release the exhaust system at the rear to allow the handbrake cable to clear as the axle is removed.
7 Undo the retaining nuts and detach the rear shock absorbers from the axle.
8 Check that the axle unit is securely supported in the centre by the jack. Get an assistant to steady the axle each side as it is lowered from the vehicle.
9 Undo the retaining nuts and remove the spring to axle U-bolts and fittings each side (photos). Check that the various axle fittings and attachments are disconnected and out of the way, then carefully lower the axle unit and withdraw it from under the vehicle.

10 Refitting the axle unit is basically a reversal of the removal procedure, but the following points should be noted.
11 Renew all self-locking nuts and lockwashers during refitting.
12 When raising the axle into position each side, engage the locating hole in the axle over the centre bolt of the spring.
13 Tighten the various nuts and bolts to their specified torque wrench settings.
14 When reconnecting the propeller shaft to the pinion flange, be sure to align the match marks made during removal.
15 Refer to Chapter 9 for details on reconnecting the brake system components. Top up and bleed the brake hydraulic system as described in that Chapter.
16 Top up the rear axle level with the correct grade of oil as described in Section 2.

Chapter 8 Rear axle

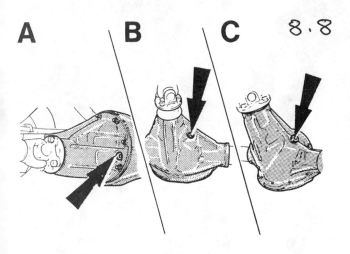

Fig. 8.8 Rear axle oil filler/level plug location (Sec 2)

A H type axle
B G type axle
C F type axle

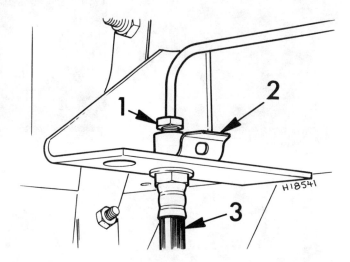

Fig. 8.9 Brake pipe to axle fitting showing pipe nut (1) spring clip (2) and flexible hose (3) (Sec 3)

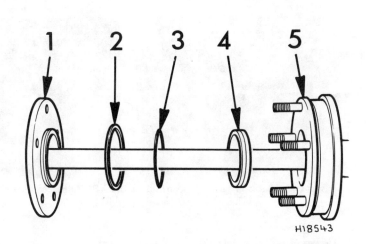

Fig. 8.10 Extract clevis pin (arrowed) to detach the LAV arm where applicable (Sec 3)

Fig. 8.11 Rear axle shaft and fittings – F and H type axles (Sec 4)

1 Axle shaft
2 Gasket
3 Seal
4 Spacer ring
5 Hub

4 Rear axleshaft (halfshaft) – removal and refitting

1 Loosen the wheel nuts. On F and H type axles, mark the position of each rear wheel relative to its drum/hub. When loosening off the nuts on six stud axle types, remember that the left-hand side has left-hand threads.
2 Chock the front wheels and jack up the rear of the vehicle supporting it firmly and adequately on stands placed at each end of the rear axle.
3 Unscrew the nuts and remove the rear wheels.
4 On F and H type axles, release the handbrake and remove the brake drum (Chapter 9) (photo).
5 Where applicable, unscrew and remove the axleshaft to hub bolts (or nuts).

6 Carefully withdraw the axleshaft from the axle housing/hub, taking care not to damage the oil seal within the hub. Release the axleshaft by carefully tapping the flange, then withdraw the shaft from the axle, being careful not to damage the splines which engage with the differential unit. Be prepared for some oil spillage (photos).
7 Recover the gasket and seal from the flange. Both the F and H type axleshafts have a spacer ring (the F type has a chamfered face and this points to it when fitted).

4.4 Withdrawing the brake drum from the H type axle

4.6A Lever the axleshaft flange away from the hub flange ...

4.6B ... and withdraw the axleshaft

4.8 O-ring seal location (arrowed) on the inboard side of the axleshaft flange (H type axle)

5.2 Type of hub unit removal tool required

5.4A Rear wheel hub and bearing (outboard side)

5.4B Rear wheel hub and oil seal (inboard side)

5.8A Refit the rear hub unit (H type axle shown) ...

5.8B ... locate the spacer ring ...

5.8C ... fit the hub nut ...

5.8D ... and tighten to the specified torque ...

5.8E ... then stake lock the nut in the axle groove

Chapter 8 Rear axle

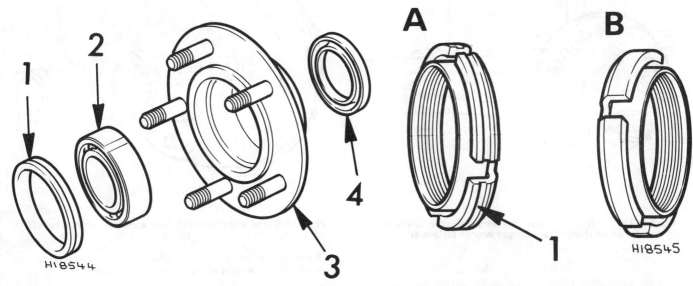

Fig. 8.12 Rear wheel hub and associate components – H type axle (Sec 5)

1 Spacer ring
2 Bearing
3 Hub
4 Oil seal

Fig. 8.13 Hub nut identification feaures – H type axle (Sec 5)

A Left-hand thread (left-hand side)
B Right-hand thread (right-hand side)
1 Identification groove (in left-hand nut)

8 Refitting the axleshaft is a reversal of the removal procedure, but note the following:

(a) Clean the mating surfaces of the axleshaft and hub before refitting and always fit a new O-ring seal and gasket (photo). On the G type axle, coat the axleshaft flange with sealer prior to fitting
(b) On the F type axle, align the marks made during removal when fitting the axleshaft to the hub, (and later the wheel)
(c) Check and top up the rear axle oil level if required as described in Section 2

---

5 Rear hub (F and H type axles) – removal, overhaul and refitting

1 Refer to the previous Section and remove the axleshaft on the side concerned.
2 Lever under the hub nut locking tab using a suitable tool to prise it clear of its slot, but take care not to damage the threads. Unscrew and remove the hub nut using Ford special tool No 15-077 or similar (photo). When removing the hub nut on the H type axle, note that it has a left-hand thread on the left- hand side (right-hand thread on the right-hand side).
3 Withdraw the hub using Ford special tool No 15-060 (if necessary). A suitable conventional slide hammer may also suffice if the special Ford tool is not readily available.
4 Wash and clean all the components in paraffin then carefully examine each item for damage and deterioration. Check the bearing wear by spinning the outer track whilst holding the inner track stationary and observing any roughness. Similarly with the bearing stationary, attempt to move the outer track laterally; excessive movement is an

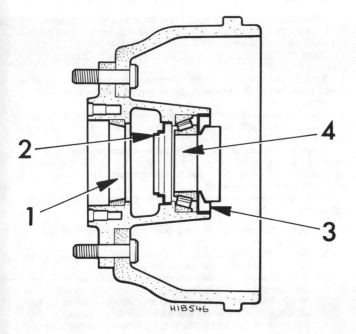

Fig. 8.14 Sectional view of G type axle hub showing (Sec 6)

1 Taper bearing cup (outer)
2 Oil seal (outer)
3 Oil seal (inner)
4 Taper bearing cup (inner)

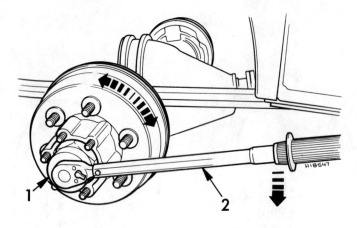

Fig. 8.15 Tighten hub nut to specified torque (G type axle) using Ford tool 15-062(1) and torque wrench (2) (Sec 6)

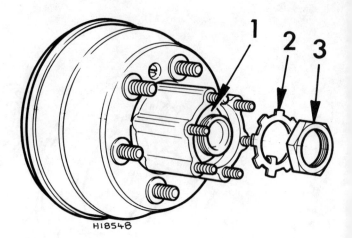

Fig. 8.16 G type axle hub nut (1), lock washer (2) and locknut (3) (Sec 6)

8 Refitting the rear hub is a reversal of the removal procedure (photos). The following special points should be noted:

(a) On the H type axle, refer to Fig. 8.13 to identify the hub nuts as they have differing threads. Stake lock the hub nuts once they are tightened to the specified torque setting
(b) Do not forget to locate the spacer ring in front of the bearing prior to refitting the axle shaft (and note that on the F type axle the spacer ring is fitted with its chamfered face out)
(c) Refit the axleshafts as described in the previous Section

6 Rear hub (G type axle) – removal, overhaul and refitting

1 Refer to Section 4 and remove the axleshaft on the side concerned.
2 Prise up the hub locknut tab, unscrew the locknut and then the hub nut. Remove the hub and brake drum (handbrake released).
3 Clean and inspect the hub as described in paragraph 4 of the previous Section.
4 If the oil seals and bearings are to be renewed, support the hub on its flange faces (not on the wheel studs) during the removal and refitting operations.
5 Drive out the outer bearing, then the inner bearing cups and oil seal using a soft metal drift, but keep each bearing inner and outer tracks together (Fig. 8.14).
6 Wash and clean all the components in paraffin then carefully examine each item for wear and deterioration. Check the bearing tapered rollers and inner and outer tracks for pitting and scoring, and if necessary obtain new bearings.
7 Refitting the rear hub is a reversal of the removal procedure, but the following points should be noted:

(a) Use suitable diameter tubing to drive the bearing cups squarely into the hub
(b) Grease the bearings with a lithium-based grease prior to assembly, and fit a new oil seal to the inner bearing
(c) When fitting the oil seals, ensure that the outer seal lip is facing the bearing
(d) When fitting the hub retaining nut, rotate the hub in each direction. Tighten the nut to the specified torque setting and then loosen it off half-a-turn (180°). Locate the lockwasher so that its central lug is under the hub nut, then fit and tighten the locknut to its specified torque wrench setting. Check the wheel bearing play using a suitable dial gauge. If necessary, further adjust the hub bearings to comply with the specified play. When the bearing play is satisfactory, bend the lockwasher tabs inwards over the locknut to secure (Figs. 8.15, 8.16 and 8.17).
(e) Refit the axle shafts as described in Section 4

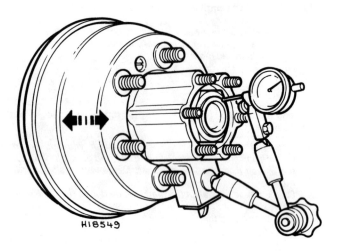

Fig. 8.17 Check the wheel bearing end play on the G type axle using a dial gauge as shown (Sec 6)

indication of wear. Any component which is unserviceable must be renewed (photos).
5 To dismantle the hub, first remove the spacer ring (if fitted), then drive out the bearing using a suitable drift whilst supporting the hub on its flange face, (not on the studs). Drive or lever out the old oil seal. Clean the hub.
6 To reassemble the hub, proceed as follows. First lubricate the bearing with grease. Support the hub with its inner flange face down. Position the bearing over the hub with its sealed face pointing up. Fit the spacer ring over the bearing, then press the bearing and spacer into the hub using a suitable tube drift. Do not press or drive against the bearing inner race. Press the bearing in until it is flush.
7 Invert the hub and support it on blocks on its outer flange face so that the studs hang free, then press the oil seal into position using a suitable tube drift. Lubricate the seal lips with grease.

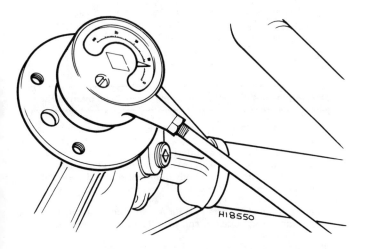

Fig. 8.18 Measure the drive pinion turning torque. Ford tool 15-041 is shown in this instance (Sec 7)

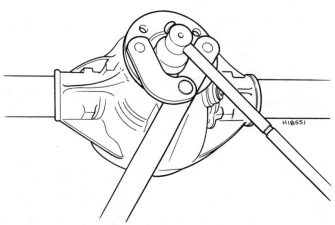

Fig. 8.19 Method of holding the drive flange when unscrewing/tightening the pinion nut. Ford tool 15-030 shown in this instance (Sec 7)

7 Drive pinion oil seal (all axle types) – renewal

All axle types are fitted with a collapsible spacer and this is shown in Figs. 8.2, 8.3 and 8.4. Renewal of the drive pinion oil seal should only be attempted by the more experienced DIY mechanic. A special torque wrench or a spring balance will be needed, and a new pinion nut must be fitted on reassembly. No difficulty should be encountered unless the pinion nut is overtightened (paragraph 1.1), in which case removal of the taper roller bearing to renew the collapsible spacer may present problems.

1 Jack up the rear of the vehicle and support it securely under the bodyframe.
2 Remove the axle shafts as described in Section 4.
3 Mark the relative positions of the propeller shaft and rear axle drive pinion flange, then unbolt and detach the propeller shaft from the flange. Position the shaft out of the way.
4 Using a spring balance and length of cord wound round the drive pinion flange, determine the torque required to turn the drive pinion and record it (Fig. 8.18).
5 Alternatively, a socket wrench fitted to the pinion nut and a suitable torque wrench may be used.
6 Mark the coupling in relation to the pinion splines for exact replacement.
7 Hold the pinion coupling flange as shown in Fig. 8.19 or by placing two 2-inch long bolts through two opposite holes and bolting them up tight. Undo the self-locking nut whilst holding a large screwdriver or tyre lever between the two bolts as a lever. Remove the flange, using a puller if necessary.
8 Carefully prise free the old oil seal using a blunt screwdriver or similar implement, then clean out the housing.
9 Fit the new oil seal first having greased the mating surfaces of the seal and the axle housing. The lips of the oil seal must face inwards. Using a piece of brass or copper tubing of suitable diameter, carefully drive the new oil seal into the axle housing recess until the face of the seal is flush with the housing. Make sure that the end of the pinion is not knocked during this operation.
10 Refit the coupling to its original position on the pinion splines.
11 Fit a new pinion nut and progressively tighten it in small increments up to the pinion turning torque value noted during removal (para. 4). Whilst tightening the nut, secure the pinion flange in the manner described in paragraph 7. When the nut is tightened to the previously recorded torque value, rotate the pinion to settle the bearing then further tighten the nut by an additional torque of 0.3 Nm (0.22 lbf ft) this

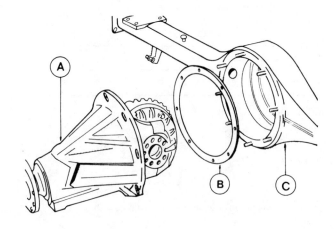

Fig. 8.20 H type axle differential unit and carrier (A), flange gasket (B) and axle case (C) (Sec 8)

being the added value required to compensate for the additional friction of the new oil seal. For example, if the original value noted during removal (para. 7) was 2.2 Nm, and the compensating value for the new oil seal of 0.3 Nm is added to it, the total tightening torque requirement in this instance will be 2.5 Nm. It is most important that the required torque is not exceeded for the reasons outlined in the introductory paragraph at the start of this Section.
12 Reconnect the propeller shaft, ensuring that the alignment marks correspond and tighten the retaining nuts to the specified torque wrench setting (Chapter 7).
13 Refit the axleshafts as described in Section 4 of this Chapter.
14 Check and top up the axle oil level (Section 2).

158   **Chapter 8 Rear axle**

8 Differential carrier (H type axle) – removal and refitting

On the H type axle the differential unit is mounted to the pinion carrier, and this is mounted on studs to the front face of the axle casing. The differential unit and carrier can therefore be removed complete with the axle casing still in position in the vehicle, but the axleshafts must first be removed. Proceed as follows.

1   Refer to Section 4 and remove both axleshafts.

2   Place a large container of at least 5 pints capacity beneath the differential carrier to catch the oil which will drain out.

3   Mark the pinion and propeller shaft flanges so that they can be refitted in their original position, then unscrew and remove the four self-locking nuts and bolts, detach the shaft, and support it with a stand.

4   Unscrew and remove the eight self-locking nuts which retain the differential carrier to the axle casing, then lift the carrier slightly to allow the oil to drain into the container.

5   Using a trolley jack or with the aid of an assistant, withdraw the differential carrier off the studs and remove it from beneath the vehicle (Fig. 8.20).

6   Peel the gasket from the axle casing studs; a new gasket will need to be fitted on reassembly.

7   Further dismantling of the differential unit is not recommended. If it is worn or damaged, its overhaul should be entrusted to your Ford dealer who will have the necessary tools required for this task. Alternatively, purchase an exchange unit.

8   Refitting the differential carrier is a reversal of the removal procedure but the following points should be noted:

(a) *Check the mating faces of the differential carrier and axle housing for burrs and file them flat; always use a new gasket*

(b) *Make sure that the differential carrier is fitted with the pinion to the bottom, and tighten the retaining nuts in diagonal sequence in three or four stages.*

(c) *The pinion and propeller shaft flanges must be aligned to the previously made marks*

(d) *Tighten all nuts and bolts to the specified torque wrench settings*

(e) *Fill the rear axle with the correct grade of oil until the level is up to the lower edge of the filler plug hole, then refit and tighten the plug*

(f) *If a new differential carrier unit has been installed, it should be run-in for 500 miles (800 km) to ensure that the new bearings bed in correctly. Change the oil after this mileage*

9 Differential unit (G and F type axles) – repair and overhaul

1   The design and layout of the G and F type axles is such that any attempt to remove the differential unit or pinion assembly from the axle housing will upset their present meshing.

2   Since special tools and skills are required to set up the crownwheel and pinion mesh, the removal, overhaul and assembly of these differential types is not recommended and should be entrusted to your Ford dealer.

3   In any case the latest trend is for rear axle components not to be supplied individually, but the complete factory-built unit only to be supplied as a replacement.

4   If required, an inspection of the differential unit can be made to assess for excessive wear or damage to its component parts. To do this, first drain the oil from the differential housing as described in Section 2, then unbolt and remove the differential housing cover.

5   Before refitting the rear cover, clean the cover and axle case mating surfaces and remove the remains of the old gasket. Also clean the threads of the retaining bolts with a wire brush, and the threaded holes in the casing with petrol or a cleaning solvent.

6   Locate the new gasket and the rear cover, smear the retaining bolt threads with sealant and fit them. Tighten the bolts in an alternate and progressive sequence to the specified torque setting.

7   Top up the rear axle oil level to complete (Section 2).

10 Fault diagnosis – rear axle

Symptom	Reason(s)
Oil leakage	Faulty pinion oil seal
	Faulty axleshaft oil seals/gasket
	Defective cover gasket
Noise	Lack of oil
	Worn bearings
	General wear
'Clonk' on taking up drive and excessive backlash	Incorrectly tightened pinion nut
	Worn components
	Worn axleshaft splines
	Elongated roadwheel bolt holes/loose wheel nuts
	Worn propeller shaft universal joints

Chapter 9 Braking system

Contents

Brake pedal – removal and refitting	21
Brake pressure control valve – description, removal and refitting	19
Fault diagnosis – Braking system	22
Front brake calliper – dismantling, overhaul and reassembly	8
Front brake calliper – removal and refitting	7
Front brake disc – examination, removal and refitting	9
Front disc pads (opposed piston calliper type) – inspection, removal and refitting	6
Front disc pads (sliding pin calliper type) – inspection, removal and refitting	5
General description	1
Handbrake – adjustment	16
Handbrake cable – removal and refitting	18
Handbrake lever and primary rod – removal and refitting	17
Hydraulic brake lines and hoses – inspection, removal and refitting	4
Hydraulic system – draining and bleeding	3
Load apportioning valve (LAV) – general	20
Master cylinder – dismantling, overhaul and reassembly	14
Master cylinder – removal and refitting	13
Rear brake shoes – inspection, removal and refitting	10
Rear brake wheel cylinder – dismantling and reassembly	12
Rear brake wheel cylinder – removal and refitting	11
Routine maintenance	2
Vacuum servo unit – removal and refitting	15

Specifications

System type .... Disc front brakes, drum rear brakes, Hydraulic operation, servo-assisted. Handbrake to rear wheels, mechanical operation

Front disc brakes

Disc diameter	254 or 270 mm (10.0 or 10.6 in), depending on model
Disc run-out (maximum)	0.13 mm (0.005 in)
Disc thickness (minimum)	12.15 mm (0.478 in)
Calliper piston diameter	38 mm (1.49 in), 41.28 mm (1.62 in) or 57.15 mm (2.25 in) depending on model
Minimum allowable pad lining thickness	1.5 mm (0.06 in)

Rear drum brakes

Maximum allowable drum diameter:

F and H type axle	229.5 mm (9.04 in)
G type axle	256 mm (10.09 in)
Minimum allowable shoe lining thickness	1.5 mm (0.06 in)

Brake shoe width:

F and H type axle	44.5 mm (1.75 in)
G type axle	55 to 70 mm (2.17 to 2.75 in)

Wheel cylinder diameter:

F and H type axle	20.3 mm (0.79 in)
Alternative for the H type axle	22.2 mm (0.87 in)
G type axle	22.2 mm (0.87 in) or 23.8 mm (0.94 in) or 25.4 mm (1.0 in)

Brake fluid type specification .... Brake fluid to Ford spec Amber SAM-6C-9103-A (Duckhams Universal Brake and Clutch Fluid)

Torque wrench settings

	Nm	lbf ft
Calliper to anchor bracket Torx bolts	21 to 29	16 to 21
Calliper anchor bracket to stub axle	95 to 123	70 to 91
Calliper to stub axle	95 to 123	70 to 91
Brake back plate to rear axle	45 to 54	33 to 40
Master cylinder to servo unit	13 to 16	10 to 12
Servo unit to bulkhead	35 to 51	26 to 38
Brake disc to hub bolts:		
IFS models	48 to 51	35 to 39
Beam axle models	60 to 78	44 to 58

1 General description

Disc brakes are fitted to the front wheels and drum brakes to the rear. All are operated under servo assistance from the brake pedal, this being connected to the master cylinder and servo assembly, mounted on the bulkhead.

A dual line brake system is fitted and will be a vertical split system or diagonal split system type according to model. The main system types used are shown in Figs. 9.1, 9.2 and 9.3.

With the vertical split system, the master cylinder primary circuit feeds the rear brakes, the secondary circuit the front brakes.

With the diagonal split system, the master cylinder primary circuit feeds the left-hand rear and the right-hand front brakes, the secondary circuit feeds the right-hand rear brake and the left-hand front brake.

The master cylinder primary circuit is at the servo end of the unit. The object of the dual line brake system is to enable the brakes to remain operative in the event of a failure in the front or rear hydraulic circuit, although the brake efficiency will of course be reduced. Servo assistance in this condition is still available.

The front disc brakes operate in a conventional manner. The brake disc is secured to the hub flange and the calliper is mounted on the stub axle, so that the disc is able to rotate in between the two halves of the calliper. The calliper type fitted is dependent on the model. On IFS models, twin piston sliding pin callipers are fitted. Beam axle models are fitted with two opposed piston or four opposed piston callipers, (according to model). The pistons in the callipers are actuated by hydraulic pressure when the foot brake is applied.

The rear drum brakes have one hydraulic cylinder operating two shoes. When the brake pedal is depressed, the hydraulic pressure expands the two pistons within the cylinder and these in turn actuate the brake shoes against the drum. The rear brakes incorporate a self-adjusting mechanism which automatically adjusts the rear brake shoes, to compensate for wear of the friction linings, whenever the footbrake is operated.

The linings are bonded into position on the shoes. The leading shoe

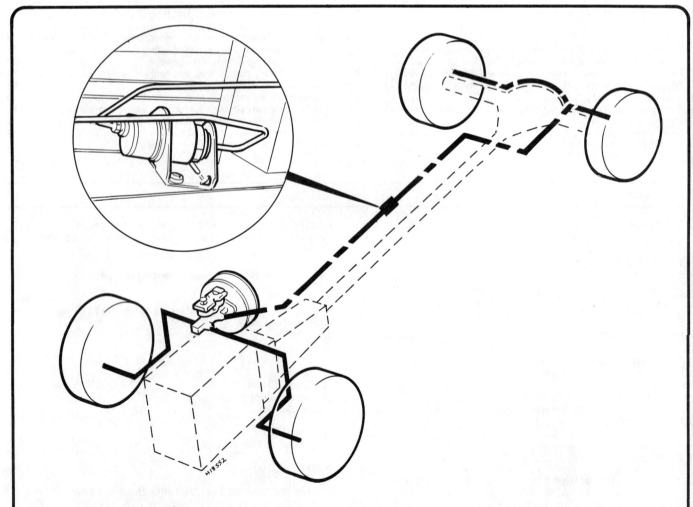

Fig. 9.1 Vertical split brake system layout with G type pressure control valve (inset) – right-hand drive shown (Sec 1)

Chapter 9 Braking system

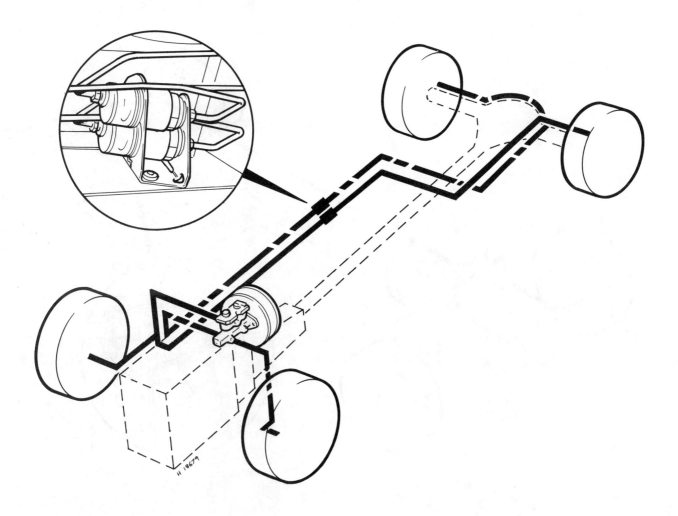

Fig. 9.2 Diagonal split brake system layout with G type pressure control valves (inset) – left-hand drive shown (Sec 1)

lining is about double the thickness of the trailing shoe lining. This ensures an even working life for both front and rear linings since the leading shoe suffers from extra wear in service.

The handbrake operates on the rear wheels only, and the centrally mounted lever is connected to the rear brake assemblies by rod and cable. Adjustment of the handbrake is possible but is not normally necessary as the rear brake adjustment is automatic.

A vacuum servo unit is fitted between the master cylinder and the bulkhead, its function being to reduce the amount of pedal pressure required to operate the brakes. In the event of the servo unit failing for any reason, the brakes will remain operational but the pedal effort required to operate them will be noticeably increased.

A brake pressure control valve (G) or load apportioning valve (LAV) is incorporated into the brake hydraulic system, the type used being dependent on model and operating territory. Their function is to adjust the hydraulic pressure to the rear brakes to prevent them from locking-up under certain operating conditions.

2 Routine maintenance

The following service checks must be made at the intervals given in the Routine maintenance *Section at the front of this manual.*

1 Check the brake hydraulic fluid level in the master cylinder reservoir. The level must be between the MAX and MIN markings on the reservoir side wall. Although it is normal for the fluid level to drop slightly due to displacement caused by the front brake pads wearing, the level should not be allowed to drop below the MIN mark at any time. If the fluid level is to be topped up, wipe clean the area around the reservoir cap to ensure that no dirt is allowed to enter the hydraulic circuit, then top up the level, but do not overfill the system. Wash off any fluid spilt onto the paintwork with cold water as it is corrosive. Always use the recommended brake fluid for topping up, as use of non-standard fluid may result in perishing or swelling of the system seals with consequent brake failure. Any sudden fall in the reservoir fluid level should be investigated immediately (photos).

2 Check the front brake pads for excessive wear. On beam axle (LCY) models, the pads can be provisionally inspected through the calliper window using a mirror and torch as shown in Fig. 9.4. This method avoids the need to remove the front roadwheels. On independent front suspension (LCX) models, a pad wear indicator tab is fitted to the calliper unit (Fig. 9.5). If the tab indicates that maximum wear has taken place, remove the front roadwheels and recheck the pad thickness through the calliper window. The pads must be renewed if any of them are worn down to or beyond the specified minimum allowable thickness.

3 The rear brake linings must also be inspected at the specified intervals for signs of excessive wear. To do this, prise free the rubber inspection plugs from the inboard side of the brake backplates and check the linings through the aperture using a torch (photo).

4 If the brake pads or linings are worn to the extent that they need

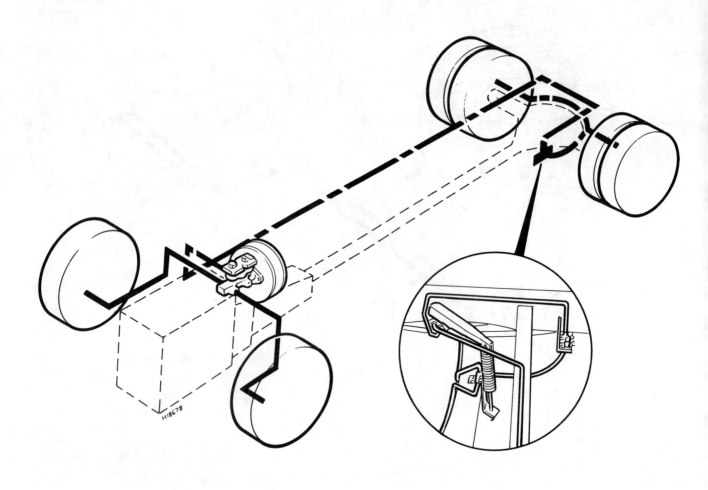

Fig. 9.3 Vertical split brake system layout with LAV pressure control valve unit (inset) – left-hand drive shown (Sec 1)

renewal, replace them as a set (front and/or rear).

5 Inspect the brake hydraulic lines and hoses for any signs of leakage, corrosion and/or chafing. Renew where necessary (photo).

6 Check the handbrake for satisfactory operation and adjust if required (not normally necessary). Lubricate the handbrake cable at the points indicated in Fig. 9.6.

7 At the specified intervals it is advisable to change the fluid in the braking system and at the same time renew all hydraulic seals and flexible hoses. This is because hydraulic fluid absorbs moisture from the air, leading to internal corrosion and reducing the boiling point of the fluid (Section 3).

3 Hydraulic system – draining and bleeding

1 If any of the hydraulic components in the braking system have been removed or disconnected, or if the fluid level in the master cylinder has been allowed to fall appreciably, it is inevitable that air will have been introduced into the system. The removal of air from the hydraulic system is essential if the brakes are to function correctly, and the process of removing it is known as bleeding.

2 There are a number of one-man, do-it-yourself, brake bleeding kits currently available from motor accessory shops. It is recommended that one of these kits should be used wherever possible as they greatly simplify the bleeding operation and also reduce the risk of expelled air and fluid being drawn back into the system.

3 If one of these kits is not available then it will be necessary to gather together a clean jar and a suitable length of clear plastic tubing which is a tight fit over the bleed screw, and also to engage the help of an assistant.

4 Before starting to bleed the brakes, check that all rigid pipes and flexible hoses are in good condition and that all hydraulic unions are tight. Take great care not to allow hydraulic fluid to come into contact with the vehicle paintwork, otherwise the finish will be seriously damaged. Wash off any spilled fluid immediately with cold water.

5 If hydraulic fluid has been lost from the master cylinder, due to a leak in the system, ensure that the cause is traced and rectified before proceeding further or a serious malfunction of the braking system may occur.

6 To bleed the system, remove the dust cap and clean the area around the bleed screw at the wheel cylinder to be bled (photo). If the hydraulic system has only been partially disconnected and suitable precautions were taken to prevent further loss of fluid, it should only be necessary to bleed that part of the system. However, if the entire system is to be bled, it is normal to start at the wheel furthest away from the master cylinder. Support the fluid bleed jar 300 mm (12 in) above the bleed nipple (see Fig. 9.7).

7 Remove the master cylinder filler cap and top up the reservoir. Periodically check the fluid level during the bleeding operation and top up as necessary, using only new brake fluid of the specified type.

8 If a one-man brake bleeding kit is being used, connect the outlet

Chapter 9 Braking system

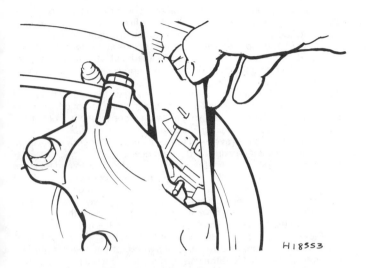

Fig. 9.4 Check the front brake pad thicknesses using a mirror held as shown in LCY models (Sec 2)

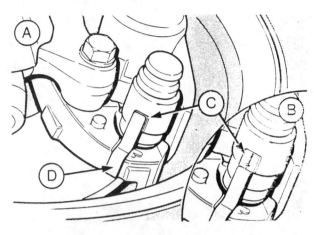

Fig. 9.5 Front brake pad wear indicator tab location on LCX models (Sec 2)

A Indicator position with new pads
B Indicator position with fully worn pads
C Calliper raised flat
D Indicator tab

tube to the bleed screw and then open the screw half a turn. If possible position the unit so that it can be viewed from the driver's seat, then depress the brake pedal to the floor and slowly release it. The one-way valve in the kit will prevent air from returning to the system at the end of each stroke. Repeat this operation until clean hydraulic fluid, free from air bubbles, can be seen coming through the tube. Now tighten the bleed screw and remove the outlet tube.
9   If a one-man brake bleeding kit is not available, connect one end of the plastic tubing to the bleed screw and immerse the other end in the

jar containing sufficient clean hydraulic fluid to keep the end of the tube submerged. Open the bleed screw half a turn and have an assistant depress the brake pedal to the floor and then slowly release it. Tighten

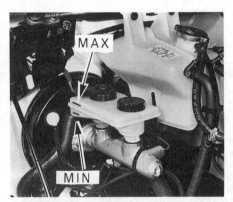

2.1A View showing master cylinder and hydraulic fluid reservoir with fluid level marks indicated

2.1B Topping up the hydraulic fluid level in the reservoir

2.3 Remove plug (arrowed) to inspect rear brake linings for excessive wear

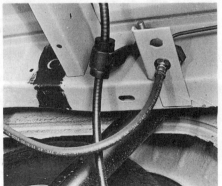

2.5 Inspect the brake hydraulic lines and hoses, also the handbrake cable for condition and security

3.6 Front brake calliper unit and bleed screw/dust cap (arrowed)

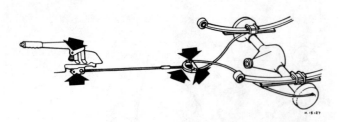

Fig. 9.6 Handbrake cable lubrication points (arrowed) (Sec 2)

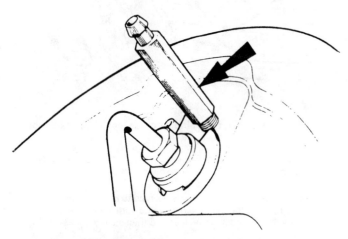

Fig. 9.7 Brake bleed valve on rear brake unit (Sec 3)

the bleed screw at the end of each downstroke to prevent expelled air and fluid from being drawn back into the system. Repeat this operation until clean hydraulic fluid, free from air bubbles, can be seen coming through the tube. Now tighten the bleed screw and remove the plastic tube.

10   If the entire system is being bled the procedures described above should now be repeated at each wheel, finishing at the wheel nearest to the master cylinder. Do not forget to recheck the fluid level in the master cylinder at regular intervals and top up as necessary. Front callipers of four-piston construction are fitted with three bleed nipples and these must be bled simultaneously by using three lengths of tubing.

11   When completed, recheck the fluid level in the master cylinder, top up if necessary and refit the cap. Check the 'feel' of the brake pedal which should be firm and free from any 'sponginess' which would indicate air still present in the system.

12   If the system is being drained as part of the fluid renewal operation, it will be necessary to depress the front brake pistons into the callipers in order to removal all the old fluid. To do this remove the disc pads as described in Section 5 or 6 (as applicable), then press the pistons fully into the callipers and retain with blocks of wood. Refit the disc pads after bleeding the system dry then fill and bleed as previously described.

13   Fluid in the rear brake circuits can be purged in the normal manner (as when bleeding).

14   Discard any expelled hydraulic fluid as it is likely to be contaminated with moisture, air and dirt which makes it unsuitable for further use.

4 Hydraulic brake lines and hoses – inspection, removal and refitting

1   Inspect the condition of the flexible hydraulic hoses leading to each of the front disc brake callipers and also the one at the front of the rear axle. If they are swollen, damaged or chafed, they must be renewed.

2   Wipe the top of the brake master cylinder reservoir and unscrew both caps. Place a piece of polythene sheet over the top of each orifice and secure with elastic bands around the filler necks. This is to stop hydraulic fluid syphoning out during subsequent operations.

3   Raise and support the vehicle on axle stands at the appropriate end to provide a suitable working height underneath the vehicle. Where applicable, remove the roadwheels to provide access to the hydraulic lines and connections to be worked on.

4   To remove a flexible brake hose, unscrew the union nuts and disconnect the rigid pipe(s). If a front hose is being removed, unscrew the banjo union bolt and collect the two copper washers. Detach the hose from its support clips or brackets and remove it from under the vehicle. Refitting is the reverse of removal but use new copper washers at the banjo union if working on the front hose.

5   The steel pipes must be thoroughly cleaned and examined for signs of dents or other percussive damage, rust and corrosion. Rust and corrosion should be scraped off and, if the depth of pitting in the pipes is significant, they will need renewal. This is most likely in those areas underneath the chassis and along the rear axle where the pipes are exposed to the full force of road and weather conditions.

6   Rigid pipe removal is usually quite straightforward. The unions at each end are undone and the pipe drawn out of the connection. The clips which may hold it to the vehicle body are bent back and it is then removed. Underneath the vehicle, exposed unions can be particularly stubborn, defying the efforts of an open-ended spanner. As few people will have the special split ring spanner required, a self-grip wrench is the only answer. If the pipe is being renewed, new unions will be provided. If not then one will have to put up with the possibility of burring over the flats on the union and use a self-grip wrench for replacement also.

7   Rigid pipes which need replacement can usually be purchased at any local garage where they have the pipe, unions and special tools to make them up. They will need to know the pipe length required and the type of flare used at the ends of the pipe. These may be different at each end of the same pipe.

8   Installation of the pipes is a reversal of the removal procedure. The pipe profile must be preset before fitting. Any acute bends must be put in by the garage on a bending machine, otherwise there is the possibility of kinking them and restricting the fluid flow.

9   On completion bleed the brake hydraulic system as described in Section 3.

5 Front disc pads (sliding pin calliper type) – inspection, removal and refitting

1   To make a provisional check of the disc pads, inspect them through the calliper window as shown in Fig. 9.4 using a mirror and torch. To make a full inspection, and to remove the pads, the calliper piston housing will need to be removed. Proceed as follows.

2   Fully apply the handbrake, loosen off the wheelnuts, then raise and support the vehicle at the front end. Remove the front wheels.

3   Undo the two Torx slide bolts securing the calliper to the anchor bracket then withdraw the calliper piston housing and position it out of the way. Support the housing so that the hydraulic hose is not distorted or stretched. Unless the calliper piston housing is to be removed, the hydraulic line can be left attached (photos).

4   Withdraw the inner and outer disc pads. If they are likely to be refitted, keep them separated and identified as they are identical and must be refitted in their original positions (photo).

5   Inspect the pads for wear. If they have worn to a lining thickness as shown in the specifications, or under, they must be renewed.

6   If it is found that the respective pads have worn unevenly, it is probable that the calliper slide sleeves are sticking, in which case they should be cleaned and lubricated with brake mechanism grease obtainable from your Ford dealer.

7   If the pads have worn to their minimum allowable thickness or unevenly, they must be renewed as a set of four (never individually).

8   Before fitting the pads into position, push the pistons back into their bores but watch for fluid spillage from the master cylinder reservoir caused by displacement. If necessary, syphon a small amount of fluid from the reservoir using a suitable pipette. Clean any fluid spilt onto the vehicle paintwork immediately with cold water, as it will damage the paint and cause corrosion.

Chapter 9 Braking system

5.3A Remove the end covers ...

5.3B ... unscrew the calliper Torx bolts ...

5.3C ... and remove them ...

5.3D ... then lift the calliper piston housing clear

5.4 Inner and outer disc pads

5.10 Inner view of piston unit prior to refitting

9   Clean the backs of the pads and brake disc surfaces of dirt and grease.
10   Refitting the disc pads is a reversal of the removal procedure, but it will be necessary to depress the footbrake pedal several times to reposition the calliper pistons and provide the disc pads with the correct adjustment. Check the brake fluid level in the reservoir and top up as necessary with fresh fluid before refitting the filler caps and wiping away any excess fluid (photo).
11   The calliper to anchor bracket Torx bolts should be tightened to the specified torque setting and the brakes operated to ensure that the slide bolts move freely in their bores.
12   Refit the roadwheels and lower the vehicle.

6 Front disc pads (opposed piston calliper type) – inspection, removal and refitting

1   Apply the handbrake firmly and remove the hub cap (when fitted). Loosen the wheel nuts and jack up the front of the vehicle, supporting it adequately on suitable stands. Remove the front wheels.
2   On inspection, if the disc pad lining thickness is found to be less than the minimum given in the Specifications, the pads must be renewed. The pads should be renewed in sets of four, and a single or single pair of pads should never be renewed alone.
3   To remove the disc pads, first open the bonnet and remove the brake fluid reservoir filler cap. Cover the fluid filler neck with a clean cloth.
4   Depress the disc pad anti-rattle clips and simultaneously grip the pad retainer pins and slide them free. Remove the clips (Fig. 9.8).
5   Extract the disc pads from the calliper housing; if they prove difficult to move by hand, a pair of long-nosed pliers can be used (Fig. 9.9).
6   The next operation will raise the level of the brake fluid in the fluid reservoir and it is advisable to remove a quantity of fluid with a pipette, although the previously positioned cloth will absorb a small amount of fluid.

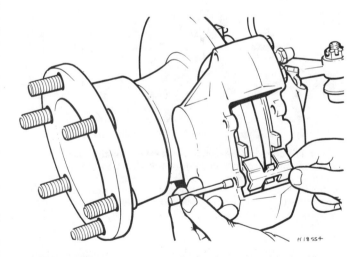

Fig. 9.8 Removing the retaining pins and anti-rattle clips from the opposed piston type calliper (Sec 6)

7   Using a flat iron bar or length of wood, carefully push the pistons fully back into their calliper bores. Be careful not to damage or distort the brake disc.
8   Clean the recesses in the calliper and the exposed faces of each piston of all traces of dirt and rust.
9   Refitting the disc pads is a reversal of the removal procedure.
10   When fitted, depress the brake pedal several times to reposition the calliper pistons and provide the disc pads with the correct adjust-

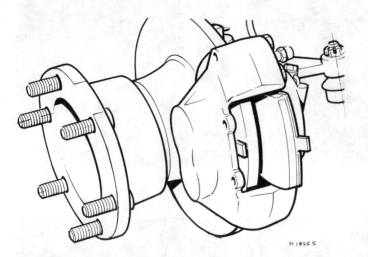

Fig. 9.9 Disc pad removal from the opposed piston type calliper (Sec 6)

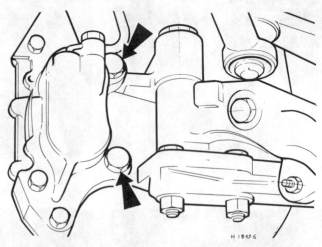

Fig. 9.10 Calliper retaining bolts (arrowed) on the opposed piston type caliper (Sec 7)

4 On the opposed piston type calliper, unscrew and remove the calliper to stub axle retaining bolts and withdraw the calliper (Fig. 9.10).
5 On the twin piston sliding calliper type unit, undo the two retaining bolts and withdraw the calliper anchor bracket and piston assembly (Fig. 9.11)
6 Refitting the front brake calliper is a reversal of the removal procedure but the following additional points should be noted:

(a) Wipe the disc contact faces clean of any split brake fluid or foreign matter
(b) Tighten all nuts and bolts to the correct torque wrench settings
(c) Refit the brake pads as described in Section 5 or 6 as applicable
(d) Bleed the brake hydraulic system as described in Section 3 and remember to remove the polythene sheeting from the fluid reservoir filler cap

8 Front brake calliper – dismantling, overhaul and reassembly

1 Remove the front brake calliper as described in Section 7. Note that under no circumstances should the two halves of the opposed piston type calliper be separated as they are specially bolted together during manufacture.
2 Position a piece of wood inside the calliper across the faces of the pistons, then use low air pressure (from a foot pump for example) applied to the fluid inlet port to force the pistons from the housing. Once the pistons have been partially ejected, they can be removed completely by hand. Suitably identify each of the pistons as to their location in the housing.
3 Using a non-metallic instrument such as a knitting needle, carefully remove the piston seals, seal retainers and dust covers as applicable.
4 Clean the pistons and housing with methylated spirit and allow to dry. Examine the surfaces of the piston and housing for wear, damage and corrosion. If the piston surface alone is unserviceable, obtain a repair kit which includes new pistons and seals, but if the housing is unserviceable renew the calliper complete. If both the pistons and housing are in satisfactory condition, obtain a repair kit of seals. Where applicable check the condition of the slide pin bolts and bushes and renew these components as necessary.
5 Dip the internal components in clean brake fluid then assemble the new seals to the calliper bores using the fingers only. Make sure the seals are fitted the correct way round.
6 Locate the seal retainers or dust covers as applicable, lubricate the calliper bores with brake fluid and carefully insert the pistons. If the original pistons are being refitted ensure that they are fitted into their original bores. With the pistons in place, engage the dust covers (where fitted) with the grooves in the calliper.
7 Refit the calliper as described in Section 7.

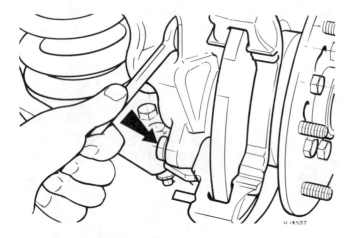

Fig. 9.11 Removing the anchor bracket bolts on the twin piston sliding calliper (Sec 7)

ment. Check the brake fluid level in the reservoir and top up as necessary with fresh fluid before refitting the filler cap and wiping away any excess fluid.

7 Front brake calliper – removal and refitting

1 Remove the front disc pads as described in Section 5 or 6 (as applicable).
2 Wipe the area around the brake fluid reservoir filler cap, remove the cap and place a piece of polythene over the filler neck. Secure the polythene with an elastic band. This will prevent an excessive amount of fluid being lost from the hydraulic system and will thus assist the final bleeding operation.
3 Using a socket or ring spanner, unscrew and remove the banjo union bolt securing the hydraulic brake pipe to the calliper. Recover the two copper washers.

Chapter 9 Braking system

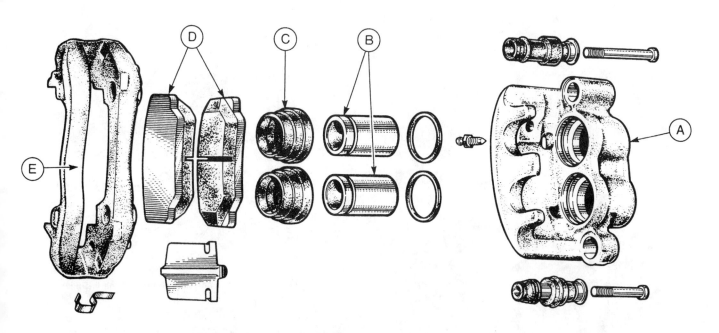

Fig. 9.12 Exploded view of the twin piston sliding calliper (Sec 8)

- A Piston housing
- B Pistons
- C Piston seal dust covers
- D Disc pads
- E Anchor bracket

9.2 Check the brake disc for thickness and condition

9.3 Checking the disc run-out using a dial gauge

10.10 Rear brake assembly shown with drum, axleshaft and wheel hub removed

9 Front brake disc – examination, removal and refitting

1 Remove the disc pads as described in Section 5 or 6 (as applicable).

2 Examine the surface of the disc for deep scoring or grooving. Light scoring is normal but anything more severe should be removed by taking the disc to be surface ground, provided the thickness of the disc is not reduced below the specified minimum; otherwise a new disc will have to be fitted (photo).

3 The disc should be checked for run-out, but before doing this, the front wheel bearings must be checked and if necessary adjusted as described in Chapter 10. The run-out is best checked with a dial gauge, although using feeler blades between the face of the disc and a fixed point will give a satisfactory result. The check should be made on both sides of the disc at a radius of 5.1 in (130 mm), the disc being slowly moved by hand (photo).

4 If the disc run-out exceeds that given in the Specifications and it is confirmed that the disc is faulty, it must be renewed.

5 To remove the brake disc, first remove the calliper unit as described in Section 7 according to type, but leave the hydraulic lines attached to the calliper. Suspend or support the calliper whilst detached to prevent the hydraulic hose from being stretched and distorted.

6 Refer to Chapter 10 and remove the front wheel hub unit complete with brake disc. If the disc and hub are to be reassembled to each other, ensure that their relative alignment marks are visible (Fig. 9.14).

7 Using a screwdriver, bend back the locktabs (where fitted) and unbolt the disc from the hub assembly in diagonal sequence.

8 Refitting the front brake disc is a reversal of the removal procedure but the following additional points should be noted:

(a) Clean the mating surfaces of the disc and hub assembly before reassembling them and ensure that the alignment marks are adjacent to each other.

(b) Always use new locktabs to secure the disc retaining bolts

(c) Adjust the wheel bearings as described in Chapter 10.

Chapter 9 Braking system

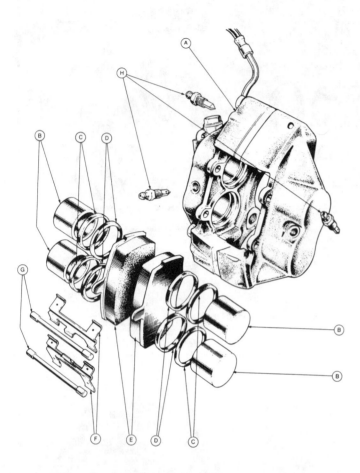

Fig. 9.13 Exploded view of the opposed piston type front brake calliper (Sec 8)

- A Calliper
- B Pistons
- C Seals
- D Retainers
- E Disc pads
- F Anti-rattle clips
- G Retaining pins
- H Bleed nipples

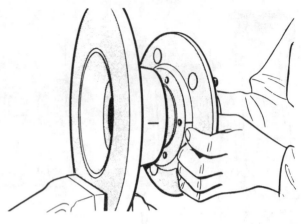

Fig. 9.14 Front hub and disc alignment mark (Sec 9)

10 Rear brake shoes – inspection, removal and refitting

A special hub nut removal tool will be required according to axle type (see text)

1 Two rubber inspection plugs are fitted to each rear backplate and these can be prised free to allow inspection of the brake linings through the apertures. This eliminates the need to remove the brake drums unless the linings are to be renewed (photo 2.3)

2 After prising out the rubber plugs, clean the edge of the shoe to ascertain how much wear has taken place and compare it with the minimum lining thickness given in the Specifications. Repeat the procedure for each shoe of both rear wheels. If it is found that any shoe is below the minimum limit, it will be necessary to renew all the rear brake shoes.

3 Remove the hub cap (when fitted) and loosen off the rear wheel nuts. Remember that six stud wheels have left-hand thread nuts on the left side, right-hand thread on the right side. Chock the front wheels and release the handbrake.

4 Raise the vehicle at the rear and support it on axle stands. On F and H type axle models remove the rear roadwheels.

5 The following instructions refer to axle types G, or F and H. Refer to Chapter 8 for the identification and description of the axle types and further details of the axleshaft removal procedure.

6 On the G type axle, unscrew and remove the axleshaft nuts in diagonal sequence and withdraw the axleshaft together with its gasket.

10.12A Upper return spring location on leading shoe

10.12B Upper return spring location on trailing shoe

10.13 Lower return spring and attachment points to the brake shoes

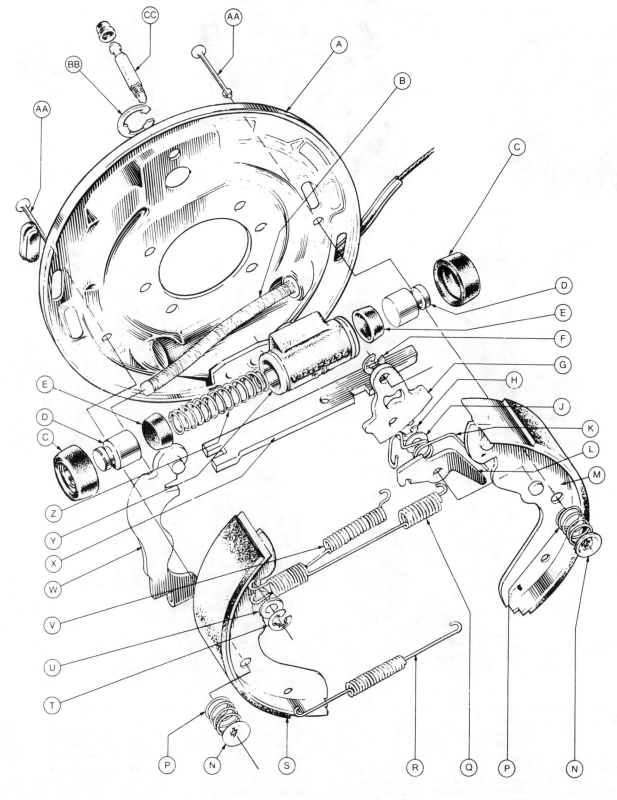

Fig. 9.15 Rear brake components (Sec 10)

A Backplate	H Spring clip	Q Upper return spring	X Spacer strut
B Handbrake cable	J Washer	R Lower return spring	X Spacer strut
C Dust cover	K Spring	S Brake shoe	Y Wheel cylinder
D Piston	L Ratchet	T Spring clip	Z Spring
E Seal	M Brake shoe	U Washer	AA Pin
F Spring clip	N Washer	V Handbrake spring	BB Circlip
G Ratchet	P Spring	W Handbrake lever	CC Bleed valve

170 Chapter 9 Braking system

10.14 Removing the leading brake shoe

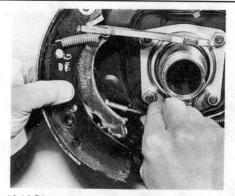
10.16 Disconnecting the handbrake cable from the trailing shoe

10.18A Trailing shoe and handbrake lever attachment – outboard side

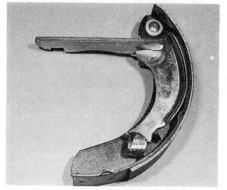
10.18B Trailing shoe and handbrake lever attachment – outboard side

10.21 General view of backplate, wheel cylinder and handbrake cable with the shoe assemblies removed

10.22A Self adjuster ratchet unit on the inboard side of the leading shoe

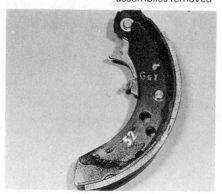
10.22B Outboard view of the leading shoe

10.23 Refitting the leading shoe – alternative method used with return springs already attached for convenience

Using a chisel or screwdriver, bend up the tabs of the lockwasher and then unscrew and remove the bearing locknut; a suitable large socket is required to loosen the bearing locknut and adjusting nut and should be obtained from the local tool hire agent or local garage if possible. Remove the lockwasher, adjusting nut, and the outer bearing cone and rollers assembly, then carefully withdraw the rear roadwheels with the hub and drum unit from the rear axle. Wipe the assembly clean of any surplus grease which may have dropped onto the brake drum inner surface.

7 On H and F axle types, mark the relative fitted positions of each drum to its hub, then withdraw the brake drums each side. Withdraw the axle shaft each side taking care not to damage the hub inner oil seal. Bend the hub locktab up and then undo the locknut using Ford special tool No 15-077.

8 With the rear brake assemblies now accessible they can be cleaned off and inspected. When cleaning, do not inhale the harmful brake asbestos dust, and take care not to allow any oil or grease to come into contact with the brake linings or brake drum friction surfaces.

9 Check the brake drums for excessive wear or signs of damage. The drums must be renewed or alternatively machined if extensive wear has taken place.

10 Check the brake shoe lining thickness and, if it is below the minimum limit given in the Specifications, or if the lining rivets are flush with the lining surface, it will be necessary to renew all four rear brake shoes. Renewal is also necessary if the linings are contaminated with oil or hydraulic fluid. In this case the source of contamination must also be rectified (photo).

11 To remove the brake shoes, first remove the holding-down spring from the leading shoe by depressing the top washer, turning it through 90° and withdrawing the washers, spring, and pin from the backplate.

12 Note the position of each brake shoe return spring, and mark the outer surface of the shoe webs to indicate their front and rear locations (photos).

13 Unhook the lower return spring from both shoes using a pair of pliers (photo).

14 Pull the lower end of the leading shoe away from the abutment and unhook it from the upper return spring; detach the return spring from the trailing shoe (photo).

Chapter 9 Braking system

11.1 Rear brake wheel cylinder unit

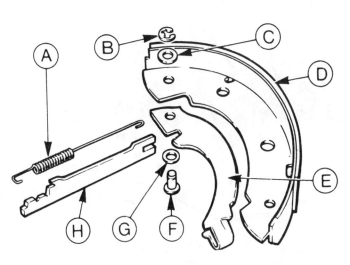

Fig. 9.17 Rear brake shoe and handbrake components (Sec 10)

- A Spring
- B Spring clip
- C Washer
- D Trailing shoe
- E Lever
- F Pivot pin
- G Washer
- H Spacer strut

18 Finally remove the handbrake lever and spacer strut from the trailing shoe by pulling the lever and twisting the strut, and then unhooking the retaining spring. Prise the pivot pin clip out of its groove and remove the handbrake lever and washers from the trailing shoe (photos).
19 With the brake shoes removed, locate an elastic band around the wheel cylinder to prevent the pistons from coming out. Take care not to depress the footbrake pedal whilst the brake shoes and drums are removed.
20 Thoroughly clean all traces of lining dust from the shoes and backplate but do not inhale the dust as it is hazardous to health.
21 Check that the wheel cylinder pistons are free to move in their bores. Check the wheel cylinder dust covers for damage and the wheel cylinders and brake pipe unions for hydraulic fluid leaks (photo).
22 Check the handbrake lever, adjuster, and self-adjusting mechanism for wear, and renew any items as necessary (photos).
23 Refitting of the rear brake shoes is a reversal of the removal procedure, but the following additional points should be noted:

(a) When refitting the brake shoes note that the leading shoe is the one with the thicker lining (photo).
(b) Ensure that the spacer strut its fitted at right-angles to the handbrake lever. The adjuster ratchets should be positioned as shown (photo 22A) prior to fitting the leading shoe
(c) When reassembling the G type axle, readjust the rear hub bearings as given in Chapter 8
(d) When reassembling the F and H type axles, tighten the hub nuts to the torque wrench setting specified in Chapter 8
(e) Depress the footbrake several times to bring the brake shoes into the correct position
(f) Adjust the handbrake as described in Section 16

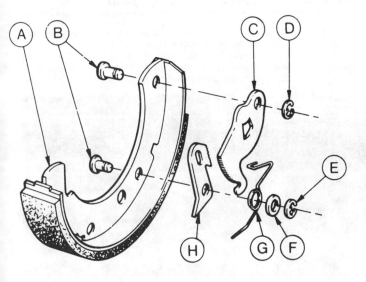

Fig. 9.16 Exploded view of the self-adjustment mechanism components (Sec 10)

- A Leading brake shoe
- B Pivot pins
- C Large ratchet
- D Spring clip
- E Spring clip
- F Washer
- G Spring
- H Small ratchet

15 Remove the holding-down spring from the trailing shoe using the procedure described in paragraph 11.
16 Pull the upper end of the trailing shoe away from the wheel cylinder, pull the handbrake cable spring back, and disconnect the handbrake cable from the handbrake lever. The trailing shoe can now be removed from the backplate (photo).
17 The self-adjusting mechanism must now be removed from the leading shoe by prising the two clips from the pivot pins. Note the correct position of the ratchet spring (Fig. 9.16).

24 Finally remove the stands, lower the vehicle to the ground, and road test the vehicle to check the operation of the brakes. If new linings have been fitted, the efficiency of the brakes may be slightly reduced until the linings have bedded-in.

Chapter 9 Braking system

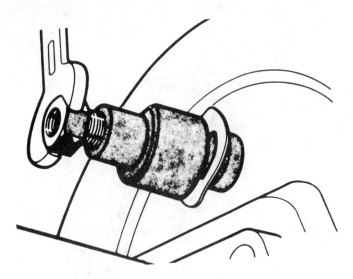

Fig. 9.18 Tool for refitting the rear wheel cylinder retaining clip (Sec 11)

11 Rear brake wheel cylinder – removal and refitting

1 If the wheel cylinder is seized or the internal seals leaking it will be necessary to overhaul or renew the unit; the removal and refitting procedure is as follows (photo).
2 Remove the rear brake shoes by referring to Section 10, then thoroughly clean the area around the wheel cylinder to prevent the ingress of foreign matter into the hydraulic system.
3 Wipe clean and then remove the brake fluid reservoir cap. Place a piece of polythene over the filler neck and secure with an elastic band to prevent all the brake fluid from draining out of the hydraulic pipes.
4 Unscrew and remove the bleed nipple from the rear of the wheel cylinder and then, using a split ring spanner if possible, unscrew the union nut securing the metal brake pipe to the wheel cylinder. Plug the end with a pencil or cover it with masking tape to prevent dirt contaminating the brake fluid (photo).
5 Using a screwdriver, prise the retaining clip from the rear of the wheel cylinder and then remove the wheel cylinder and gasket from the brake backplate.
6 Clean the wheel cylinder location on the backplate on both sides.
7 Refitting the rear wheel cylinder is a reversal of the removal procedure but the following additional points should be noted:

(a) Fit a new gasket and retaining clip to the wheel cylinder each time it is removed
(b) A special tool is required to fit the retaining clip to the wheel cylinder and this is shown in Fig. 9.18. If the tool is not readily available, a suitable length of tubing, a threaded stud, a nut, and washer can be made into a tool quite easily by the home mechanic
(c) It is important to ensure that the location peg on the wheel cylinder locates with the hole in the backplate
(d) Bleed the braking system as described in Section 3 and then depress the brake pedal several times to centralise and adjust the brake shoes by actuation of the automatic adjuster. Don't forget to remove the polythene sheet from the brake fluid reservoir.
(e) Adjust the handbrake as described in Section 16.

12 Rear brake wheel cylinder – dismantling and reassembly

1 Remove the wheel cylinder as described in the previous Section, then brush and clean away all external dirt.
2 Prise each rubber boot from its groove in the cylinder body. Extract each piston together with its seal from the cylinder bore; if difficulty is experienced apply air pressure from a tyre pump to the fluid inlet, and

11.4 Brake line connection and bleed nipple (arrowed) to rear wheel cylinder

13.1 Detaching the fluid level indicator lead connector

16.3 View showing cable adjuster and locknut (1), the equalizer (2) and return spring (3)

17.4 Handbrake lever retaining bolts (arrowed)

17.7 Handbrake primary rod to relay lever connection

Chapter 9 Braking system

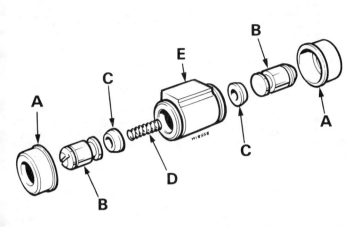

Fig. 9.19 Exploded view of the rear wheel cylinder unit (Sec 12)

    A   Dust caps   D   Spring
    B   Pistons   E   Cylinder
    C   Seals

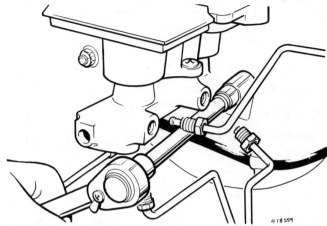

Fig. 9.20 Master cylinder removal from the servo unit (Sec 13)

wrap the assembly in a clean cloth to catch the parts in as they emerge. The bleed nipple must be fitted or the injected air will escape through its hole.

3   The cylinder components and their order of fitting are shown in Fig. 9.19.

4   Inspect the inside of the cylinder for score marks caused by impurities in the hydraulic fluid. If any are found, the cylinder and pistons will require renewal. If the wheel cylinder requires renewal always ensure that the replacement is exactly similar to the one removed.

5   If the cylinder is sound, thoroughly clean it out with fresh hydraulic fluid.

6   The old rubber seals will probably be swollen and visibly worn. Smear the new rubber seals with hydraulic fluid and refit to the pistons, making sure they are the correct way round.

7   Wet the cylinder bore with clean hydraulic fluid and then insert the return spring, followed by the seal and piston at each end. Ensure that the seal is correctly located and not distorted or damaged during fitting into the cylinder bore.

8   Relocate the rubber boots on each end of the cylinder and check that they are fully engaged in the cylinder grooves to complete.

9   Refit the cylinder as described in the previous Section.

13 Master cylinder – removal and refitting

1   Raise and support the bonnet. Detach the brake fluid level warning indicator wire connector (photo).

2   Wipe the area around the hydraulic fluid reservoir filler cap, unscrew the cap, place a piece of polythene over the filler neck and secure with an elastic band. This will help retain the fluid in the reservoir while the master cylinder is being removed.

3   Referring to Fig. 9.20 disconnect the fluid lines at their connections to the master cylinder. Note the pipe locations. Plug the pipe ends to prevent spillage and the ingress of dirt.

4   Undo and remove the two nuts and spring washers that secure the master cylinder to the rear of the servo unit. Lift away the master cylinder taking care that no hydraulic fluid is allowed to drip onto the paintwork. Wash any fluid spillage from paintwork or fittings in the engine compartment using cold water as the fluid will damage paintwork.

5   Refit in the reverse sequence to removal. Refer to Section 14 if fitting a new unit. When reattaching the fluid line unions, they should be loosely assembled until all are located. The master cylinder should also be loosely attached to assist if necessary in allowing the pipe connections to be engaged easily and without the danger of cross-threading.

With all pipes connected, tighten the fittings and the master cylinder retaining nuts.

6   Top up the hydraulic reservoir fluid level and bleed the system as described in Section 3. Don't forget to remove the polythene.

14 Master cylinder – dismantling, overhaul and reassembly

If a replacement master cylinder is to be fitted, it will be necessary to lubricate the seals before fitting to the vehicle as they have a protective coating when originally assembled. Remove the blanking plugs from the hydraulic pipe union seatings. Inject clean hydraulic fluid into the master cylinder and operate the primary piston several times so that the fluid spreads over all the internal working surfaces. If the master cylinder is to be dismantled after removal, proceed as follows:

1   First remove the reservoir filler cap and invert the cylinder to allow the fluid to drain from the cylinder. Detach the reservoir from the cylinder by unscrewing the two retaining screws. On some models the reservoir is removed by simply pulling upwards from the cylinder and removing the rubber seals. Do not re-use fluid emptied from the cylinder. (Figs. 9.21 and 9.22).

2   Mount the cylinder in a soft-jawed vice, then use a length of dowel rod to depress the primary piston and, with it held down, locate and remove the stop pin from the front reservoir inlet (Fig. 9.23).

3   Extract the circlip from the mouth of the cylinder and then remove the internal components, placing them on a clean surface in the exact order of removal; tap the cylinder on a wooden block to remove the secondary piston assembly.

4   Remove the primary piston spring by unscrewing and removing the retaining screw and sleeve.

5   Prise the seals from the pistons as necessary and then wash all components in clean hydraulic fluid and methylated spirit. Examine the pistons and cylinder bore surfaces for scoring, scratches, or bright wear areas and if any are observed, renew the master cylinder as a complete unit. If the surfaces are in good condition, discard the old seals and obtain a repair kit comprising primary piston and new seals.

6   Dip the internal components in clean hydraulic fluid and then fit them into the master cylinder bore in the reverse order to removal, making sure that the new seals are fitted the correct way round. Refer to Figs. 9.24 and 9.25 for the correct position of the primary and secondary piston components. When handling the seals use only the fingers to manipulate them into position.

7   The remaining reassembly procedure is a reversal of the dismantling procedure. As a safety precaution, when the cylinder has been refitted to the vehicle, have an assistant depress the footbrake pedal hard for a minimum of ten seconds and observe the cylinder for signs of fluid leakage.

15 Vacuum servo unit – removal and refitting

1 The vacuum servo unit is located between the master cylinder and the engine bulkhead. First remove the master cylinder as described in Section 13.
2 Prise free and release the check valve with vacuum hose from the servo unit. Note that if either of these items require renewal (hose or check valve) they must be replaced as a pair, not individually.
3 Working from inside the vehicle, disconnect the servo operating rod from the brake pedal by extracting the spring clip and withdrawing the clevis pin.
4 Detach the wiring connector from the brake stop light switch on the brake pedal bracket, then twist the switch through 90° and remove it from the bracket.
5 Unscrew and remove the two nuts retaining the servo unit to the bulkhead and remove the servo unit from the vehicle (Fig. 9.26).
6 The servo unit is not repairable and therefore it must be renewed if it is defective.
7 Refitting the vacuum servo unit is a reversal of the removal procedure, but the following points should be noted:
  (a) Tighten the retaining nuts to the specified torque wrench setting
  (b) Ensure that the brake light switch is securely clipped into position and the wiring connection securely made
  (c) Refer to Section 13 for details on refitting the master cylinder and then top up and bleed the hydraulic system (Section 3)

16 Handbrake – adjustment

1 Normally the automatic adjustment of the rear brakes will also remove any excess handbrake lever movement, but where the handbrake cable has stretched it will be necessary to carry out the following adjustment.
2 Jack up the rear of the vehicle, chock the front wheels and support it adequately with suitable stands.
3 Fully release the handbrake lever and then working underneath the vehicle, loosen off the cable adjuster locknuts. (One locknut has a left-hand thread). Turn the adjuster until all tension is removed from the operating rods and cable, and then clean and lubricate the adjuster threads thoroughly (photo). Operate the footbrake pedal several times to ensure that the rear brakes are adjusted, then pull the handbrake lever up 3 notches.
4 Retighten the adjuster until it is hand tight but no more and then tighten the locknuts against the adjuster; the threads each side of the adjuster should be of equal length and the threads should be visible through the adjuster holes.
5 Under normal operation, the handbrake should be effective when applied 3 to 5 notches. Release the handbrake fully and check that the rear wheels are free to turn, then reapply the handbrake and lower the vehicle to the ground.

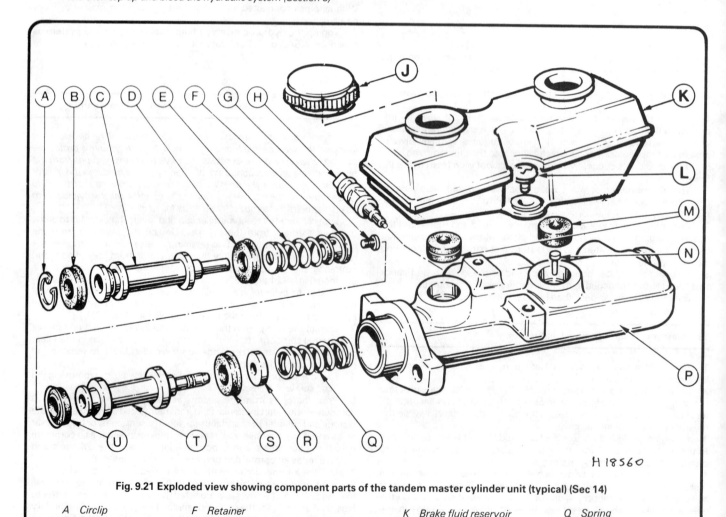

Fig. 9.21 Exploded view showing component parts of the tandem master cylinder unit (typical) (Sec 14)

A Circlip	F Retainer	K Brake fluid reservoir	Q Spring
B Seal	G Screw	L Screw	R Retainer
C Primary piston	H Brake failure warning light switch where fitted	M Seal	S Seal
D Seal		N Piston stop pin	T Secondary piston
E Spring	J Reservoir cap	P Master cylinder	U Seal

Chapter 9 Braking system

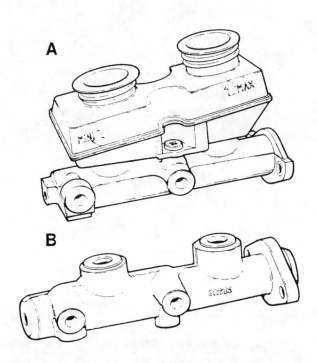

Fig. 9.22 Master cylinder and reservoir type identification (Sec 14)

A Bendix type with screw fixing reservoir
B Automotive products with clip on reservoir

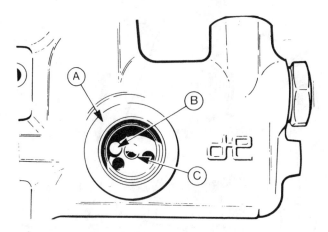

Fig. 9.23 Master cylinder (A) showing the piston stop pin (B) and secondary inlet port (C) (Sec 14)

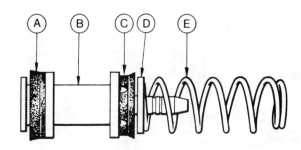

Fig. 9.25 Secondary piston components – assembled position (Sec 14)

A Seal
B Piston
C Seal
D Retainer
E Spring

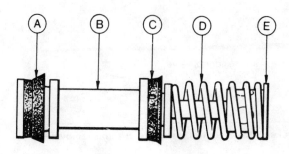

Fig. 9.24 Primary piston components – assembled position (Sec 14)

A Seal
B Piston
C Seal
D Spring
E Retainer

17 Handbrake lever and primary rod – removal and refitting

Handbrake lever

1   Chock the front roadwheels and then fully release the handbrake.
2   Prise free the handbrake lever gaiter from the floor and ease it up the lever out of the way (Fig. 9.28).
3   Extract the spring clip then withdraw the clevis pin from the lever to relay link rod connection.
4   Undo the lever to floor mounting bolts and remove the lever (photo).
5   Refit in the reverse order of removal. Renew the clevis pin if it is excessively worn (also the spring clip if necessary) and lubricate the pin as it is fitted. On completion, check and if necessary adjust the handbrake as described in Section 16.

Primary rod

6   Chock the front roadwheels, release the handbrake, then raise and support the vehicle at the rear on axle stands.
7   With the handbrake fully released, disconnect the primary rod from the relay lever by extracting the spring clip and withdraw the clevis pin (photo).
8   Detach the primary rod from the cable equaliser unit by removing the spring clip and clevis pin. Remove the primary rod.
9   Refit in reverse order of removal. Renew the clevis pins if they are excessively worn, also the spring clips if necessary. Lubricate the clevis pins as they are fitted.
10   On completion, check and if necessary adjust the handbrake as described in Section 16.

Fig. 9.26 Servo unit retaining nut locations (arrowed) (Sec 15)

18.6 Handbrake cable and backplate

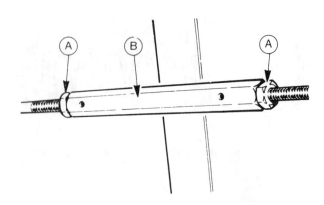

Fig. 9.27 Handbrake cable adjustment locknuts (A) and adjuster (B) (Sec 16)

19.2 Brake pressure control valve

18 Handbrake cable – removal and refitting

1 Chock the front wheels, slacken the rear wheel nuts, raise the vehicle at the rear and support it on axle stands.
2 Release the handbrake, then working from underneath the vehicle, detach the handbrake primary rod to equaliser by removing the clevis pin and clip.
3 Detach the return spring from the equaliser.
4 Remove the rear brake drums as described in Section 10.
5 Detach the brake cable from the brake shoe mechanism operating lever by pulling on the cable return spring so that the cable can be disengaged from the lever.
6 Detach the cable from the location bracket on the backplate and withdraw the cable (photo).
7 Refit in the reverse order of removal. Refer to Section 10 to refit the brake drum. On short wheelbase models, engage the cables in the clips at their crossing points (Fig. 9.29).
8 Before lowering the vehicle, check and if necessary adjust the handbrake as described in Section 16

19 Brake pressure control valve – description, removal and refitting

1 Some models are fitted with a load conscious brake pressure control valve, the purpose of which is to regulate the pressure applied to the rear brakes regardless of vehicle loading. This then prevents the wheels 'locking up' when heavy brake pressure is applied.
2 The valve assembly is mounted on the underside of the vehicle, attached to the inboard side of the right-hand chassis member opposite the fuel tank (photo).
3 The valve operation is fully automatic and it cannot be adjusted or repaired. If defective, it must therefore be renewed.
4 To remove the pressure control valve, first raise and support the vehicle at the rear end on axle stands. Allow a suitable working clearance underneath.
5 Wipe clean and then remove the brake fluid reservoir filler cap then position a clean piece of polythene sheet over the filler neck and secure with an elastic band. This will prevent the brake fluid from draining from the hydraulic pipes when they are detached from the valve.
6 Clean the hydraulic line connections at the control valve, then

Chapter 9 Braking system 177

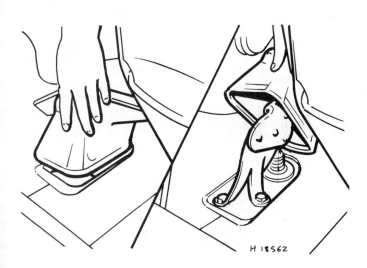

Fig. 9.28 Handbrake lever gaiter removal (Sec 17)

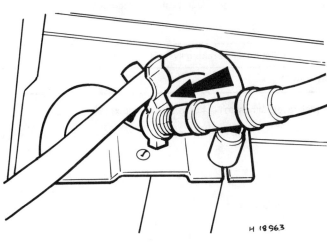

Fig. 9.29 Handbrake cable retaining clip (arrowed) fitted to short wheelbase models (Sec 18)

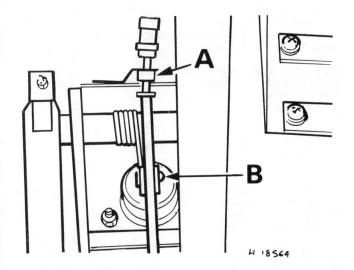

Fig. 9.30 Brake pedal to pushrod clevis pin (B) and stop lamp switch (A) (Sec 21)

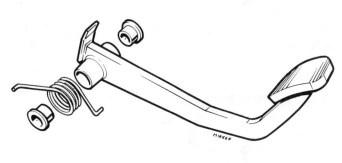

Fig. 9.31 Brake pedal and associate fittings (Sec 21)

unscrew the union nuts and detach the rigid hydraulic lines from the valve. Plug the lines to prevent fluid loss and the ingress of dirt.
7   Withdraw the valve retaining clip and withdraw the valve from its bracket.
8   Refit in the reverse order of removal. As the valve is fitted to the bracket, align the cut out. When the valve is refitted, top up the hydraulic fluid level and bleed the rear brakes as described in Section 3. Once the rear brakes are bled, get an assistant to slowly depress the brake pedal whilst you unscrew the bleed screw on the pressure control valve, but only a little to allow the fluid past. As soon as the fluid is seen to exit from the valve retighten the bleed screw.
9   Lower the vehicle to complete.

20 Load apportioning valve (LAV) – general

1   The function of this device is similar in certain aspects to that of the brake pressure control valve described in the previous Section, but it is only fitted to certain models and differs in design and location.

2   The LAV unit is mounted on the chassis in front of the rear axle and is operated by a lever and load sensing spring interconnected between the valve and the axle (Fig. 9.3).
3   The only time that the LAV unit will need removal is if the valve is defective or the axle is to be removed, in which case the valve unit can be left in position but the coil spring will have to be detached at its top end. If the spring is detached at its bottom end the valve will have to be re-adjusted. Refer to Chapter 8 for details on removing/refitting the rear axle.
4   The valve will also have to be re-adjusted if the assembly is removed. This requires the use of specialised equipment and reference to several graphs depending on the vehicle body type. The removal and refitting of this unit should therefore be entrusted to a Ford dealer.

21 Brake pedal – removal and refitting

1   Detach the brake pedal to servo pushrod by extracting the spring clip and withdrawing the clevis pin (Fig. 9.30).
2   Detach the lead connector from the brake stop light switch then twist the switch through 90° and withdraw it from the bracket.
3   Unhook and remove the spring clip securing the brake pedal shaft

Chapter 9 Braking system

on the brake side, then press the pedal towards the clutch so that the brake pedal is released (Fig. 9.31).

4   Press or drive out the pedal bushes and remove the spring.

5   Refit in the reverse order of removal. Align the cut out in the shaft with the pedal box as they are reassembled.

22   Fault diagnosis – braking system

Symptom	Reason(s)
Pedal travels almost to floor before brakes operate	Wheel cylinder leaking Master cylinder leaking (bubbles in master cylinder fluid) Brake flexible hose leaking Brake line fractured Brake system unions loose Air in hydraulic system Defective automatic adjuster mechanism
Brake pedal feels 'springy'	New pads or linings not yet bedded-in Excessive brake disc run-out Brake drums badly worn and weak or cracked Brake discs badly worn or loose Master cylinder securing nuts loose
Brake pedal feels 'spongy' and 'soggy'	Wheel cylinder leaking Master cylinder leaking (bubbles in master cylinder reservoir) Brake pipe line or flexible hose leaking Unions in brake system loose Air in hydraulic system
Excessive effort required to brake vehicle	Linings or pads badly worn New linings recently fitted – not yet bedded-in Harder linings or pads fitted than standard causing increase in pedal pressure Primary or secondary hydraulic circuit failure Linings, brake drums or discs contaminated with oil, grease or hydraulic fluid Leaking vacuum hose Servo unit worn internally
Brakes uneven and pulling to one side	Linings, pads and brake drums or discs contaminated with oil, grease, or hydraulic fluid Tyre pressures unequal Brake backplate loose Brake shoes of pads fitted incorrectly Different type of linings fitted at each wheel Anchorages for front suspension or rear axle loose Brake drums or disc badly worn, cracked or distorted
Brakes tend to bind, drag, or lock-on	Faulty automatic adjuster causing brake shoes to be adjusted too tightly Handbrake cable over-tightened Brake calliper or pistons seized Reservoir vent hole in cap blocked with dirt Master cylinder trap valves restricted – brakes seize in 'on' position Wheel cylinder seized in 'on' position Brake shoe pull-off springs broken, stretched or loose Brake shoe pull-off springs fitted wrong way round, omitted, or wrong type used Handbrake system rusted or seized in the 'on' position

Chapter 10 Suspension and steering

Contents

Anti-roll bar (beam axle type) – removal and refitting	15
Anti-roll bar (IFS) – removal and refitting	9
Coil spring and lower suspension arm (IFS) – removal and refitting	6
Fault diagnosis – suspension and steering	39
Front beam axle – removal and refitting	11
Front beam axle leaf spring – removal and refitting	13
Front shock absorber (beam axle type) – removal and refitting	16
Front shock absorber (IFS) – removal and refitting	10
Front suspension bracket bush (IFS) – removal	7
Front wheel alignment	37
Front wheel hub – dismantling and reassembly	4
Front wheel hub – removal, refitting and adjustment	3
General description	1
Lower arm balljoint (IFS) – removal and refitting	8
Power-assisted steering system – fluid level check and system bleeding	33
Power steering gear unit – removal and refitting	35
Power steering hoses – removal and refitting	36
Power steering pump – removal and refitting	34
Rack and pinion steering gear unit – dismantling, overhaul and reassembly	25
Rack and pinion steering gear unit – removal and refitting	24
Rear axle leaf spring – removal and refitting	17
Rear shock absorber – removal and refitting	18
Routine maintenance	2
Steering column – dismantling, overhaul and reassembly	21
Steering column – removal and refitting	20
Steering column lower universal coupling – removal and refitting.	23
Steering column upper bearing – renewal	22
Steering gear rubber bellows – renewal	27
Steering wheel – removal and refitting	19
Stub axle (beam axle type) – removal, overhaul and refitting	12
Suspension bump stop (beam axle type) – removal and refitting	14
Suspension bump stop (IFS) – removal and refitting	5
Track rod end – removal and refitting	26
Worm and nut steering drag link – removal and refitting	32
Worm and nut steering drop arm – removal and refitting	31
Worm and nut steering gear – dismantling, overhaul and reassembly	29
Worm and nut steering gear – removal and refitting	28
Worm and nut steering gear – rocker shaft pre-load adjustment	30
Wheels and tyres – general care and maintenance	38

Specifications

Front suspension

Short wheelbase models

Type	Independent 'hybrid' MacPherson strut, coil springs, double acting telescopic shock absorbers and anti-rollbar
Caster angle	0° 15′ to 4° 30′
Maximum caster angle variation, left-to-right-hand	1°
Camber angle	0°30′ to 2° 30′
Maximum camber angle variation, left-to-right-hand	1° 15′
Toe-in (measured at wheel rims)	0.001 to 1.60 mm (0.00 to 0.063 in)
Toe-in angle	0° to 0° 16′
Front wheel bearing endfloat	0.025 to 0.13 mm (0.001 to 0.005 in)

Long wheelbase models

Type	Beam axle with semi-elliptic leaf springs and telescopic shock absorbers
Caster angle	2° 30′ to 6° 15′
Maximum caster angle variation, left-to-right-hand	1°
Camber angle	0° 30′ to 1° 30′
Maximum camber angle variation, left-to-right-hand	1°
Toe-in (measured at wheel rims)	0.001 to 1.60 mm (0.00 to 0.063 in)
Toe-in angle	0° 0′ to 0° 16′
Front wheel bearing end float	0.025 to 0.13 mm (0.001 to 0.005 in)
Stub axle to axle beam shim:	
Maximum clearance	0.1 mm (0.004 in)
Minimum clearance	0.025 mm (0.001 in)
Shim sizes	3.3 to 4.6 mm (0.13 to 0.18 in) in 0.1 mm (0.004 in) increments

180   Chapter 10   Suspension and steering

Rear suspension
Type .... Rigid rear axle suspended from semi-elliptic leaf springs. Telescopic shock absorbers

Roadwheels
Type .... Five or six stud fixing, steel
Size .... 14 x 5.5 or 14 x 5

Tyres

Model	Body	Tyre size	Pressure – bar (lbf/in^2)	
			front	rear
80	Van/Combi	185R 14 REIN	2.4 (35)	2.9 (42)
100	Van/Combi	185R 14C PR6	2.6 (38)	3.4 (50)
100	Bus	185R 14C PR6	2.4 (35)	2.9 (42)
100L	Van/Combi/C. Cab	195R 14C PR8	2.5 (36)	4.1 (59)
115	Bus	195R 14C PR6	2.3 (34)	3.0 (44)
120	All models	195R 14C PR6	2.3 (34)	3.7 (54)
130	All models	185R 14 REIN	2.4 (35)	2.3 (34)
130	All models	185R 14C PR6	2.6 (38)	2.5 (36)
160	All models	185R 14C REIN	2.4 (35)	2.7 (39)
160	All models	185R 14C PR6	2.6 (38)	2.9 (42)
190	Van/Combi/C. Cab	185R 14C PR6	2.6 (38)	3.2 (47)

Steering
Type .... Rack and pinion, (short wheelbase models), or worm and nut. Power-assisted steering available on some models

Rack and pinion steering
Adjustment .... Yoke plug and spring
Turns lock-to-lock 5.3
Pinion turning torque .... 0.9 to 1.8 Nm (0.6 to 1.3 lbf ft)
Steering gear lubricant .... To Ford spec. SLM-1C-9110A grease
Lubricant capacity .... 180 grams

Worm and nut steering
Ratio 24:1
Rocker shaft pre-load (through centre of travel) .... 2.5 to 3.0 Nm (1.8 to 2.2 lbf ft)
Upper bearing pre-load .... 0.3 to 0.8 Nm (0.2 to 0.6 lbf ft)
Steering gear lubricant .... To Ford spec. SLM-1C-9110A grease
Lubricant capacity .... 200 grams

Power-assisted steering
Steering gear type .... Bendix
Pump type .... Hobourn Eaton roller vane
Power steering fluid type .... ATF to Ford spec SQM-2C-9010-A (Duckhams D-Matic)

Torque wrench settings
Front suspension (IFS models)

	Nm	lbf ft
Front wheel bearing adjuster nut:		
Stage 1	23 to 33	17 to 25
Stage 2	Slacken to give specified endfloat	
Crossmember to chassis	70 to 90	52 to 66
Front bracket to chassis nuts	148 to 201	109 to 148
Front bracket to crossmember bolts	26 to 32	19 to 24
Lower balljoint nut	130	96
Lower balljoint to lower arm nuts	40 to 51	30 to 38
Anti-roll bar to chassis bolts	20 to 28	15 to 21
Anti-roll bar link nut	20 to 28	15 to 21
Shock absorber mounting nuts	40 to 51	30 to 38
Shock absorber mounting bolts	40 to 57	30 to 42
Brake disc to hub bolts	48 to 51	35 to 38
Lower arm through bolt nut (see text):		
Clamping torque	100	74
Fully slacken	-	-
Snug torque	70	52
Torque angle	90°	90°

Front suspension (beam axle models)

	Nm	lbf ft
Front wheel bearing adjuster nut:		
Stage 1	23 to 33	17 to 25
Stage 2	Slacken to give specified endfloat	

Chapter 10 Suspension and steering

Torque wrench settings (continued)   Nm   lbf ft

Front suspension (beam axle models)

	Nm	lbf ft
Spring U-bolt nuts:		
Stage 1	35 to 45	26 to 33
Stage 2	75 to 85	55 to 63
Stage 3	114 to 146	84 to 108
Spring shackle nuts	70 to 90	52 to 66
Spring front eye bolt nut	156 to 196	115 to 145
Spring bump stop nut	40 to 51	30 to 38
Shock absorber upper nuts	40 to 51	30 to 38
Shock absorber lower nuts	70 to 90	52 to 66
Anti-roll bar link nut	70 to 97	52 to 72
Anti-roll bar anchor plate nuts	70 to 97	52 to 72

Rear suspension

	Nm	lbf ft
Spring U-bolt nuts:		
F and H type axles	88 to 100	65 to 74
G type axle	120 to 130	89 to 96
Spring front eye bolt nut	159 to 196	117 to 145
Spring shackle bolts:		
F and H type axles	70 to 90	52 to 66
G type axle	175 to 220	129 to 162

Rack and pinion steering

	Nm	lbf ft
Steering wheel nut	45	33
Column tube to bulkhead	23	17
Column tube upper mounting	23	17
Coupling shaft pinch bolt	30	22
Inner track rod balljoint	103	76
Track rod end locknut	80	59
Track rod end to steering arm	60	44
Steering gear unit to crossmember	50	37
Pinion bearing cover	85	63
Preload yoke slipper plug:		
Stage 1	10	7
Stage 2	Slacken by 30°	Slacken by 30°

Worm and nut steering

	Nm	lbf ft
Steering wheel nut	45	33
Column tube to bulkhead	23	17
Column tube upper mounting	23	17
Coupling shaft pinch bolt	30	22
Drop arm locknut	270	199
Rocker arm and adjuster locknut	35	26
Rocker arm side cover	21	16
Steering wormshaft adjuster locknut	135	100
Steering nut clamp screws	6	4
Drag link and track rod end castle nuts	50 to 70	37 to 52
Steering unit to body	84	62
Steering arm to stub axle	58	43
Drag link end locknuts	80	59
Lock stop nut	20 to 28	15 to 21

Power steering and associated systems (where different from above)

	Nm	lbf ft
Pump mounting bolts (M10)	41 to 58	30 to 43
Pump mounting bolts (M12)	70 to 97	52 to 72
Pump idler pulley bolts	21 to 28	16 to 21
Fluid hose connection nut	27 to 34	20 to 25
Fluid hose clamps	2.5 to 3.5	2 to 3
Coupling shaft/column pinch bolt	26 to 32	19 to 24
Drag link/track rod castle nut:		
Stage 1	115	85
Stage 2	Tighten further to next split pin hole	

Roadwheels

	Nm	lbf ft
Five stud wheel	75 to 95	55 to 70
Six stud wheel	155 to 180	114 to 133

1 General description

The front suspension system will be of either independent (IFS) type, fitted to short wheelbase models, or beam axle type, fitted to long wheelbase and 100L models (Figs. 10.1 and 10.2).

The independent front suspension system comprises a coil spring lower arm and a telescopic double acting shock absorber each side. The beam axle suspension comprises a single taper leaf spring and a double-acting telescopic shock absorber each side.

An anti-roll bar is fitted to the front suspension system on some models.

The rear suspension system on all models is much the same as that fitted on earlier Transit models, and comprises single or multiple

Chapter 10 Suspension and steering

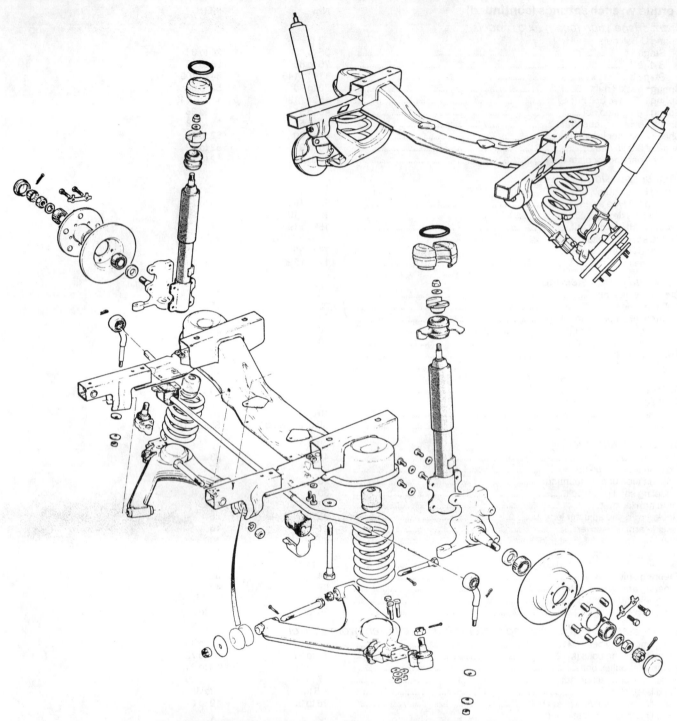

Fig. 10.1 Exploded view of the independent front suspension system (Sec 1)

semi-elliptic leaf springs and telescopic double-acting shock absorbers.

The steering system fitted will be either rack and pinion type (IFS) models, or worm and nut (beam axle models). The steering column used is common to all models. Power-assisted steering is available on the worm and nut steering models.

2 Routine maintenance

A regular routine inspection and service of the steering and suspension is essential, particularly where the vehicle operates in arduous conditions with regular maximum capacity loading. Carry out the following service procedures at the intervals specified in the *Routine maintenance* Section at the start of this manual.

All models

1 Check and if necessary adjust the tyre pressures. Inspect the tyres for signs of uneven or excessive wear, or any signs of damage. If abnormal wear is found, a thorough check of the steering and suspension components should be made to find the cause. Refer to Section 38 for further details.

2 Check the track rod balljoints and covers for excessive wear (Fig. 10.3). With the vehicle free standing on the ground, get an assistant to actuate the steering wheel back and forth whilst you watch the balljoint

Chapter 10 Suspension and steering

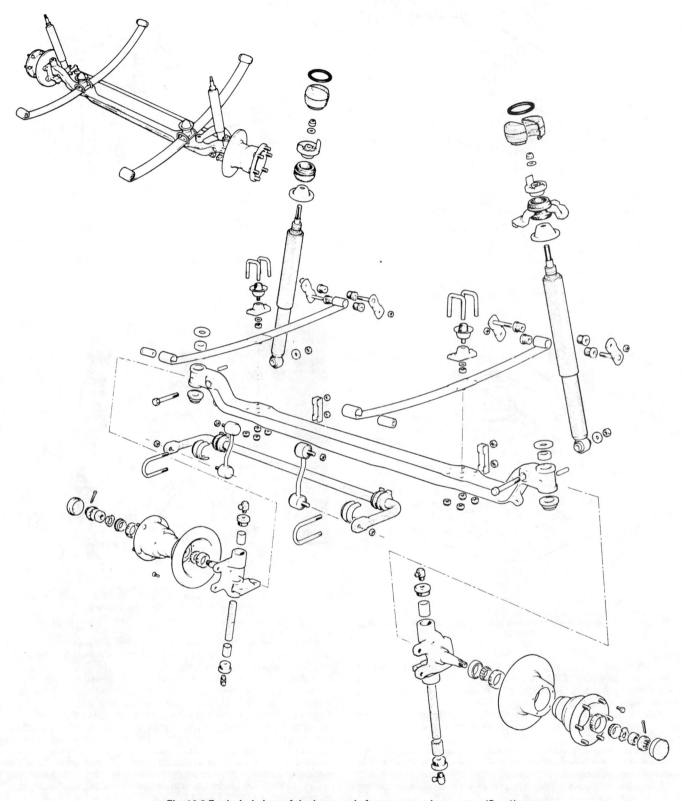

Fig. 10.2 Exploded view of the beam axle front suspension system (Sec 1)

for signs of excessive movement. Renew the balljoint(s) if necessary (Section 26).

3 Check the steering and suspension fastenings for security, the U-bolt nuts in particular (Fig. 10.4). Also check the roadwheel nuts for tightness.

4 Check the wheel hub bearings for excessive wear and play. Refer to Section 3 for details on the front hubs, and Section 5 or 6 in Chapter 8 for details on the rear wheel hubs.

IFS models

5 Check the condition of the steering balljoint linkages and seal rubbers. To check for excessive wear, locate a suitable lever between

Chapter 10 Suspension and steering

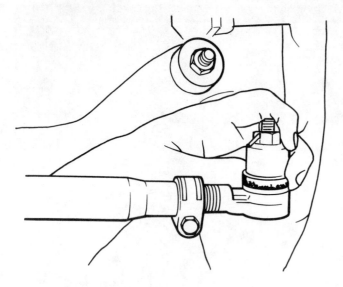

Fig. 10.3 Check the balljoint covers (Sec 2)

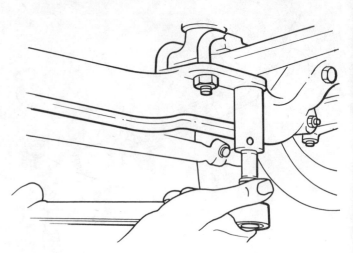

Fig. 10.4 Check the suspension U-bolt nuts for tightness (Sec 2)

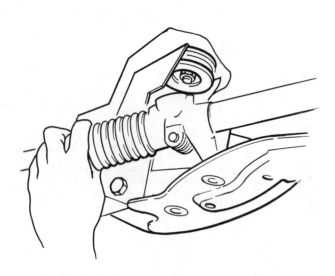

Fig. 10.5 Checking an inner track rod joint track and pinion steering (Sec 2)

Fig. 10.6 Lubricate the beam axle king pins and bushes via the upper and lower nipples (arrowed) (Sec 2)

the roadwheel inner rim and the lower arm, and carefully apply an upward pressure. On LCX Series models, the joints are not spring loaded and there should be no noticeable movement. On other models the balljoints are spring loaded and a small amount of movement is acceptable but this should not exceed 0.5 mm (0.020 in). Renew the joints if they are excessively worn or the rubber seals are in poor condition (Section 8).

6 Check the steering rack bush for excessive wear by moving the rack up and down at the non-pinion end. Check the rack bellows for damage and renew as necessary (Section 26).

7 Check the inner rack joints for excessive wear by squeezing them through the bellows as shown (Fig. 10.5).

Beam axle models

8 Check the front axle kingpins and bushes for excessive wear. This check must be made before the kingpins are greased. Raise the vehicle at the front end so that the front roadwheels are clear of the ground, then grip each front roadwheel in turn at the top and bottom and with the wheel in the straight-ahead position, rock it in and out by pulling on the top whilst pushing on the bottom and vice-versa. Any excessive play in the axle will be felt. A small amount of play is acceptable but if in doubt, get your Ford dealer to check and assess the wear. Renew the kingpins if necessary (Section 12).

9 Lubricate the kingpins and bushes using a suitable grease gun. Raise and support the vehicle at the front end so that the front roadwheels are clear of the ground. Lubricate the axle via the grease nipples until clean grease is seen to exit from the joints (Fig. 10.6).

10 Check the steering free-play at the steering wheel. With the vehicle free standing on level ground and the correct tyre pressures, position the wheels in the straight-ahead position, then lightly holding the steering wheel, rock it back and forth. At the same time check the amount of steering wheel movement before the front roadwheels start to turn. An assistant will be useful in observing this. The total movement must not exceed 35 mm (1.37 in), at the steering wheel rim. If it does exceed this margin, check and adjust the steering rocker (sector) shaft pre-load setting (Section 30).

Power steering

11 Check the fluid level in the power steering fluid reservoir and top up the level if required (Section 33).

12 Check the power steering drivebelt tension and condition, and adjust/renew as necessary (Chapter 2).

Chapter 10 Suspension and steering

3.5 Removing the hub dust cap

6.6 Steering balljoint to suspension arm retaining nuts

6.8 Suspension arm to front bracket pivot bolt, nut and bushes

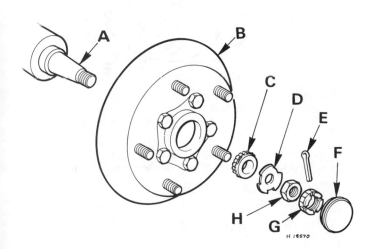

Fig. 10.7 Exploded view of the front hub (Sec 3)

A Stub axle
B Hub/disc
C Bearing (outer)
D Washer
E Split pin
F Dust cap
G Retainer
H Nut

Fig. 10.8 Checking the front wheel bearings for endfloat (Sec 3)

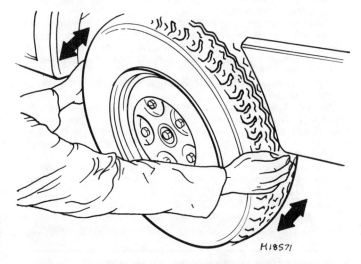

Fig. 10.9 Checking the front wheel bearing endfloat using a dial gauge (Sec 3)

3 Front wheel hub – removal, refitting and adjustment

1 Loosen off the wheel nuts (left-hand thread on left-hand side on beam axle variants), then raise and support the vehicle at the front on axle stands.
2 Remove the nuts and remove the roadwheels.
3 On independent front suspension (IFS) models, unbolt and detach the front brake pipe location bracket from the shock absorber.
4 Unbolt and remove the brake calliper unit from the stub axle unit, referring to Chapter 9 for details. The hydraulic hose can be left attached, but support the weight of the calliper to prevent the hose from being strained. Note that the upper calliper bolt on IFS models will not fully withdraw.
5 Prise free the dust cap from the hub (photo), wipe away the grease, then extract the split pin and remove the nut retainer. Unscrew the hub nut, then remove the washer and withdraw the hub unit complete with its outer bearing from the stub axle. Note that the hub nut on the left-hand side has a left-hand thread.
6 Clean the hub and bearings for inspection.
7 Examine the bearing surfaces for signs of excessive wear and pitting. If new bearings are required, renew the bearing cones and cups as a pair. If the oil seal is defective it must be renewed. To renew the hub bearings and oil seal, proceed as described in Section 4.

8 Lubricate the hub bearings and oil seal lips with a suitable grease prior to refitting the hub onto the stub axle.
9 Take care when refitting the hub unit not to damage the oil seal lips. Slide the unit onto position on the stub axle and then fit the outer bearing cone, washer and adjuster nut, but do not tighten the nut fully at this stage.
10 Refit the brake calliper unit (Chapter 9) and on IFS models, reconnect the brake hose to the location clip on the shock absorber.
11 Refit the front roadwheel(s) and tighten the retaining nuts to the specified torque wrench setting.
12 Wheel bearing adjustment: Tighten the wheel bearing adjuster nut to the specified torque wrench setting whilst simultaneously spinning the roadwheel in each direction to ensure that the wheel bearings are fully seated.
13 Loosen off the adjuster nut by one flat, then rock the roadwheel by hand to further seat the bearings. Now hold the wheel at diagonally opposite points and then pull and push the wheel in and out to feel for a small amount of endfloat which should be present. If the required endfloat is not felt, unscrew the adjuster nut a further flat, then try again. Do not apply a rocking motion to the wheel when testing for endfloat or a false impression may possibly be gained.
14 An accurate wheel bearing endfloat check should now be made using a dial indicator gauge attached as shown in Fig. 10.9. Turn the bearing adjuster nut as required to achieve the specified endfloat then remove the gauge, fit the nut retainer over the adjuster nut and insert a new split pin to secure the nut in the set position.
15 Smear some wheel bearing grease over the nut, then refit the dust cap. Lower the vehicle to the ground.

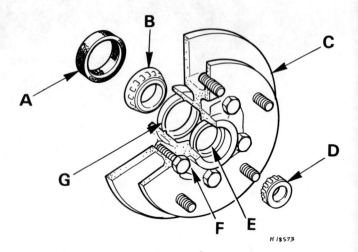

4 Front wheel hub – dismantling and reassembly

1 Remove the wheel hub from the vehicle as described in the previous Section.
2 Withdraw the inner oil seal from the hub by levering it out with a suitable screwdriver or similar tool.
3 Extract the inner and outer bearing cones (Figs. 10.10 and 10.11).
4 Remove the bearing cups by driving them out of the hub using a suitable soft metal drift. Take care not to damage the brake disc during this operation.
5 To remove the brake disc from the hub, mark the disc and hub for alignment so that they can be reassembled in their original positions, then bend back the retaining bolt lock tabs (IFS only), and undo the retaining bolts. Separate the disc from the hub. Clean their mating surfaces prior to reassembly.
6 New lock washers must be fitted when reassembling the disc and hub on IFS models. Ensure that the disc to hub alignment marks correspond then insert and tighten the retaining bolts to the specified torque wrench setting. Bend the lock washer tabs over the nut to secure on IFS models.
7 If the bearings are being renewed, replace the cups and cones as they are matched pairs.
8 Drive the bearing cups into their housings using a suitable tube drift, whilst supporting the hub unit.
9 Lubricate the bearing cones and cups with wheel bearing grease, then fit the cones. The outboard cone can be left until later for fitting if required (after the hub is refitted onto the stub axle).
10 Drive a new oil seal into position in the hub on the inboard side.
11 Refit the wheel hub and disc unit as described in the previous Section and adjust the endfloat.

5 Suspension bump stop (IFS) – removal and refitting

1 Prise or cut free the bump stop from its location peg on the crossmember. Clean the peg thoroughly before fitting the new bump stop (Fig. 10.12).
2 Lubricate the new bump stop and the new location peg with a soapy solution to ease its fitting, then press the stop onto the peg using

Fig. 10.10 Exploded view of front hub bearings (IFS type) (Sec 4)

A Grease seal
B Inner bearing cone
C Hub
D Outer bearing cone
E Outer bearing cup
F Locking tab
G Inner bearing cup

a jack. When correctly fitted, it should be possible to rotate the stop on the peg.

6 Coil spring and lower suspension arm (IFS) – removal and refitting

1 Loosen off the wheel nuts, raise and support the vehicle at the front end on axle stands, allowing a minimum clearance between the crossmember and the ground of 500 mm (20 in).
2 Remove the front roadwheels.
3 Disconnect the track rod end from the stub axle as described in Section 26. Move the track rod out of the way.
4 Where applicable, disconnect the anti-roll bar from the suspension arm (Section 9).
5 Position a jack under the suspension arm and raise it to just support its weight.
6 Unscrew the three retaining nuts securing the steering balljoint to the suspension arm. Remove the bolts and detach the suspension arm (photo).
7 Slowly lower the jack under the arm and allow the arm to drop. Remove the coil spring and its upper rubber pad.
8 Straighten the retaining split pins in the bolts of the suspension arm to crossmember and front bracket, then withdraw the pins. Unscrew

Chapter 10 Suspension and steering

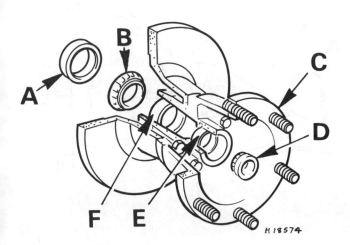

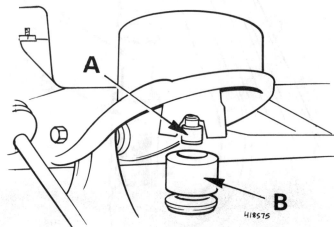

Fig. 10.12 View showing bump stop (B) and peg (A) (Sec 5)

Fig. 10.11 Exploded view of front hub bearings (beam axle type) (Sec 4)

A Grease seal
B Inner bearing cone
C Hub
D Outer bearing cone
E Outer bearing cup
F Inner bearing cup

the retaining nuts, remove the washers, and withdraw the pivot bolts (photo). Lower the suspension arm from its location and remove it.
9 If the suspension arm bush is worn or perished and in need of replacement, it can be withdrawn from the suspension arm eye using a simple puller comprising a metal tube of suitable diameter, (just under that of the arm eye inside diameter), washers, a long bolt and nut. Alternatively (and preferably), if a press is at hand, the bush can be pushed out using a suitable rod or a length of tube of suitable diameter.
10 If the suspension arm has been damaged or distorted in anyway, it must be renewed.
11 To ease the fitting of the new bush into the suspension arm, lubricate the bush and the eye in the arm with engine oil, then press or draw (as applicable) the bush into the arm so that the bush lips are seated correctly. To ensure that the bush is fully seated, the tube used to press (or draw) the bush into position should have a cut-out section at one end to allow it to pass over the welded seam on the suspension arm (Figs. 10.13 and 10.14).
12 To refit the suspension arm, locate it with the crossmember and front bracket and insert the pivot bolts. Fit the dished washer so that its convex face is towards the bush of the front bracket. Do not fully tighten the retaining nuts at this stage. They must be fully tightened when the vehicle is free standing.
13 Locate the rubber pad onto the top end of the coil spring and locate the spring into position between the crossmember and the suspension arm. If the rubber pad is in poor condition, it must be renewed. Engage the spring end into the slot in the pad. Ensure that the lower end of the spring engages in the recess in the suspension arm.
14 Raise the suspension arm using a suitable jack and engage the outer end with the stub of the stub axle balljoint. Align the three retaining bolt holes, insert the bolts, fit the retaining nuts and tighten them to the specified torque wrench setting. Take care when fitting the stub to the suspension arm not to engage more than is necessary, as the balljoint bellows can be easily damaged.
15 Reconnect the steering track rod end balljoint with the stub axle as described in Section 26.
16 If applicable, reconnect the anti-roll bar link to the suspension arm (Section 9).

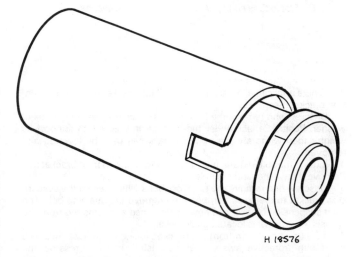

Fig. 10.13 Lower arm bush and removal tool (Sec 6)

17 Refit the roadwheel(s), raise the vehicle and remove the axle stands then lower the vehicle so that it is free standing.
18 The lower suspension arm retaining nuts can now be tightened to the specified torque wrench setting and new split pins fitted to secure. When tightening the nut securing the rear pivot bolt to the crossmember the following tightening sequence must be adopted. First tighten the nut to the clamping torque figure given in the Specifications then slacken the nut fully. Tighten it again to the 'snug' torque figure then tighten further through one quarter turn (90°).
19 Tighten the wheel nuts to the specified torque to complete.

7 Front suspension bracket bush (IFS) – renewal

Before starting it should be noted that if the right and left-hand side front suspension bracket bushes are to be renewed, they must only be renewed individually. If they are both renewed at the same time, the suspension geometry will be disturbed and will require checking and possibly adjustment by a Ford garage on completion.

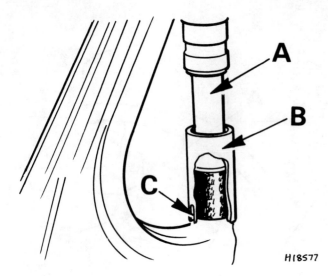

Fig. 10.14 Inserting the bush into the lower suspension arm (Sec 6)

    A Bush press tool
    B Bush and outer tool (tube)
    C Slot in tube engaged over weld seam of lower arm

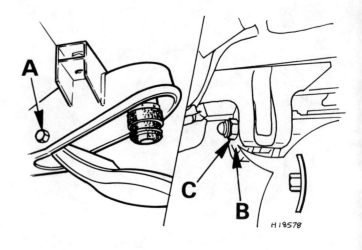

Fig. 10.15 Lower suspension arm (Sec 6)

    A Pivot bolt
    B Dished washer (inset shows orientation)
    C Front bracket/lower arm nut

1 Raise and support the front end of the vehicle on axle stands.
2 Where applicable, detach the anti-roll bar bracket from the suspension bracket (Section 9).
3 Unscrew and remove the front bracket to chassis securing nut and washer. Position a jack under the bracket and raise it to support its weight, then withdraw the bolt. The jack can then be lowered and removed.
4 Extract the split pin and unscrew the lower arm to front bracket nut. Remove the washer and the bolt.
5 Unscrew and remove the two Torx bolts which secure the suspension bracket to the crossmember, then remove the retaining bolt. The front suspension bracket can now be detached and removed from the chassis (on which it is located on dowels).
6 To extract the bush from the bracket, locate the bracket in a vice fitted with protective jaws so that the bush is up, then press out the bush using a two-leg puller or alternatively, draw the bush out using a long bolt, spacer tube, washers and nut (Fig.10.17).
7 Clean out the eye in the bush before inserting the new bush. Lubricate the eye and bush with engine oil, then press or draw the bush into position by reversing the withdrawal process (Fig.10.18). Ensure that when fitted, the bush lips protrude each side of the bracket.
8 Refit the front bracket to the chassis and reconnect the associate items in the reverse order of removal, but note the following special points:

(a) Ensure that the bracket is fully engaged with the location dowels before tightening the retaining nuts and bolts
(b) The suspension arm pivot washer must be fitted so that its convex side is facing the suspension bush. Do not tighten the retaining nut to the specified torque setting until after the vehicle is lowered and is free standing on its wheels. A new split pin can then be inserted to secure
(c) The crossmember to chassis bolt and the two Torx bolts should only be hand tightened initially. The bracket to chassis bolts are then fitted (jack under the bracket to align the bolt holes), and tightened. Remove the jack and tighten the crossmember to chassis bolts to the specified torque wrench setting
(d) Refer to Section 9 to reconnect the anti-roll bar (if applicable)

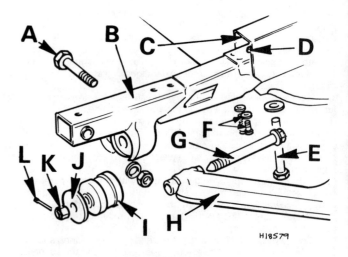

Fig. 10.16 Front suspension lower arm and front bracket components (Sec 7)

    A Bolt (chassis/bracket)   G Lower arm/bracket bolt
    B Front bracket   H Lower arm
    C Crossmember   I Bracket bush
    D Dowel   J Dished washer
    E Crossmember/chassis bolt   K Nut
    F Torx bolts   L Split pin

Chapter 10 Suspension and steering 189

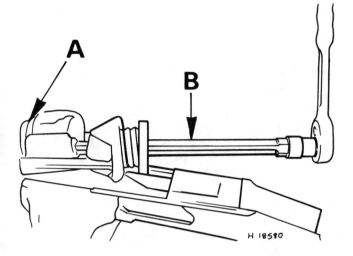

Fig. 10.17 Bush (A) removal from bracket using a two-legged puller (B) (Sec 7)

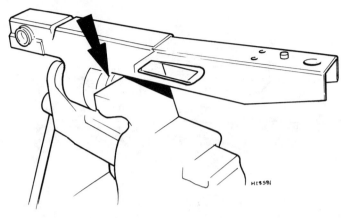

Fig. 10.18 Bush installation into the bracket using a vice as a press (Sec 7)

8 Lower arm balljoint (IFS) – removal and refitting

1 Loosen off the front roadwheel nuts then raise and support the vehicle at the front end on axle stands. Remove the front roadwheels.
2 Position a jack under the suspension arm and raise it to just support the weight of the arm.
3 Undo the retaining bolts and disconnect the shock absorber from the stub axle unit.
4 Extract the split pin and unscrew the castle nut from the balljoint. If the stud rotates when unscrewing the nut, lever upon the arm while unscrewing to prevent the stud from turning.
5 To detach the balljoint from the stub axle, fit and tighten a balljoint separator tool to release the tapered joint.
6 Unscrew the three retaining nuts, remove the bolts and withdraw the balljoint stub from the suspension arm.
7 Refitting is a reversal of the removal procedure but note the following points:

   (a) When fitting the balljoint unit, align the split pin hole in the stud in a transverse plane to the balljoint stub so that the pin can be more easily fitted
   (b) Take care not to damage the balljoint gaiter during reassembly, particularly when fitting the stub to the suspension arm. If it is inserted too far, the gaiter can easily be damaged
   (c) Tighten the castle nut to the specified torque and then if the split pin hole is not suitably aligned to allow the pin to be inserted, further tighten the nut to suit. Use a new split pin
   (d) Tighten the shock absorber retaining nuts to the specified torque setting

9 Anti-roll bar (IFS) – removal and refitting

1 The anti-roll bar and its connecting links are best removed and refitted with the vehicle free standing, therefore if possible, position it over an inspection pit or run it onto ramps at the front end.
2 To remove the anti-roll bar, undo the retaining bolts and detach the bar from the crossmember on each side then undo the retaining nuts and detach it from the suspension arms (complete with link rods) (Figs. 10.20 and 10.21).
3 Extract the spring clip at each end and remove the link rods from the anti-roll bar, using a puller to withdraw them or alternatively by drifting them off.

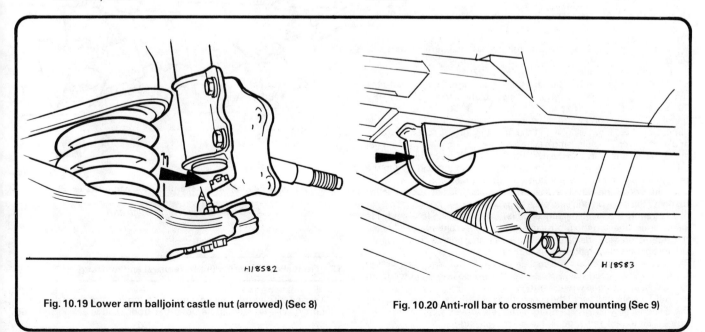

Fig. 10.19 Lower arm balljoint castle nut (arrowed) (Sec 8)

Fig. 10.20 Anti-roll bar to crossmember mounting (Sec 9)

Chapter 10 Suspension and steering

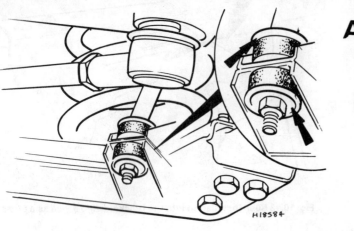

Fig. 10.21 Anti-roll bar link rod and bushes. Note orientation of the washers (arrowed) (Sec 9)

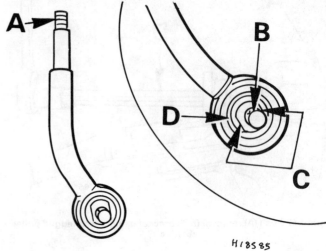

Fig. 10.22 Anti-roll bar link bush arrangement (Sec 9)

A Link thread   C Flat faces
B Bar shank   D Metal inner bush

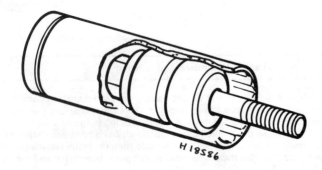

Fig. 10.23 Link bush assembly located in fitting tube (Sec 9)

4 Inspect the bushes of the anti-roll bar and link arm bushes and renew them if they are perished or worn. Cut or prise free the link arm bush from the lower suspension arm (Fig. 10.22).
5 To fit the link arm bush to the suspension arm, first lubricate the bush and its port in the link arm with soapy solution, then insert an M8 x 70 mm bolt fitted with a flat washer into the bush. Locate the bush and bolt into the bore of a 25 mm extra long socket or a tube of suitable dimensions (see Fig. 10.23). Insert the bolt through the hole in the arm, and then pull on the bolt using a suitable self-gripping wrench or similar and draw the bush through the eye and into the arm. Keep the socket (or tube) flush against the arm whilst pulling the bush through. When the bush is inserted, remove the bolt and then renew the bush on the opposite side of the vehicle in the same manner (it is not good practice to renew the bush on one side only) (Fig. 10.24).
6 To renew the insulator rubbers, first check that the left-hand side insulator engages on the protruding locator on the anti-roll bar. Locate the right-hand side insulator onto the bar and ensure that both insulators are fitted with their joints to the rear.
7 Locate the anti-roll bar and engage the bracket lugs into their slots in the front bracket each side. Locate the retaining bolts and tighten them to the specified torque wrench setting.
8 Refit the anti-roll bar to the connecting links each side, ensuring that the dished link washers are fitted with their convex side towards the bushes. Fit the link to bar retaining clip at each end.

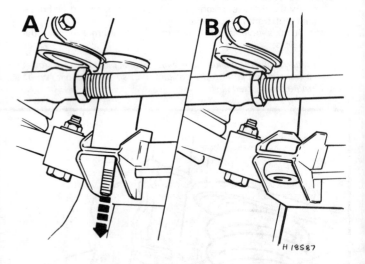

Fig. 10.24 Link bush assembly to lower arm (Sec 9)

A Draw bush dowel through arm
B Bush fitted position

10 Front shock absorber (IFS) – removal and refitting

1 Loosen off the front wheel nuts, raise the front of the vehicle and support it on axle stands. Remove the front roadwheel on the side concerned.

Chapter 10 Suspension and steering

10.2 View showing shock absorber, suspension arm, steering stub axle and associate items:
1 Brake hydraulic line location bracket to shock absorber
2 Brake calliper
3 Lower arm
4 Shock absorber to stub axle bolts

10.6 Front shock absorber upper mounting and retaining nut

2 Undo the retaining bolt and release the brake line location bracket from the front shock absorber (photo).
3 Position a jack under the suspension arm and raise the jack to support the weight of the arm (not lift it).
4 Support the weight of the wheel hub unit with a block or suitable stand. This will prevent possible damage to the lower balljoint during subsequent operations.
5 Working in the vehicle cab area, prise free the rubber retaining ring from the shock absorber top cover, then separate and remove the cover half sections.
6 Unscrew and remove the shock absorber upper retaining nut, then withdraw the washer and retainer (photo).
7 Unscrew the four shock absorber to stub axle retaining bolts, and then lower and withdraw the shock absorber unit from the vehicle.
8 If the shock absorber unit is leaking or defective in any way it must be renewed as it cannot be repaired. To check the action of the shock absorber, mount it in a vice and then operate the piston rod a few times through its full stroke. If the action is noticeably weak or there is an uneven resistance, the unit is in need of replacement. It is advisable to renew both front shock absorbers at the same time or the handling characteristics of the vehicle could well be adversely affected.
9 Before refitting the shock absorber, check the condition of the insulator bush in the body. If perished or worn it should be renewed. To remove it from the panel, prise it free or failing this, carefully cut it out. Clean the aperture in the body panel prior to fitting the replacement.
10 Lubricate the insulator and the location bore with a soapy solution, then insert the insulator by pressing and twisting it into position.
11 Smear the inside of the insulator and the shock absorber upper end with a soapy solution to ease fitting, then insert it up through the insulator and fit the retainer, washer and nut.
12 Align the lower mounting holes of the shock absorber with the corresponding holes in the stub axle and then insert the retaining bolts.
13 Tighten the upper retaining nut and the lower retaining bolts to the specified torque wrench settings. Refit the upper cover and secure with the retaining ring.
14 Reconnect the brake line to the shock absorber, then refit the roadwheel. Raise the vehicle, remove the axle stands and supports under the stub axle and suspension arm, then lower the vehicle to the ground.

11 Front beam axle – removal and refitting

1 The front axle beam can be removed in a fully assembled state or partially dismantled; if the kingpins are seized it will be advantageous to remove the axle beam first and press them out with a hydraulic press.
2 Loosen off the front wheel nuts, check that the handbrake is fully applied, then chock the rear wheels and jack up the front of the vehicle (unladen) supporting it with stands placed beneath the chassis members. Support the axle beam with a trolley jack and stands.
3 Unscrew the nuts and remove the front roadwheels.
4 Unscrew and remove the nuts securing each front shock absorber to the axle beam. Note that the bolts are withdrawn from the front.
5 Detach the track rod end balljoints from the stub axle (Section 26).
6 Disconnect the steering drag link from the drop arm using a suitable balljoint separator as shown in Fig. 10.53 (see Section 31).
7 Unscrew the brake calliper unit retaining bolts, withdraw the calliper from the stub axle assembly. The hydraulic line can be left attached to the calliper, but suspend or support it so that the hydraulic line is not

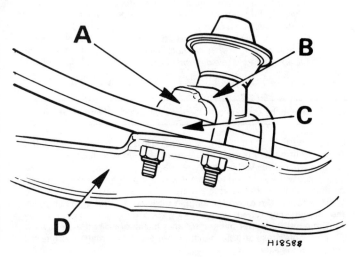

Fig. 10.25 Beam axle to spring assembly (Sec 11)

A Top plate
B U-bolt
C Leaf spring
D Axle

Chapter 10 Suspension and steering

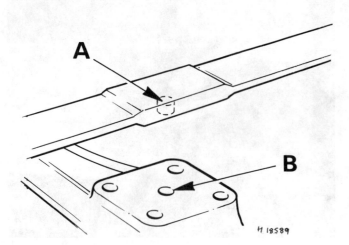

Fig. 10.26 Spring dowel (A) locates in hole in axle (B) (Sec 11)

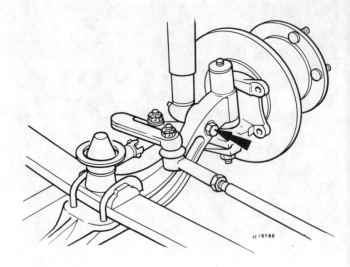

Fig. 10.27 Beam axle shock absorber retaining bolt (arrowed) (Sec 11)

stretched or deformed. Repeat the procedure with the opposing calliper unit.

8 Where applicable, unscrew the retaining nuts and remove the bar anchor plates, clamps and U-bolts.

9 Unscrew the retaining nuts and remove the axle to spring U-bolts and top plates. Ensure that the axle is well supported each side before detaching the U-bolts (Fig. 10.25).

10 Enlist the aid of an assistant to help steady the axle beam as it is lowered and withdrawn from the vehicle.

11 Lower and withdraw the axle from under front of the vehicle.

12 If distortion of the axle beam is suspected it should be checked by a suitably equipped garage and renewed if necessary.

13 Refitting the front axle beam is a reversal of the removal procedure but the following additional points should be noted:

(a) When raising the axle into position, engage it with the dowel location peg in the spring, then fit and tighten the U-bolts and nuts to the specified torque wrench setting (the nuts being tightened in a progressive sequence) (Fig. 10.26)

(b) Ensure that the shock absorber retaining bolts are fitted with their heads to the front

(c) If an anti-roll bar is fitted, set the insulators at a distance of 620 mm (24.5 in) apart, and an equal distance from the anti-roll bar cranked ends (Fig. 10.28)

(d) Tighten all fastenings to their specified torque wrench settings or to the nearest split pin hole alignment position (where applicable). Always use new split pins to secure

12 Stub axle (beam axle type) – removal, overhaul and refitting

1 Remove the front hub unit as described in Section 3. Ensure that the vehicle is firmly supported at the front end on axle stands set in a position where they will not be in the way.

2 Unscrew the nuts and remove the steering arm to stub axle retaining bolts. Remove the steering arm from the stub axle (Fig.10.29).

3 Unscrew and remove the kingpin upper and lower grease nipples and cap nuts (Fig. 10.30).

4 Drive out the cotter pin and remove the kingpin. You may have to use a suitable press or drift to extract the pin, which may in turn entail removal of the axle beam.

5 Remove the stub axle and shim.

6 Assuming that the stub axle is being overhauled, remove the grease seal from the upper bore and press out the stub axle bottom bearing (or drive it out using a suitable drift).

7 Remove the opposite stub axle in the same manner but keep the respective right and left-hand components separate.

8 Do not remove the steering lock stop bolts unless absolutely

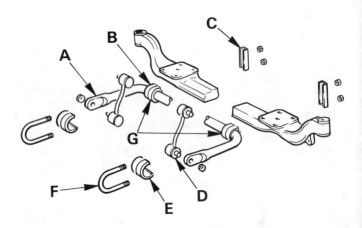

Fig. 10.28 Beam axle and anti-roll bar assembly components (Sec 11)

A Anti-roll bar
B Insulator
C Anchor plate
D Link
E Mounting clamp
F U-bolt
G Insulator inner faces

necessary, as these are adjusted for position and will require special tooling to reset them when they are refitted. This will have to be done by a Ford dealer.

9 Thoroughly clean the stub axle beam with paraffin and dry them with a lint-free cloth. Examine the stub axle and axle beam for signs of distortion or damage. If in doubt regarding the condition of the axle beam or stub axle, have your Ford dealer make an inspection and renew as necessary.

10 Assuming that the kingpin is to be renewed, the stub axle pivot bushes will also have to be renewed. As the bush removal and refitting procedure requires the use of specialised equipment, their replacement is best entrusted to your Ford dealer who will also ream the newly fitted bushes to suit the kingpin.

11 With the new bushes fitted and the lower bearing and grease seal in place, locate the stub axle unit to the axle beam and then select and insert a new shim of the required thickness between the upper face of the axle eye and the bottom face of the upper stub axle eye (Fig. 10.31).

Chapter 10 Suspension and steering

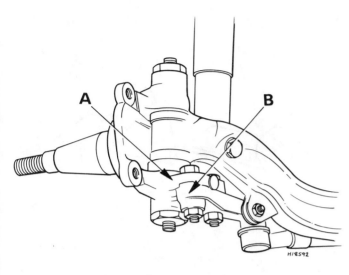

Fig. 10.29 Stub axle (A) and steering arm (B) showing retaining bolts (Sec 12)

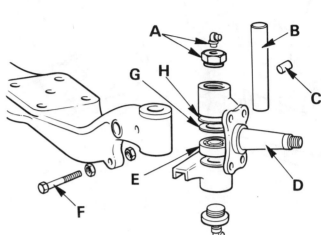

Fig. 10.30 Exploded view of the stub axle (Sec 12)

- A Grease nipple and cap-nut
- B King pin
- C Cotter pin
- D Stub axle
- E Lower bearing
- F Lock stop bolt
- G Shim
- H Grease seal

Fig. 10.31 Select a shim of the required thickness (Sec 12)

Measure the required shim thickness using a feeler gauge, and select a shim which will provide the specified stub axle to axle beam clearance.
12 Lightly smear the kingpin with grease, then push it downwards just past the shim location, insert the shim and then push the kingpin back up and align the cut-away section with the cotter pin hole in the axle.
13 Insert a new cotter pin and drive it home so that it is flush with the axle case.
14 Refit the upper and lower cap nuts and grease nipples. Lubricate the kingpin at each end, and then check its turning action. If it is excessively tight, a thinner shim is probably required. If it is excessively loose in its feel it may need a thicker shim. Recheck the clearance between the stub axle and the axle.
15 Refit the steering arm to the stub axle and tighten the retaining nuts to the specified torque setting. If detached, reconnect the steering track rod as described in Section 26.
16 Refit the hub unit as described in Section 3 and check/adjust the endfloat as given.

13 Front beam axle leaf spring – removal and refitting

1 Apply the handbrake firmly and jack up the front of the vehicle (unladen), supporting it adequately with stands beneath the chassis members. Support the axle beam on a trolley jack.
2 Unscrew and remove the U-bolt nuts then slowly lower the jack under the axle beam and then disengage the U-bolts and top plate from the spring. The jack should be lowered to the point where the axle/spring location dowel is clear.
3 Unscrew and remove the self-locking nuts from the spring shackles, then, using a soft metal drift, drive out the shackle pins and plates and extract the rubber bushes (Fig. 10.32).
4 Unscrew and remove the front mounting nut and washer, and drive out the bolt whilst supporting the spring.
5 Lift the spring away from the axle and mountings.
6 Examine the front and rear mounting bushes and the condition of the shackle, U-bolts and bump stop, and renew any faulty components.
7 The spring eye bush can be renewed by pressing, drawing or cutting out the old bush, according to the facilities available. Whatever method is employed to remove and refit the bush, take care not to damage the spring eye.
8 When fitting the new bush, ensure that when installed, there is an equal amount of bush protruding each side of the spring eye. Support the spring to avoid the possibility of damage and distortion.
9 Refitting of the front spring is a reversal of the removal procedure, but the following special points must be noted:

(a) Assemble the spring to the chassis, but do not fully tighten the shackle bolts until after the vehicle is lowered to the ground and is free standing. Then tighten them to the specified torque setting
(b) When assembling the spring to the axle, ensure that the location dowel engages in the hole in the axle platform (Fig.10.26)
(c) Centralise the top plate on the spring before fitting the U-bolts and tightening the retaining nuts to the specified torque setting

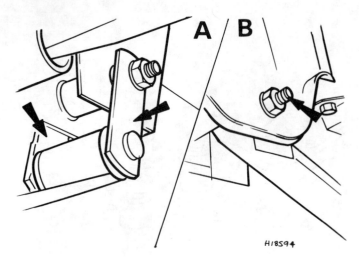

Fig. 10.32 Leaf spring and shackle plates at rear end (A) and eye bolt (B) at front end (Sec 13)

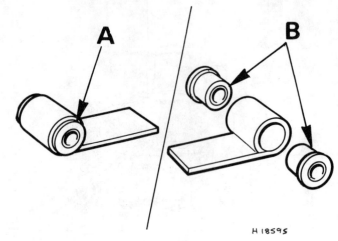

Fig. 10.33 Showing spring front end bush (A) and rear end bushes (B) (Sec 13)

14 Suspension bump stop (beam axle type) – removal and refitting

1 Proceed as described in paragraphs 1 and 2 in the previous Section and remove the top plate nut from the spring.
2 Unscrew and remove the bump stop retaining nut and washer, then remove the bump stop from the plate (Fig. 10.34).
3 Refit in the reverse order of removal. Tighten the nut to the specified torque.

15 Anti-roll bar (beam axle type) – removal and refitting

1 Raise the vehicle at the front end and support it on axle stands. Ensure that the handbrake is fully applied.
2 Unscrew the anti-roll bar link to chassis mounting nuts.
3 Unscrew the retaining nuts and detach the anchor plates, the insulator clamps and the U-bolts. Disconnect the anti-roll bar and remove it from the vehicle.
4 Renew the insulators and any other items which are worn or damaged (Fig. 10.28).
5 Refit in the reverse order of removal but note the following special points:

  (a) When refitting the anti-roll bar insulators, they must be positioned 620 mm (24.5 in) apart. Measure this distance from the inner face of each insulator. When set, the insulators must be equally distanced each side from the cranked ends of the stabilizer bar.
  (b) Tighten the retaining nuts to their specified torque wrench settings

16 Front shock absorber (beam axle type) – removal and refitting

1 Working within the vehicle cab area, prise free the rubber retaining ring from the shock absorber top cover, then separate and remove the cover half sections (Fig. 10.35).
2 Unscrew and remove the shock absorber upper retaining nut and washer, then remove the upper retainer (and insulator if necessary).

Fig. 10.34 Bump stop and top plate (Sec 14)

3 Working underneath the front end of the vehicle, unscrew and remove the shock absorber bottom end retaining nut and washer. Withdraw the bolt and then lower and remove the shock absorber unit. (Fig. 10.36).
4 If the shock absorber unit is leaking or is defective in anyway it must be renewed. To check the action of a shock absorber for operational efficiency, mount it in a vice and operate the piston rod a few times. If the action is weak or there is an uneven resistance, it is in need of renewal. It is advisable to renew both front shock absorbers at the same time. The renewal of one only can cause detrimental steering and handling effects.
5 The insulator bush should be inspected for wear in the bottom eye and if required this can be pressed or driven out for renewal. Smear the new bush with a soapy solution to ease fitting, then press it into position so that it protrudes an equal amount each side of the eye.
6 If the upper insulator is still in position in the body, inspect it for signs of perishing or excessive wear. To remove it from the panel, prise or if necessary, cut it free.
7 Clean the upper insulator aperture in the body, then lubricate the new insulator with a soapy solution, and insert it by pressing and twisting it into position.
8 Locate the lower retainer over the top of the shock absorber unit engaging the 'D' shaped hole over the flat of the piston rod.

Chapter 10 Suspension and steering

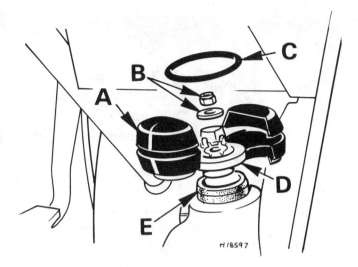

Fig. 10.35 Front shock absorber top mounting and cover (Sec 16)

- A Half covers
- B Nut and washer
- C Rubber ring
- D Retainer (upper)
- E Insulator

Fig. 10.36 Shock absorber lower mounting to beam axle (Sec 16)

- A Shock absorber
- B Nut, washer and bolt

9   Insert the shock absorber upwards through the body panel and fit the upper retainer, washer and securing nut. Get an assistant to help guide it up through the body, and lubricate the upper retainer with a soapy solution to ease fitting. It should be noted that the upper retainer can only be fitted in one direction due to the 'D' shaped location hole, and the flat on the shock absorber stud. When fitting the retaining nut, hand tighten it only at this stage.
10 Turn the upper retainer and piston rod to align the arrow mark on the top face of the retainer next to the R (right-hand) or L (left-hand) mark (according to side), so that it points towards the front of the vehicle.
11 This is necessary to ensure that the upper and lower angled retainer bores and the shock absorber is correctly aligned (Fig.10.37).
12 With the upper retainer and piston rod correctly aligned, tighten the upper retainer nut to the specified torque wrench setting.
13 Refit the cover half sections and securing ring.
14 Reconnect the lower mounting and tighten its retaining nut to the specified torque.

17 Rear axle leaf spring – removal and refitting

1   Chock the front wheels and jack up the rear of the vehicle (unladen), supporting it adequately with stands placed beneath the chassis members. Remove the roadwheel(s).
2   Place a jack beneath the rear axle, and raise to support it.
3   Unscrew and remove the spring shackle retaining nuts and withdraw the plate, shackle pins, and rubber bushes using a soft metal drift as necessary (photo).
4   Unscrew and remove the front mounting retaining nut and washer and drive out the mounting bolt with a soft metal drift(photo).
5   Unscrew and remove the U-bolt nuts and detach the U-bolts together with the clamp plate; the rear spring can now be lifted from the rear axle and withdrawn from beneath the vehicle (Fig.10.38).
6   Examine the front and rear (where fitted) mounting bushes and the condition of the shackle U-bolts, and bump stop, and renew any faulty components. Always renew the self-locking nuts once they are removed.
7   Refitting the rear spring is a reversal of the removal procedure, but note the following additional points:

(a) The shackle pins and nuts must not be fully tightened until after the vehicle is lowered to the ground.
(b) When fitting the spring to the axle, engage the centre bolt fully before fitting the U-bolts and clamp plates
(c) Tighten all nuts and bolts to their specified torque wrench settings
(d) On models equipped with a load apportioning valve in the brake system, have the setting checked by a Ford dealer at the earliest opportunity

17.3 Rear spring and shackle

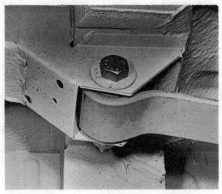

17.4 Rear spring and front end eye bolt mounting

18.2 Rear shock absorber

Chapter 10 Suspension and steering

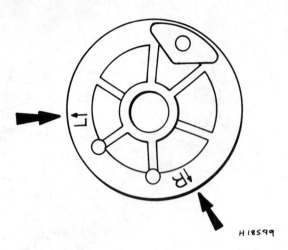

Fig. 10.37 Upper retainer positional marks (L and R) (Sec 16)

18 Rear shock absorber – removal and refitting

1 Raise the vehicle at the rear and support it on axle stands.
2 Undo the retaining nut, and withdraw the shock absorber lower mounting bolt (photo).
3 Unscrew the upper mounting bolt to the underbody and then remove the shock absorber (photo).
4 To test the shock absorber for efficiency, grip it at the upper or lower mounting eye in a vice and then pump the piston repeatedly through its full stroke. If the resistance is weak or is felt to be uneven, the shock absorber is defective and must be renewed. It must also be renewed if it is leaking fluid. The shock absorber(s) cannot be repaired. It is advisable to renew both rear shock absorbers at the same time or the handling characteristics of the vehicle could be adversely affected.
5 Refit the shock absorber in the reverse order of removal. Tighten the mounting bolts to their specified torque wrench settings when the vehicle is lowered to the ground and is freestanding.

19 Steering wheel – removal and refitting

1 Disconnect the battery earth lead.
2 Centralise the front roadwheels so that they are in the straight-ahead position. Prise free the centre pad from the steering wheel (photo).
3 Unscrew the steering wheel retaining nut, then pull free the steering wheel from the shaft. The wheel is located on the shaft by a master spline.
4 Refit the steering wheel in the reverse order of removal. Align the master spline in the steering wheel hub with the shaft master spline and push it firmly into position. Tighten the retaining nut to the specified torque wrench setting. Refit the centre pad. Reconnect the battery earth lead.

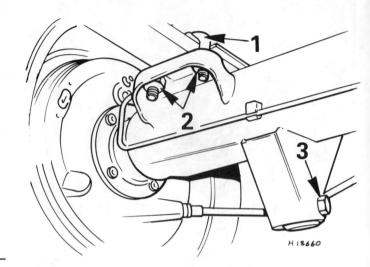

Fig. 10.38 Rear leaf spring to axle fixings (Sec 17)

1 U-bolt
2 U-bolt nuts
3 Shock absorber/axle mounting bolt

18.3 Rear shock absorber to underbody mounting

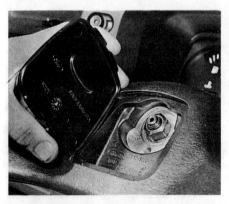

19.2 Removing the steering wheel centre pad

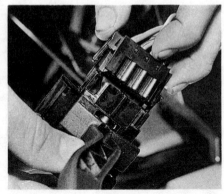

20.4 Disconnect the steering switch multi-connectors

20.7 Steering column upper retaining nuts

197

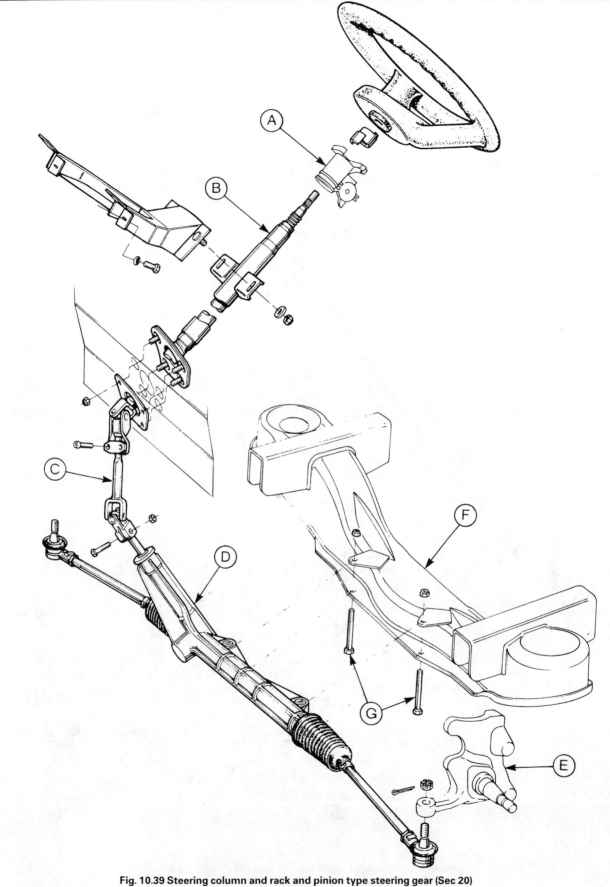

Fig. 10.39 Steering column and rack and pinion type steering gear (Sec 20)

A Column lock unit
B Column tube and shaft
C Universal joint coupling unit
D Rack and pinion unit
E Stub axle
F Crossmember
G Steering gear unit mounting bolts

198

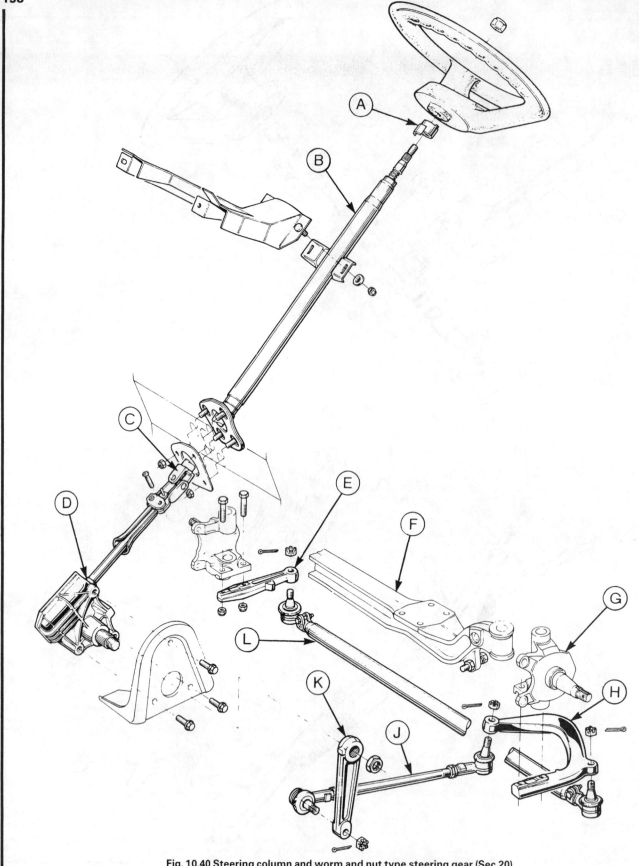

Fig. 10.40 Steering column and worm and nut type steering gear (Sec 20)

A Indicator cam
B Column tube/shaft unit
C Universal joint
D Steering gear unit
E Steering arm
F Axle beam
G Stub axle
H Steering arm
J Drag link
K Drop arm
L Track rod

Chapter 10 Suspension and steering

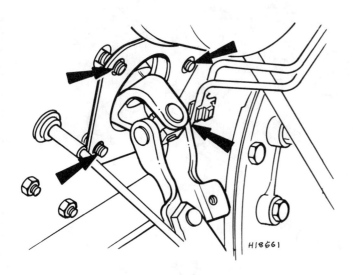

Fig. 10.41 Steering column to bulkhead attachment points (arrowed) (Sec 20)

20 Steering column – removal and refitting

1 Disconnect the battery earth lead.
2 Centralise the steering so that the front roadwheels are in the straight-ahead position. Remove the steering wheel as described in Section 19.
3 Undo the retaining screws and remove the upper and lower column shrouds.
4 Disconnect the wiring loom multi-connectors from the column switches and release the loom to column retaining straps (photo).
5 Working from the engine compartment side, unscrew and withdraw the lower column to universal joint pinch-bolt and move the clamp plate aside.
6 Unscrew the four nuts which secure the steering column tube to the bulkhead (Fig. 10.41).
7 From inside the vehicle, support the weight of the column and undo the two upper retaining nuts. Withdraw the column unit from within the vehicle (photo).
8 Refitting is a reversal of the removal procedure but note the following special points:

(a) Ensure that both the steering column and roadwheels are centralised when refitting the column to the universal joint
(b) Tighten all fastenings to their specified torque wrench settings but loosely tighten the column to bulkhead nuts, followed by the upper column nuts, then tighten the four nuts to their torque setting, followed by the two upper column nuts
(c) Refit the steering wheel with reference to Section 19
(d) Ensure that all wiring loom connectors are securely made and fit new retaining straps. Check the operation of the various switches before refitting the shrouds

21 Steering column – dismantling, overhaul and reassembly

1 Remove the steering column from the vehicle as described in the previous Section, then support the column in a vice fitted with soft jaw protectors.
2 If not already removed, withdraw the steering wheel from the column as described in Section 19.
3 Remove the indicator cam and then prise free the star washer. Withdraw the spring and upper thrustwasher (Fig. 10.42).
4 Fit the ignition key into its switch and turn it to position II. The steering column shaft can now be withdrawn.
5 Undo the screws securing the column multi-function switches and remove the switches.
6 To remove the steering column lock unit, drill out the tube to lock shear bolt using a $\frac{3}{16}$ in diameter drill and then carefully tap the lock unit free (Fig. 10.43).
7 The upper column support bearing is removed by prising it out.
8 Undo the two grub screws to remove the ignition switch loom plate from the lock housing.
9 To remove the ignition switch lock barrel, turn the key to position I, then insert a small diameter punch into the hole in the lock housing and release the lock barrel (photo).
10 Renew any defective or obviously worn components. If the column has been damaged in any way, it must be renewed.
11 Commence reassembly by refitting the lock unit. With the ignition key set in position I, then insert the lock into its housing and push it home until the lock pin is felt to engage in the column lock sleeve.
12 Turn the key to position O then fit the loom plate, securing with the two grub screws.
13 Push the lock housing onto the column tube until it is abutting the tube. Screw the new shear bolt into position and tighten it to the point where the head of the bolt shears off.
14 Locate the upper column shaft thrust bearing to the outer tube.
15 Turn the ignition key to position II then slide the column through the tube, locate the lower bearing and fit the upper thrust washer and spring.
16 Locate the star washer over the shaft and drive it onto its seat using a suitable tube drift so that it secures the spring. Refit the indicator cam.
17 Refit the multi-function column switches then refit the column and steering wheel as described in the previous Sections.
18 On completion, check the operation of the multi-switches and the steering lock for satisfactory operation.

22 Steering column upper bearing – renewal

1 The steering column upper bearing can be removed from the column with the column in position in the vehicle. No special tools are required.
2 Disconnect the battery and remove the steering wheel as described in Section 19.
3 Undo the retaining screws and remove the upper and lower column shrouds.
4 Detach and remove the multi-function switches from the steering column.
5 Withdraw the indicator cam from the steering column shaft then prise free the star washer from its locating groove in the shaft using a suitable screwdriver. Remove the star washer and the coil spring.
6 Withdraw the column upper bearing thrust washer and then prise out the upper bearing from the column tube.
7 To fit the new bearing, drive it into the column tube using a suitable tube drift.
8 Fit the bearing/thrust washer and coil spring, then secure them with a star washer. Press the star washer down the shaft and into its groove using a suitable tube or a socket. Relocate the indicator cam.
9 Refit the multi-function switches and the steering wheel and shrouds. Refer to Section 19 when refitting the steering wheel. Reconnect the battery.
10 Check the operation of the steering column switches and the steering action on completion.

23 Steering column lower universal coupling (IFS) – removal and refitting

1 Disconnect the battery earth lead and centralise the front roadwheels so that they are in the straight-ahead position.
2 Unscrew and remove the steering column to universal joint clamp bolt then pivot the clamp plate to one side and let the coupling shaft hang free.

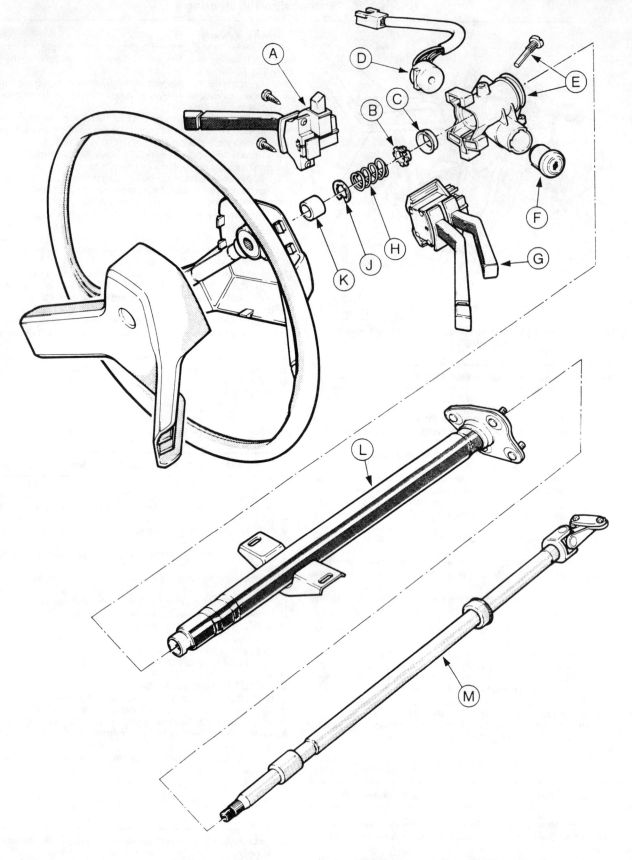

Fig. 10.42 Exploded view of the steering column and associate components (Sec 21)

- A Multi-function switch
- B Thrustwasher
- C Thrust bearing
- D Wiring loom
- E Lock housing (and bolt)
- F Lock barrel
- G Multi-function switch
- H Pre-load spring
- J Circlip
- K Indicator cam
- L Tube
- M Shaft

Chapter 10 Suspension and steering

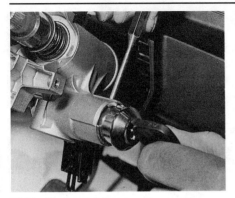
21.9 Ignition switch lock barrel removal

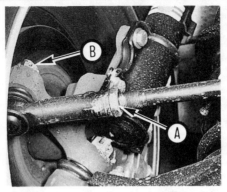
26.2 Showing track rod end locknut (A) and balljoint (B) – rack and pinion steering type

27.4 Correctly fitted steering gear bellows

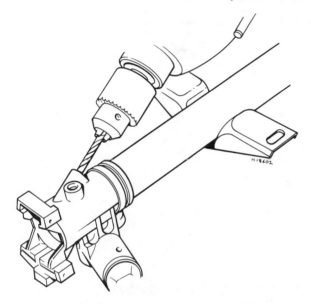
Fig. 10.43 Drilling out the column lock shear bolt (Sec 21)

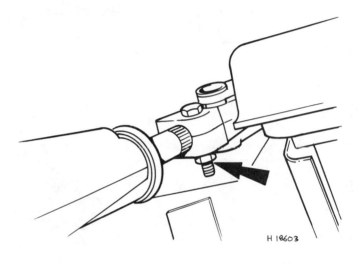
Fig. 10.44 Steering column lower coupling pinch bolt (Sec 23)

3 Unscrew and remove the pinch-bolt and nut from the lower universal joint then withdraw the coupling unit from the pinion shaft (Fig. 10.44).
4 Refitting is a reversal of the removal procedure but note the following special points:

(a) Ensure that the block splines are correctly positioned when engaging the universal joint coupling shaft onto the steering gear pinion shaft (Fig. 10.45)
(b) Tighten the retaining nuts and bolts to the specified torque wrench setting

24 Rack and pinion steering gear unit – removal and refitting

1 Disconnect the battery earth lead.
2 Centralize the steering gear so that the roadwheels are in the straight-ahead position and lock the column in this position.
3 Unscrew the clamp plate pinch-bolt from the column to universal joint coupling. Slide the clamp plate to one side.
4 Loosen off the front roadwheel bolts then raise and support the vehicle at the front end on axle stands. Remove the front roadwheels.
5 Refer to Section 26 and disconnect the track rod end ball joints from the steering arms.
6 Unscrew and remove the steering gear unit to crossmember retaining bolts and then withdraw the steering gear unit, withdrawing it downwards to disengage the universal joint from the column.

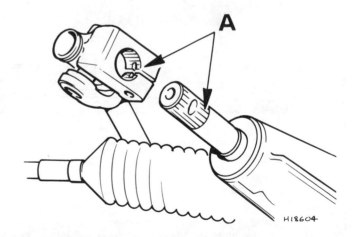
Fig. 10.45 Block splines (A) of the coupling and pinion shaft must align (Sec 23)

7 Unscrew the pinch-bolt and remove the universal joint from the pinion shaft.
8 To remove the mounting bushes, press or drift them out using a suitable tube or socket.
9 Refitting is a reversal of the removal procedure, but before

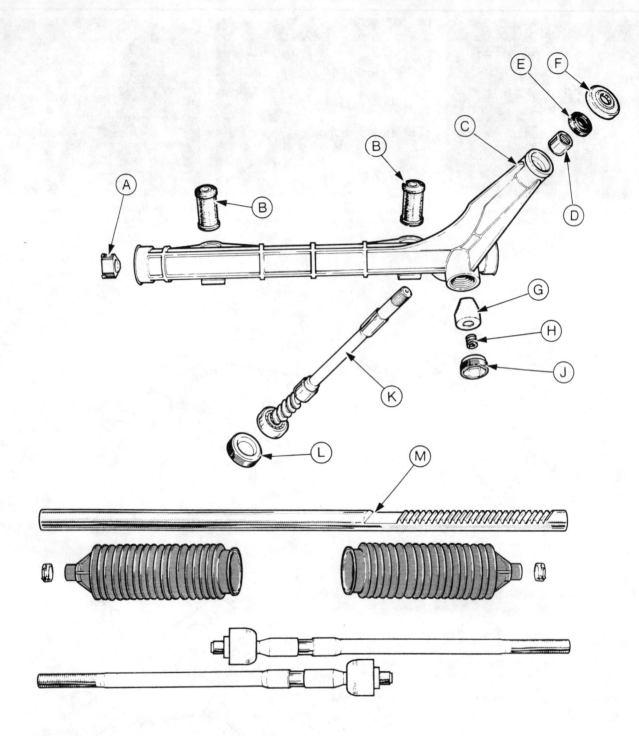

Fig. 10.46 Exploded view of the rack and pinion steering gear unit and associated components (Sec 24)

- A Support bush
- B Mounting bushes
- C Housing
- D Upper pinion bearing
- E Upper seal
- F Dust cover
- G Yoke
- H Yoke spring
- J Plug
- K Pinion
- L Pinion bearing cover
- M Rack

Chapter 10 Suspension and steering

Fig. 10.48 Ford tool 13-011 for removal of yoke plug (Sec 25)

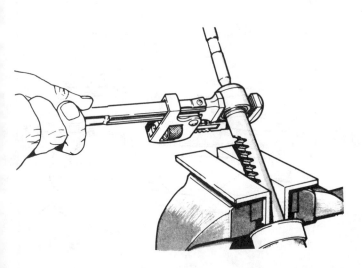

Fig. 10.47 Track rod removal from the steering rack (Sec 25)

reconnecting the column universal joint coupling, set the steering wheel and front wheel in the straight-ahead position. If a new steering gear is being fitted the pinion position can be ascertained by halving the number of turns necessary to move the rack from lock-to-lock. Tighten the nuts and bolts to the specified torque and fit a new split pin. Finally check and if necessary adjust the front wheel alignment as described in Section 37.

25 Rack and pinion steering gear unit – dismantling, overhaul and reassembly

1 Remove the steering gear as described in Section 24 then clean the exterior of the unit with paraffin and wipe dry.
2 Mount the steering gear unit in a vice fitted with protective jaws. Cut free the bellows retaining clips and slide the bellows down the track rods. Remove the pinion dust cover.
3 Move the rack fully to the left and grip the rack in a soft jawed vice (Fig. 10.47).
4 If the original track rods are fitted use a pipe wrench to unscrew the balljoint from the rack and remove the track rod. If service replacement track rods are fitted use a spanner on the machined flats.
5 Remove the right-hand track rod in the same way.
6 Using a hexagon key, unscrew and remove the yoke plug and remove the spring and yoke. If available, use Ford tool No 13-011 to unscrew the yoke plug (Fig. 10.48).
7 Remove the lower pinion bearing cover, again using a hexagon key or by reversing tool No 13-011 (if available).
8 Move the rack to one side then extract the pinion unit from the housing. As it is withdrawn, note the orientation of the master spline of the pinion shaft. If the pinion and/or the bearings are damaged or badly worn, they must be renewed as a combined unit, so do not dismantle them.
9 Withdraw the rack from the housing. Carefully lever out the rack support bearing from the housing.
10 The pinion bearings and seal can be removed from the gear housing by driving them out using a suitable punch.
11 Clean, inspect and renew any defective components. Clean old sealant from the yoke plug and housing. Clean off and dress the rack tube staking before reassembly.
12 Insert the rack support bush into the steering rack housing, but ensure that the lugs are correctly engaged.
13 Drift the pinion seal and upper bearing into position, with the bearing positioned flush to the seal shoulder (Fig. 10.49).
14 Lubricate the bearings, bushes and seals with the specified grease, then slide the rack into its housing and position it so that all of its teeth are visible.

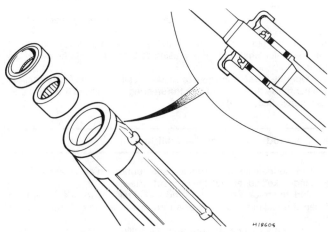

Fig. 10.49 Refit the upper bearing and seal (Sec 25)

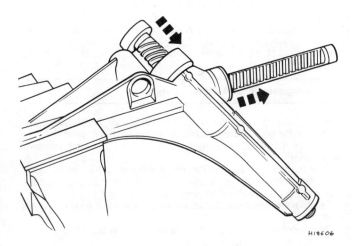

Fig. 10.50 Refit the pinion and bearing (Sec 25)

15 Refit the pinion and bearing unit. Press the rack along its housing and engage the pinion so that when the rack is in its central location in the housing, the pinion master spline is positioned as noted during removal (Fig. 10.50).
16 Fill the pinion and rack housing with the specified grease, refit the pinion bearing cover and tighten it to the specified torque setting. Stake the cover and housing using a suitable punch or chisel.
17 Centralise the steering rack (halve the pinion rotations lock to lock) and clean any grease from the housing and plug threads.
18 Refit the rack yoke and spring into the housing, smear the plug thread with a sealant solution then fit and tighten the plug to the specified torque setting. Move the rack through five full operating movements in each direction to settle it, then back off the plug by 30°.
19 Using a piece of cord and a spring balance, check the pinion turning

torque. Connect the cord and spring balance to the pinion, centralise the rack, turn the pinion one turn in an anti-clockwise rotation then clockwise two turns and note the turning torque. Further turn it one turn anti-clockwise. The turning torque through the centre of travel must be between 0.9 and 1.8 Nm (0.664 to 1.3 lbf ft). If required the yoke plug can be turned a further 5° in either direction (from the minus 30° position). When loading the rack against the spring, no free play between the back of the rack and the yoke should be felt.

20 Stake the yoke plug and the pinion plug to the housing to secure in three equidistant positions.

21 Slide the rack to one side to expose the teeth, then relocate the gear unit in the vice, clamping the teeth securely in protective jaws.

22 Screw the inner rack balljoint units to the rack and tighten them to the specified torque. Stake the joint flange into the rack groove to lock. If a pipe wrench was used to tighten the units, remove any burrs by filing them smooth.

23 Relocate the steering gear bellows over the steering gear tube and secure with clips.

24 Refit the pinion dust cover with the specified grease, then slide it over the pinion shaft onto the housing so that it is in contact with the upper seal lip.

25 Operate the steering gear to ensure that it has a satisfactory action, then centralise the rack and refit the steering gear unit to the vehicle as described in Section 24.

26 Track rod end – removal and refitting

1 Jack up the front of the vehicle and support on axle stands. Apply the handbrake and remove the relevant wheel.

2 Mark the track rod and track rod end in relation to each other then loosen the locknut a quarter of a turn or slacken the clamp bolt nut (photo).

3 Extract the split pin and unscrew the balljoint nut.

4 Using a balljoint separator tool release the track rod end from the steering arm (Figs. 10.51 and 10.52).

5 Unscrew the track rod end from the track and noting the number of turns necessary to remove it. Note that on beam axle models the track rod ends have left or right-hand threads depending on the side concerned.

6 Refitting is a reversal of removal, but tighten the nuts to the specified torque and fit a new split pin. On completion check and if necessary adjust the front wheel alignment as described in Section 37.

27 Steering gear rubber bellows – renewal

1 Remove the track rod end as described in Section 26. Also unscrew and remove the track rod and locknut (having noted its position).

2 Remove the clips and slide the bellows from the track rod and steering gear.

3 Slide the new bellows over the track rod and onto the steering gear. Where applicable make sure that the bellows seat in the cut-outs provided in the track rod and housing (support end only).

4 Fit and tighten the clips, but ensure that the bellows are not twisted (photo).

5 Reconnect the track rod ends as described in Section 26.

28 Worm and nut steering gear – removal and refitting

1 Raise the vehicle at the front end and support it on axle stands. Centralise the steering (wheels in straight-ahead position).

2 Extract the split pin, and loosen off the nut securing the drag link to the drop arm balljoint. Detach the two by releasing the tapered joint using a balljoint separator (Fig. 10.53).

3 Unscrew and remove the drop arm to steering gear retaining nut. If no master spline is visible, mark the fitted position of the drop arm on the steering shaft, by scribing an alignment mark across the two end faces then withdraw the drop arm from the shaft using a suitable puller as shown (Fig. 10.54).

4 Unscrew and remove the steering column lower coupling pinch-bolt from the clamp plate (Fig. 10.55) and move the clamp plate to one side.

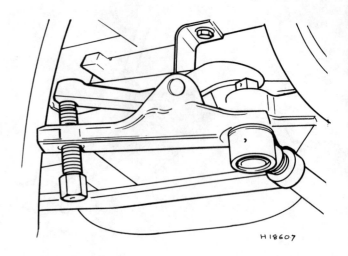

Fig. 10.51 Using a balljoint separator tool to release the track rod end from the steering arm (Sec 26)

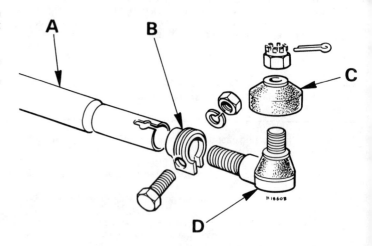

Fig. 10.52 Track rod end and track rod – worm and nut type steering (Sec 26)

A Track rod
B Clamp
C Cap
D Track rod end

5 Unscrew and remove the steering gear to chassis bolts and remove the steering gear unit (Fig. 10.56).

6 Refitting is a reversal of the removal procedure but note the following special points:

(a) When fitting the drop arm to the steering gear rocker shaft ensure that the alignment marks and/or master splines are in alignment

(b) Tighten all fastenings to their specified torque wrench settings

29 Worm and nut steering gear – dismantling, overhaul and reassembly

1 Clean the exterior of the steering gear with paraffin and thoroughly dry it.

2 Loosen the locknut which ensures the rocker shaft adjuster screw,

Chapter 10 Suspension and steering

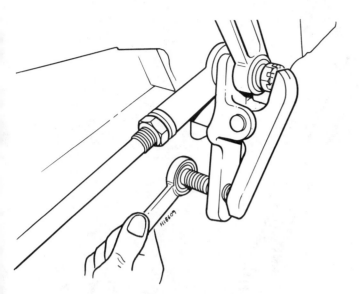

Fig. 10.53 Detaching the drop arm from the drag link using a separator tool (Sec 28)

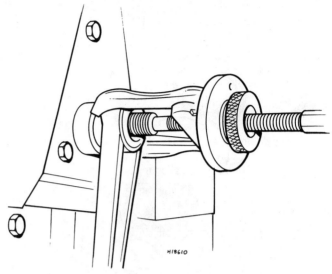

Fig. 10.54 Withdrawing the drop arm from the steering rocker shaft (Sec 28)

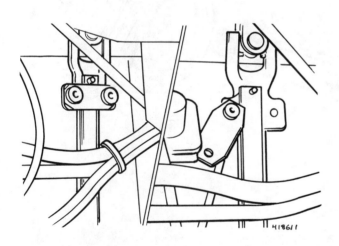

Fig. 10.55 Steering column lower coupling pinch-bolt removal (Sec 28)

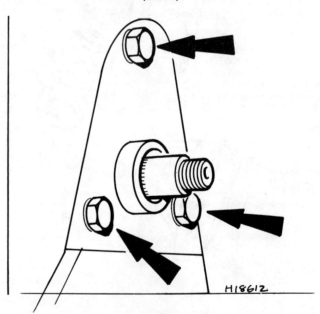

Fig. 10.56 Unscrew the three retaining bolts (arrowed) – manual steering models (Sec 28)

then unscrew and remove the three bolts which retain the rocker shaft housing side cover in position. Remove the side cover, gasket and rocker shaft assembly (Fig. 10.58).

3 Remove the adjustment locknut and unscrew the side cover from the adjusting screw, then slide the adjusting and spacer out of the rocker shaft location.

4 Straighten the tabs of the lock washer retaining the upper bearing housing locknut and unscrew and remove the locknut; a special C-spanner is required to do this and should be borrowed from a tool agent (Fig. 10.59).

5 Unscrew and remove the upper bearing housing and carefully withdraw the steering shaft together with the upper and lower bearings.

6 Using a suitable drift, drive the upper and lower bearing cups out of the upper housing and steering box.

7 Hold the worm shaft in a soft-jawed vice and detach the transfer tubes and balls from the nut by unscrewing and removing the clamp retaining screws (Fig. 10.60).

8 Slide off the nut assembly together with the 62 steel balls.

9 Prise the oil seal from the drop arm end of the steering box.
10 Thoroughly wash all components in paraffin and dry them with a lint-free cloth.
11 Examine all the components for damage, fractures, and excessive wear. Fit the rocker shaft temporarily in the steering box bush and check that there is no excessive clearance, then inspect the teeth of the sector and nut assembly for wear. Check the bearing races and balls for pitting and signs of wear and inspect the bush in the side cover for wear. New bushes should be drifted into position where necessary, but if the rocker and steering shafts need renewal it will probably be more economical to obtain a reconditioned steering gear.
12 Obtain a new oil seal, side cover gasket and bearing housing lock washer.
13 To reassemble the steering gear first drive the oil seal and lower bearing cup into the steering box making sure that they are fitted squarely. Similarly, using suitable diameter tubing, drive the upper bearing cup into the upper housing.

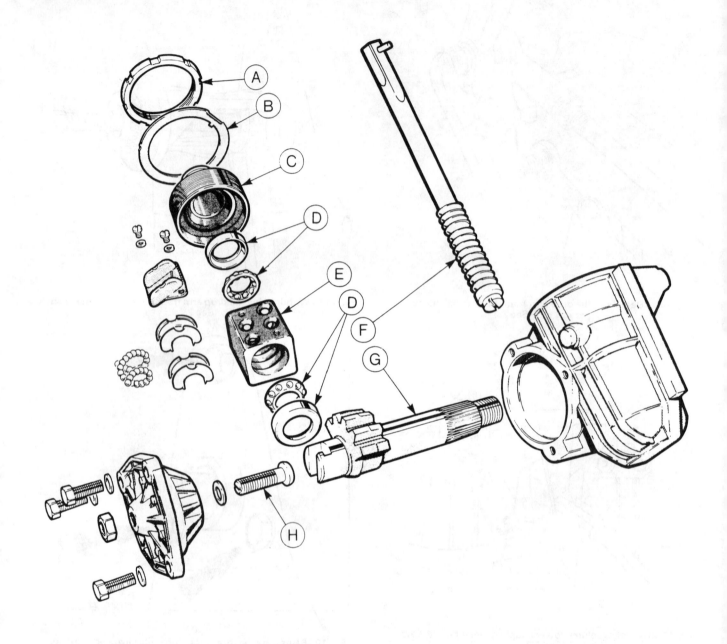

Fig. 10.57 Worm and nut steering gear components (Sec 28)

A Adjustment locknut
B Lock washer
C Adjuster screw
D Thrust bearings
E Worm nut
F Steering shaft worm
G Rocker shaft
H Rocker shaft adjustment screw

14 Smear the worm, nut and transfer tubes with the specified grease and press the steel balls into the transfer tubes.
15 Slide the nut onto the worm and press the steel balls into each of the four holes until all 62 are in position; it will be necessary to shake the assembly to settle the balls in their grooves, but take care not to allow the balls to fall between the worm, nut and ball tracks.
16 Refit the transfer tubes, align the transfer holes and tighten the retaining screws.
17 Grease the caged ball races and position them in the bearing cups, then screw the upper bearing housing onto the worm shaft and tighten to achieve the specified pre-load. (Use a spring balance and some string around the shaft to measure this.)
18 Refit the lock washer and lock ring and tighten it with the special tool. Bend one tab into the locknut slot and the remaining tab over the housing (Fig. 10.61).
19 Fit the adjuster screw to the rocker shaft and select a spacer to give the adjuster screw 0.05 mm (0.002 in) clearance in the slot. Remove the screw, fit the spacer, and refit the screw.
20 Screw on the side cover and locknut and position the new joint on the side cover with grease.
21 Insert the rocker shaft and side cover, making sure that the centre teeth of both shaft and nut are engaged, then tighten the three side cover retaining bolts.
22 Adjust the steering gear as described in Section 30.
23 Fill the steering box with the correct amount of grease as given in the Specifications.

Chapter 10 Suspension and steering

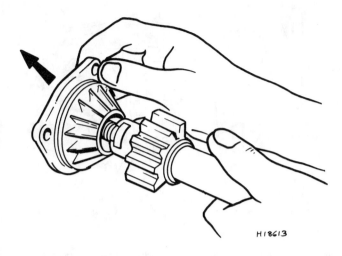

Fig. 10.58 Removing the rocker shaft and side cover (Sec 29)

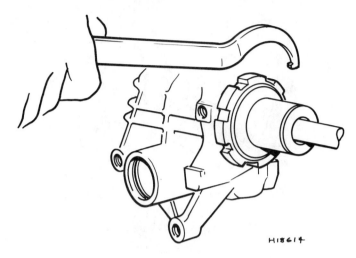

Fig. 10.59 Removing the locknut using a 'C' spanner (Sec 29)

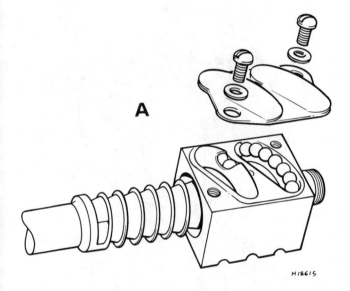

Fig. 10.60 Transfer tube unit showing clamp (A) (Sec 29)

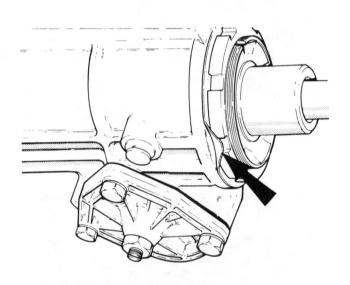

Fig. 10.61 Bend lockwasher tab (arrowed) to secure the locknut (Sec 29)

30 Worm and nut steering gear – rocker shaft pre-load adjustment

1   Raise and support the vehicle at the front end on axle stands.
2   Extract the split pin, unscrew the retaining nut and detach the drag link from the drop arm using a suitable balljoint separator.
3   Release the steering lock, then move the steering wheel onto its full lock position. From this point count the number of turns of the wheel which are necessary to move to the opposite lock. Now centralise the wheel by moving it back half the lock-to-lock turns. A 'high spot' should be felt on the steering wheel at this point.
4   From the centre position, turn the steering wheel two full turns in an anti-clockwise direction, then proceed as follows.
5   Loosen off the rocker shaft adjuster screw locknut and screw.
6   Prise free the steering wheel centre pad. Connect up a socket extension and a torque gauge (or cord and spring balance) and then turn the steering clockwise four complete turns to measure the turning torque required. The turning torque should be as given in the Specifications (rocker shaft pre-load) (Fig. 10.62).
7   If adjustment is necessary, adjust the rocker shaft screw as required to meet this pre-load, then tighten the locknut (Fig. 10.63).
8   Reassemble in the reverse order of renewal. Use a new split pin to secure the drop arm balljoint nut and tighten the nut to the specified torque setting.
9   Check that the steering action is fully satisfactory by turning it from lock-to-lock and then lower the vehicle to the ground.

31 Worm and nut steering drop arm – removal and refitting

1   Unscrew and remove the nut securing the drop arm to the steering box shaft. If no master spline is visible, make an alignment mark across the face of the shaft and arm to indicate their relative positions during refitting, then withdraw the drop arm from the shaft using a suitable puller (Fig. 10.54).
2   Extract the split pin, unscrew the retaining nut and then detach the drop arm from the drag link using a balljoint separator tool (Fig. 10.53).
3   Refitting is a reversal of the removal procedure but note the following points:

(a) Centralise the steering before reconnecting the drop arm, and ensure that the master spline or alignment marks made during removal align

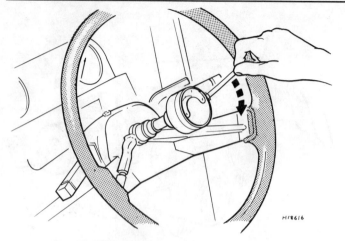

Fig. 10.62 Checking the steering torque (Sec 30)

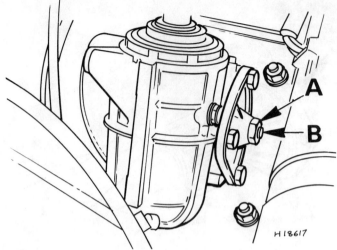

Fig. 10.63 Steering gear adjuster screw (B) and locknut (A) (Sec 30)

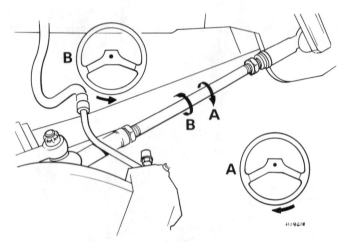

Fig. 10.64 Adjust drag link as required to centralise the steering (Sec 32)

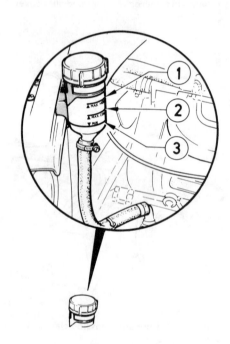

Fig. 10.65 Power steering fluid reservoir showing level marks (Sec 33)

1 Maximum level (with engine hot)
2 Maximum level (with engine cold)
3 Minimum level (with engine cold)

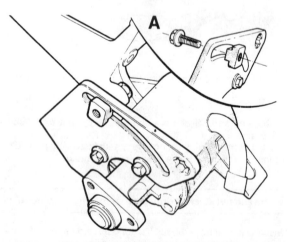

Fig. 10.66 Power steering pump and mounting bracket showing adjuster bolt and nut (A) (Sec 34)

(b) Tighten the retaining nuts to their specified torque settings and insert a new split pin to secure the castellated nut
(c) On completion, check that the steering action is satisfactory (lock-to-lock) and also that the front wheels are in the straight-ahead position when the steering wheel is centralised. If required the drag link can be adjusted as described in Section 32

32 Worm and nut steering drag link – removal and refitting

1 Straighten and extract the split pins from the ball pin nuts at each end of the drag link, then unscrew and remove the nuts.
2 Using a universal balljoint separator, free the ballpins and withdraw the drag link.
3 To remove the drag link ends, loosen the locknuts and unscrew the ends, noting that they have left- and right-hand threads.
4 Screw the new ends onto the drag link an equal number of threads so that the ballpin centres are 457.0 mm (18.0 in) apart.
5 Refit the drag link, tighten the ballpin nuts and fit new split pins.
6 With the wheels in the straight-ahead position, the steering gear

Chapter 10 Suspension and steering

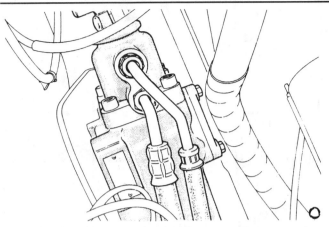

Fig. 10.67 Power steering hydraulic line connections to steering gear unit (Sec 35)

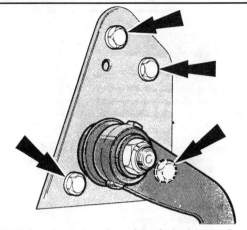

Fig. 10.68 Steering gear unit retaining bolts (arrowed) – power steering gear models (Sec 35)

must be at the centre of its travel; adjust the drag link accordingly and then tighten the clamp bolts.
7   Check the steering angles and alignment as described in Section 37 to complete.

33 Power-assisted steering system – fluid level check and system bleeding

Fluid level
1   The fluid level reservoir has three level marking on its transport casing, these indicate (1) the maximum level when the engine is hot, (2) the maximum level when the engine is cold and (3) the minimum level when the engine is cold. The correct fluid level must be maintained at all times and should be checked at the intervals given in the *Routine maintenance* Section at the start of this manual. Only the fluid type specified must be used in the system. Do not overfill the system (Fig. 10.65).

Bleeding the system
2   If the power steering system has been emptied of fluid, top up the fluid level, then raise and support the vehicle at the front end so that the front wheels are clear of the ground. Turn the steering wheel from lock-to-lock twice (engine switched off). As the wheels are being turned, top up the reservoir fluid level as necessary. Lower the vehicle at the front end.
3   Start up the engine and allow it to run at its normal idle speed whilst the above procedure is repeated, again keeping the reservoir topped up to the required level.

34 Power-steering pump – removal and refitting

1   Place a suitable container under the power-steering pump, disconnect the fluid pipes, and drain the fluid.
2   Remove the drive belts as described in Chapter 2. Unbolt the pulley if necessary to ease removal.
3   Unscrew the nuts and bolts and remove the pump (Fig. 10.66).
4   Refitting is a reversal of removal, but tighten the nuts and bolts to the specified torque. Tension the drivebelts as described in Chapter 2. Refill the power-steering system with fluid and bleed it as described in Section 33.

35 Power steering gear unit – removal and refitting

1   The removal procedure is as given for the manual steering gear unit in Section 28, but in addition, the hydraulic lines will need to be detached. To do this, place a container beneath the steering gear then unscrew the pressure and return pipe unions and drain the power steering fluid. Cover the pipe ends and steering gear apertures with masking tape to prevent the ingress of foreign matter (Fig. 10.67). Note

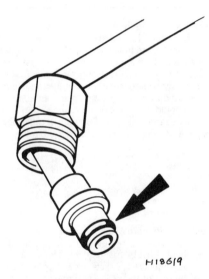

Fig. 10.69 O-ring seal on high pressure fluid hose connector (Sec 36)

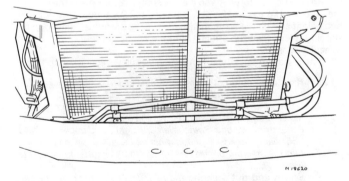

Fig. 10.70 Power steering hoses and clamps to front panel (Sec 36)

also that there are four steering gear retaining bolts rather than three on manual types (Fig. 10.68).
2   The refitting procedure is the reversal of the removal procedure, but note the following points:

(a) When reconnecting the steering gear hydraulic connections, take care not to over tighten the unions
(b) Refill and bleed the power steering system as described in Section 33
(c) Check and if necessary adjust the front wheel alignment as described in Section 37

36 Power steering hoses – removal and refitting

1 Disconnect the battery earth lead.
2 Detach and remove the radiator grille (Chapter 11).
3 Unscrew and remove the hydraulic hose clamp bolts from the front panel, also the single screw from the power steering pump unit.
4 Locate suitable container(s) under the steering gear and pump units, then disconnect the hoses from each. Clean the area around the hose unions and plug them to prevent further fluid leakage and the possible ingress of dirt.
5 If the hose(s) and/or their union O-ring seals are perished or damaged they must be renewed. The O-ring seal is only available with a new hose, so treat it with care if they are to be re-used (Fig. 10.69).
6 Refit in the reverse order of removal. Ensure that all connections are clean and do not over tighten the unions (Fig.10.70).
7 On completion, top up the system fluid level (do not re-use the old fluid) and bleed the system (Section 33).

37 Front wheel alignment

1 Accurate front wheel alignment is essential to provide good steering and handling characteristics and to prevent excessive tyre wear. The camber and castor angles are built into the axle and spindle bodies and are not adjustable. Only the front wheel toe-in is adjustable.
2 Checking of all wheel alignment angles and adjustments is best carried out by a suitably equipped garage, but the toe-in can be checked by the home mechanic by obtaining or making an adjustable tracking gauge. The gauge should have two pointers, one adjustable which can be positioned between the inner or outer faces of the wheels.
3 Before making any adjustments check that the following are within limits:

   (a) Tyre pressures
   (b) Wheel run-out
   (c) Front wheel bearing adjustment
   (d) Front axle bushes
   (e) Steering balljoints

The vehicle must be at its normal kerb weight (unladen) when making the check and the front suspension settled by 'bouncing' the body a few times. Finally the check must be made on a flat level surface.
4 Using the gauge, measure the distance between the wheel rims at the hub height at the rear of the wheel and mark the tyre with chalk to indicate where the measurement was taken.
5 Roll the vehicle forwards so that the chalk mark is now at the front of the wheel and measure the distance between the wheel rims again at hub height and on the same measuring points; the latter measurement should be less than the original by the amount of toe-in given in the Specifications.
6 If necessary, adjust the toe-in setting as follows according to type:

   (a) **Rack and pinion steering models:** *Loosen off the track rod to balljoint locknut and also loosen off the rack bellows outer end on the track rod. Grip the track rod and turn it in the appropriate direction to adjust as required. As the rod is turned, prevent the gaiter from twisting. Both the left and right-hand track rods have right-hand threads. Each rod must be equally adjusted to maintain the steering parallel. Retighten the locknut(s) to complete (Photo 26.2)*
   (b) **Worm and nut steering models:** *The track rod/balljoint unit threads have left and right-hand threads according to side. Loosen off the rod to balljoint clamp then turn the rod in the required direction to adjust the setting. Each rod must be equally adjusted to maintain the steering parallel. Retighten the clamps to complete (Fig. 10.52)*

38 Wheels and tyres – general care and maintenance

Wheels and tyres should give no real problems in use provided that a close eye is kept on them with regard to excessive wear or damage. To this end, the following points should be noted.

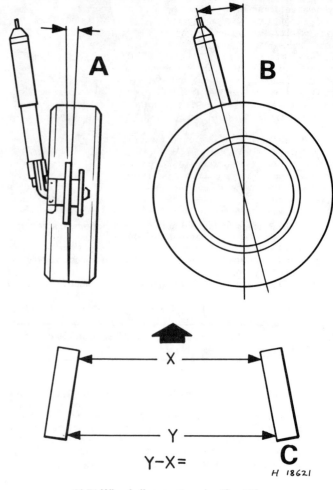

10.71 Wheel alignment angles (Sec 37)

A *Camber angle*
B *Castor angle*
C *Toe-in setting (Y-X)*

Ensure that tyre pressures are checked regularly and maintained correctly. Checking should be carried out with the tyres cold and not immediately after the vehicle has been in use. If the pressures are checked with the tyres hot, an apparently high reading will be obtained owing to heat expansion. Under no circumstances should an attempt be made to reduce the pressures to the quoted cold reading in this instance, or effective underinflation will result.

Underinflation will cause overheating of the tyre owing to excessive flexing of the casing, and the tread will not sit correctly on the road surface. This will cause a consequent loss of adhesion and excessive wear, not to mention the danger of sudden tyre failure due to heat build-up.

Overinflation will cause rapid wear of the centre part of the tyre tread coupled with reduced adhesion, harsher ride, and the danger of shock damage occurring in the tyre casing.

Regularly check the tyres for damage in the form of cuts or bulges, especially in the sidewalls. Remove any nails or stones embedded in the tread before they penetrate the tyre to cause deflation. If removal of a nail *does* reveal that the tyre has been punctured, refit the nail so that its point of penetration is marked. Then immediately change the wheel and have the tyre repaired by a tyre dealer. Do *not* drive on a tyre in such a condition. In many cases a puncture can be simply repaired by the use of an inner tube of the correct size and type. If in any doubt as to the possible consequences of any damage found, consult your local tyre dealer for advice.

Periodically remove the wheels and clean any dirt or mud from the inside and outside surfaces. Examine the wheel rims for signs of rusting, corrosion or other damage. Wheels are easily damaged by 'kerbing' whilst parking, and may become dented or buckled. Renewal of the

Chapter 10 Suspension and steering

wheel is very often the only course of remedial action possible.

The balance of each wheel and tyre assembly should be maintained to avoid excessive wear, not only to the tyres but also to the steering and suspension components. Wheel imbalance is normally signified by vibration through the vehicle's bodyshell, although in many cases it is particularly noticeable through the steering wheel. Conversely, it should be noted that wear or damage in suspension or steering components may cause excessive tyre wear. Out-of-round or out-of-true tyres, damaged wheels and wheel bearing wear/maladjustment also fall into this category. Balancing will not usually cure vibration caused by such wear.

Wheel balancing may be carried out with the wheel either on or off the vehicle. If balanced on the vehicle, ensure that the wheel-to-hub relationship is marked in some way prior to subsequent wheel removal so that it may be refitted in its original position.

General tyre wear is influenced to a large degree by driving style – harsh braking and acceleration or fast cornering will all produce more rapid tyre wear. Interchanging of tyres may result in more even wear, but this should only be carried out where there is no mix of tyre types on the vehicle. However, it is worth bearing in mind that if this is completely effective, the added expense of replacing a complete set of tyres simultaneously is incurred, which may prove financially restrictive for many owners.

Front tyres may wear unevenly as a result of wheel misalignment. The front wheels should always be correctly aligned according to the settings specified by the vehicle manufacturer.

Legal restrictions apply to the mixing of tyre types on a vehicle. Basically this means that a vehicle must not have tyres of differing construction on the same axle. Although it is not recommended to mix tyre types between front axle and rear axle, the only legally permissible combination is crossply at the front and radial at the rear. When mixing radial ply tyres, textile braced radials must always go on the front axle, with steel braced radials at the rear. An obvious disadvantage of such mixing is the necessity to carry two spare tyres to avoid contravening the law in the event of a puncture.

In the UK, the Motor Vehicles Construction and Use Regulations apply to many aspects of tyre fitting and usage. It is suggested that a copy of these regulations is obtained from your local police if in doubt as to the current legal requirements with regard to tyre condition, minimum tread depth, etc.

Before diagnosing faults from the following chart, check that any irregularities are not caused by:

(a) *Binding brakes*
(b) *Incorrect 'mix' of radial and crossply tyres*
(c) *Incorrect tyre pressures*
(d) *Misalignment of the bodyframe*

39 Fault diagnosis – suspension and steering

Symptom	Reason(s)
Steering feels vague, vehicle wanders and floats at speed	Tyre pressures uneven Shock absorbers worn Spring U-bolts broken (beam axle models) Steering balljoints badly worn Suspension geometry incorrect Kingpins and bushes worn (beam axle models) Worn steering gear or steering gear bushes Weak or broken front coil spring (IFS models) Steering gear adjustment incorrect (worm and nut type) Chassis underframe out of alignment Vehicle overladen
Stiff and heavy steering	Tyre pressures too low Kingpins need greasing Steering gear needs topping up or is incorrectly adjusted Steering balljoints seizing Wheel alignment incorrect Steering column misaligned Suspension geometry incorrect Power steering pump drivebelt slipping (where applicable)
Wheel wobble and vibration	Wheel nuts loose Front wheels and tyres out of balance Steering balljoints badly worn Hub bearings badly worn Steering gear free play excessive (worm and nut type) Front springs loose, weak or broken Front shock absorbers worn Kingpins and bushes worn (beam axle models)
Excessive pitching and rolling on corners and during braking	Shock absorbers worn Spring leaf broken (beam axle models) Vehicle overladen Weak or broken front coil spring (IFS models)

Chapter 11 Bodywork

Contents

Bonnet – removal, refitting and adjustment	7
Bonnet release cable and latch – renewal	8
Bonnet safety catch – removal and refitting	9
Bumpers – removal and refitting	25
Door mirror and glass – removal and refitting	28
Door trim panels – removal and refitting	10
Facia – removal and refitting	29
Front door – removal and refitting	12
Front door fittings – removal and refitting	11
Front door quarter glass – removal and refitting	20
Front door sliding window glass and frame – dismantling and reassembly	22
Front door sliding window glass and frame – removal and refitting	21
Front door window and regulator – removal and refitting	19
Front grille – removal and refitting	26
General description	1
Headlining – removal and refitting	30
Maintenance – bodywork and under frame	2
Maintenance – hinges and locks	6
Maintenance – upholstery and carpets	3
Major body damage – repair	5
Minor body damage – repair	4
Opening rear quarter window – removal and refitting	23
Rear doors – removal and refitting	15
Rear door fittings – removal and refitting	16
Seats – removal and refitting	31
Seat belt and stalk – removal and refitting	32
Sliding side door – removal and refitting	13
Sliding side door fittings – removal and refitting	14
Tailgate – removal and refitting	18
Tailgate fittings – removal and refitting	17
Windscreen and fixed windows – removal and refitting	24
Windscreen grille – removal and refitting	27

Specifications

Torque wrench settings

	Nm	lbf ft
Bench seat pivot bolt	6 to 8	4 to 6
Seat track to floor bolts	21 to 25	16 to 19
Seat frame to track nuts	21 to 25	16 to 19
Seat belt stalk (to floor)	38 to 53	28 to 39
Seat belt inertia reel fixings	38 to 53	28 to 39
Bonnet safety catch release lever screw	3.5 to 4.5	2.6 to 3.3

1 General description

The body and chassis on all versions of the Transit is of all steel construction. Three basic chassis types are available, being short wheelbase, long wheelbase and extended wheelbase models.

The three main body types are van, bus and chassis cab. Twin opening rear doors or a tailgate is fitted and, on some models, side opening door(s) are available. The bodyshell is aerodynamic in shape to promote economy and reduce wind noise levels.

During manufacture, each body is carefully prepared for painting, given a zinc phosphate treatment, sprayed with a polyester primer, which is then oven baked and finally, two top finisher coats of enamel are applied. The body cavities are wax-injected to prevent corrosion.

Due to the large number of specialist applications of this vehicle range, information contained in this Chapter is given on parts found to be common on the popular factory produced version. No information is given on special body versions.

2 Maintenance – bodywork and underframe

The general condition of a vehicle's bodywork is the one thing that significantly affects its value. Maintenance is easy but needs to be regular. Neglect, particularly after minor damage, can lead quickly to further deterioration and costly repair bills. It is important also to keep watch on those parts of the vehicle not immediately visible, for instance the underside, inside all the wheel arches and the lower part of the engine compartment.

The basic maintenance routine for the bodywork is washing – preferably with a lot of water, from a hose. This will remove all the loose solids which may have stuck to the vehicle. It is important to flush these off in such a way as to prevent grit from scratching the finish. The wheel arches and underframe need washing in the same way to remove any accumulated mud which will retain moisture and tend to encourage rust. Paradoxically enough, the best time to clean the underframe and wheel arches is in wet weather when the mud is thoroughly wet and soft. In very wet weather the underframe is usually cleaned of large accumulations automatically and this is a good time for inspection.

Periodically, except on vehicles with a wax-based underbody protective coating, it is a good idea to have the whole of the underframe of the vehicle steam cleaned, engine compartment included, so that a thorough inspection can be carried out to see what minor repairs and renovations are necessary. Steam cleaning is available at many garages and is necessary for removal of the accumulation of oily grime which sometimes is allowed to become thick in certain areas. If steam cleaning facilities are not available, there are one or two excellent grease solvents available, such as Holts Engine Cleaner or Holts Foambrite, which can be brush applied. The dirt can then be simply hosed off. Note that these methods should not be used on vehicles with wax-based underbody protective coating or the coating will be removed. Such vehicles should be inspected annually, preferably just prior to winter, when the underbody should be washed down and any damage to the wax coating repaired using Holts Undershield. Ideally, a completely fresh coat should be applied. It would also be worth considering the use of such

Chapter 11 Bodywork

wax-based protection for injection into door panels, sills, box sections, etc, as an additional safeguard against rust damage where such protection is not provided by the vehicle manufacturer.

After washing paintwork, wipe off with a chamois leather to give an unspotted clear finish. A coat of clear protective wax polish, like the many excellent Turtle Wax polishes, will give added protection against chemical pollutants in the air. If the paintwork sheen has dulled or oxidised, use a cleaner/polisher combination such as Turtle Extra to restore the brilliance of the shine. This requires a little effort, but such dulling is usually caused because regular washing has been neglected. Care needs to be taken with metallic paintwork, as special non-abrasive cleaner/polisher is required to avoid damage to the finish. Always check that the door and ventilator opening drain holes and pipes are completely clear so that water can be drained out. Bright work should be treated in the same way as paint work. Windscreens and windows can be kept clear of the smeary film which often appears by the use of a proprietary glass cleaner like Holts Mixra. Never use any form of wax or other body or chromium polish on glass.

3 Maintenance – upholstery and carpets

Mats and carpets should be brushed or vacuum cleaned regularly to keep them free of grit. If they are badly stained remove them from the vehicle for scrubbing or sponging and make quite sure they are dry before refitting. Seats and interior trim panels can be kept clean by wiping with a damp cloth and Turtle Wax Carisma. If they do become stained (which can be more apparent on light coloured upholstery) use a little liquid detergent and a soft nail brush to scour the grime out of the grain of the material. Do not forget to keep the headlining clean in the same way as the upholstery. When using liquid cleaners inside the vehicle do not over-wet the surfaces being cleaned. Excessive damp could get into the seams and padded interior causing stains, offensive odours or even rot. If the inside of the vehicle gets wet accidentally it is worthwhile taking some trouble to dry it out properly, particularly where carpets are involved. *Do not leave oil or electric heaters inside the vehicle. for this purpose.*

4 Minor body damage – repair

Note: *For more detailed information about bodywork repair, the Haynes Publishing Group publish a book by Lindsay Porter called The Car Bodywork Repair Manual. This incorporates information on such aspects as rust treatment, painting and glass fibre repairs, as well as details on more ambitious repairs involving welding and panel beating.*

Repair of minor scratches in bodywork

If the scratch is very superficial, and does not penetrate to the metal of the bodywork, repair is very simple. Lightly rub the area of the scratch with a paintwork renovator like Turtle Wax New Color Back, or a very fine cutting paste like Holts Body + Plus Rubbing Compound to remove loose paint from the scratch and to clear the surrounding bodywork of wax polish. Rinse the area with clean water.

Apply touch-up paint, such as Holts Dupli-Color Color Touch or a paint film like Holts Autofilm, to the scratch using a fine paint brush; continue to apply fine layers of paint until the surface of the paint in the scratch is level with the surrounding paintwork. Allow the new paint at least two weeks to harden: then blend it into the surrounding paintwork by rubbing the scratch area with a paintwork renovator or a very fine cutting paste, such as Holts Body + Plus Rubbing Compound or Turtle Wax New Color Back. Finally, apply wax polish from one of the Turtle Wax range of wax polishes.

Where the scratch has penetrated right through to the metal of the bodywork, causing the metal to rust, a different repair technique is required. Remove any loose rust from the bottom of the scratch with a penknife, then apply rust inhibiting paint, such as Turtle Wax Rust Master, to prevent the formation of rust in the future. Using a rubber or nylon applicator fill the scratch with bodystopper paste like Holts Body + Plus Knifing Putty. If required, this paste can be mixed with cellulose thinners, such as Holts Body + Plus Cellulose Thinners, to provide a very thin paste which is ideal for filling narrow scratches. Before the stopper-paste in the scratch hardens, wrap a piece of smooth

cotton rag around the top of a finger. Dip the finger in cellulose thinners, such as Holts Body + Plus Cellulose Thinners, and then quickly sweep it across the surface of the stopper-paste in the scratch; this will ensure that the surface of the stopper-paste is slightly hollowed. The scratch can now be painted over as described earlier in this Section.

Repair of dents in bodywork

When deep denting of the vehicle's bodywork has taken place, the first task is to pull the dent out, until the affected bodywork almost attains its original shape. There is little point in trying to restore the original shape completely, as the metal in the damaged area will have stretched on impact and cannot be reshaped fully to its original contour. It is better to bring the level of the dent up to a point which is about $\frac{1}{8}$ in (3 mm) below the level of the surrounding bodywork. In cases where the dent is very shallow anyway, it is not worth trying to pull it out at all. If the underside of the dent is accessible, it can be hammered out gently from behind, using a mallet with a wooden or plastic head. Whilst doing this, hold a suitable block of wood firmly against the outside of the panel to absorb the impact from the hammer blows and thus prevent a large area of the bodywork from being 'belled-out'.

Should the dent be in a section of the bodywork which has a double skin or some other factor making it inaccessible from behind, a different technique is called for. Drill several small holes through the metal inside the area – particularly in the deeper section. Then screw long self-tapping screws into the holes just sufficiently for them to gain a good purchase in the metal. Now the dent can be pulled out by pulling on the protruding heads of the screws with a pair of pliers.

The next stage of the repair is the removal of the paint from the damaged area, and from an inch or so of the surrounding 'sound' bodywork. This is accomplished most easily by using a wire brush or abrasive pad on a power drill, although it can be done just as effectively by hand using sheets of abrasive paper. To complete the preparation for filling, score the surface of the bare metal with a screwdriver or the tang of a file, or alternatively, drill small holes in the affected area. This will provide a really good 'key' for the filler paste.

To complete the repair see the Section on filling and re-spraying.

Repair of rust holes or gashes in bodywork

Remove all paint from the affected area and from an inch or so of the surrounding 'sound' bodywork, using an abrasive pad or a wire brush on a power drill. If these are not available a few sheets of abrasive paper will do the job just as effectively. With the paint removed you will be able to gauge the severity of the corrosion and therefore decide whether to renew the whole panel (if this is possible) or to repair the affected area. New body panels are not as expensive as most people think and it is often quicker and more satisfactory to fit a new panel than to attempt to repair large areas of corrosion.

Remove all fittings from the affected area except those which will act as a guide to the original shape of the damaged bodywork (eg headlamp shells, etc). Then, using tin snips or a hacksaw blade, remove all loose metal and any other metal badly affected by corrosion. Hammer the edges of the hole inwards in order to create a slight depression for the filler paste.

Wire brush the affected area to remove the powdery rust from the surface of the remaining metal. Paint the affected area with rust inhibiting paint like Turtle Rust Master; if the back of the rusted area is accessible treat this also.

Before filling can take place it will be necessary to block the hole in some way. This can be achieved by the use of aluminium or plastic mesh, or aluminium tape.

Aluminium or plastic mesh or glass fibre matting, such as the Holts Body + Plus Glass Fibre Matting, is probably the best material to use for a large hole. Cut a piece to the approximate size and shape of the hole to be filled, then position it in the hole so that its edges are below the level of the surrounding bodywork. It can be retained in position by several blobs of filler paste around its periphery.

Aluminium tape should be used for small or very narrow holes. Pull a piece off the roll and trim it to the approximate size and shape required, then pull off the backing paper (if used) and stick the tape over the hole; it can be overlapped if the thickness of one piece is insufficient. Burnish down the edges of the tape with the handle of a screwdriver or similar, to ensure that the tape is securely attached to the metal underneath.

Bodywork repairs – filling and re-spraying

Before using this Section, see the Sections on dent, deep scratch, rust holes and gash repairs.

214   Chapter 11   Bodywork

Many types of bodyfiller are available, but generally speaking those proprietary kits which contain a tin of filler paste and a tube of resin hardener are best for this type of repair, like Holts Body + Plus or Holts No Mix which can be used directly from the tube. A wide, flexible plastic or nylon applicator will be found invaluable for imparting a smooth and well contoured finish to the surface of the filler.

Mix up a little filler on a clean piece of card or board – measure the hardener carefully (follow the maker's instructions on the pack) otherwise the filler will set too rapidly or too slowly. Alternatively, Holts No Mix can be used straight from the tube without mixing, but daylight is required to cure it. Using the applicator apply the filler paste to the prepared area; draw the applicator across the surface of the filler to achieve the correct contour and to level the filler surface. As soon as a contour that approximates to the correct one is achieved, stop working the paste – if you carry on too long the paste will become sticky and begin to 'pick up' on the applicator. Continue to add thin layers of filler paste at twenty-minute intervals until the level of the filler is just proud of the surrounding bodywork.

Once the filler has hardened, excess can be removed using a metal plane or file. From then on, progressively finer grades of abrasive paper should be used, starting with a 40 grade production paper and finishing with 400 grade wet-and-dry paper. Always wrap the abrasive paper around a flat rubber, cork, or wooden block – otherwise the surface of the filler will not be completely flat. During the smoothing of the filler surface the wet-and-dry paper should be periodically rinsed in water. This will ensure that a very smooth finish is imparted to the filler at the final stage.

At this stage the 'dent' should be surrounded by a ring of bare metal, which in turn should be encircled by the finely 'feathered' edge of the good paintwork. Rinse the repair area with clean water, until all of the dust produced by the rubbing-down operation has gone.

Spray the whole repair area with a light coat of primer, either Holts Body + Plus Grey or Red Oxide Primer – this will show up any imperfections in the surface of the filler. Repair these imperfections with fresh filler paste or bodystopper, and once more smooth the surface with abrasive paper. If bodystopper is used, it can be mixed with cellulose thinners to form a really thin paste which is ideal for filling small holes. Repeat this spray and repair procedure until you are satisfied that the surface of the filler, and the feathered edge of the paintwork are perfect. Clean the repair area with clean water and allow to dry fully.

The repair area is now ready for final spraying. Paint spraying must be carried out in a warm, dry, windless and dust free atmosphere. This condition can be created artificially if you have access to a large indoor working area, but if you are forced to work in the open, you will have to pick your day very carefully. If you are working indoors, dousing the floor in the work area with water will help to settle the dust which would otherwise be in the atmosphere. If the repair area is confined to one body panel, mask off the surrounding panels; this will help to minimise the effects of a slight mis-match in paint colours. Bodywork fittings (eg chrome strips, door handles etc) will also need to be masked off. Use genuine masking tape and several thicknesses of newspaper for the masking operations.

Before commencing to spray, agitate the aerosol can thoroughly, then spray a test area (an old tin, or similar) until the technique is mastered. Cover the repair area with a thick coat of primer; the thickness should be built up using several thin layers of paint rather than one thick one. Using 400 grade wet-and-dry paper, rub down the surface of the primer until it is really smooth. While doing this, the work area should be thoroughly doused with water, and the wet-and-dry paper periodically rinsed in water. Allow to dry before spraying on more paint.

Spray on the top coat using Holts Dupli-Color Autospray, again building up the thickness by using several thin layers of paint. Start spraying in the centre of the repair area and then, with a side-to-side motion, work outwards until the whole repair area and about 2 inches of the surrounding original paintwork is covered. Remove all masking material 10 to 15 minutes after spraying on the final coat of paint.

Allow the new paint at least two weeks to harden, then, using a paintwork renovator or a very fine cutting paste such as Turtle Wax New Color Back or Holts Body + Plus Rubbing Compound, blend the edges of the paint into the existing paintwork. Finally, apply wax polish.

Plastic components

With the use of more and more plastic body components by the vehicle manufacturers (eg bumpers, spoilers, and in some cases major body panels), rectification of more serious damage to such items has become a matter of either entrusting repair work to a specialist in this field, or renewing complete components. Repair of such damage by the DIY owner is not really feasible owing to the cost of the equipment and materials required for effecting such repairs. The basic technique involves making a groove along the line of the crack in the plastic using a rotary burr in a power drill. The damaged part is then welded back together by using a hot air gun to heat up and fuse a plastic filler rod into the groove. Any excess plastic is then removed and the area rubbed down to a smooth finish. It is important that a filler rod of the correct plastic is used, as body components can be made of a variety of different types (eg polycarbonate, ABS, polypropylene).

Damage of a less serious nature (abrasions, minor cracks etc) can be repaired by the DIY owner using a two-part epoxy filler repair material like Holts Body + Plus or Holts No Mix which can be used directly from the tube. Once mixed in equal proportions (or applied direct from the tube in the case of Holts No Mix), this is used in similar fashion to the bodywork filler used on metal panels. The filler is usually cured in twenty to thirty minutes, ready for sanding and painting.

If the owner is renewing a complete component himself, or if he has repaired it with epoxy filler, he will be left with the problem of finding a suitable paint for finishing which is compatible with the type of plastic used. At one time the use of a universal paint was not possible owing to the complex range of plastics encountered in body component applications. Standard paints, generally speaking, will not bond to plastic or rubber satisfactorily, but Holts Professional Spraymatch paints to match any plastic or rubber finish can be obtained from dealers. However, it is now possible to obtain a plastic body parts finishing kit which consists of a pre-primer treatment, a primer and coloured top coat. Full instructions are normally supplied with a kit, but basically the method of use is to first apply the pre-primer to the component concerned and allow it to dry for up to 30 minutes. Then the primer is applied and left to dry for about an hour before finally applying the special coloured top coat. The result is a correctly coloured component where the paint will flex with the plastic or rubber, a property that standard paint does not normally possess.

---

5   Major body damage – repair

1   With the exception of chassis cab versions, the chassis members are spot welded to the underbody, and in this respect can be termed of monocoque or unit construction. Major damage repairs to this type of body combination must of necessity be carried out by body shops equipped with welding and hydraulic straightening facilities.

2   Extensive damage to the body may distort the chassis and result in unstable and dangerous handling as well as excessive wear to tyres and suspension or steering components. It is recommended that checking of the chassis alignment be entrusted to a Ford agent with specialist checking jigs.

---

6   Maintenance – hinges and locks

1   Every six months or 6000 miles (9600 km), the bonnet catch, door hinges, door check straps, sliding door, and sliding step (where fitted), should be oiled with a few drops of engine oil from an oil can, or a multi-purpose grease (photo).

2   At the same interval the door striker plates should be given a thin smear of grease.

3   The door locks should be lubricated prior to each winter using a suitable lock lubricant or multi-purpose grease.

---

7   Bonnet – removal, refitting and adjustment

1   Open the bonnet and support it with its stay rod.

2   Detach the windscreen washer tubes and the engine compartment light lead.

3   Mark around the bonnet hinges to show the outline of their fitted positions for correct realignment on assembly (photo).

4   Get an assistant to support the bonnet whilst you unscrew and remove the hinge retaining bolts. Then lift the bonnet clear.

5   Refit in reverse sequence and only tighten the hinge bolts fully when bonnet alignment is satisfactory.

Chapter 11 Bodywork

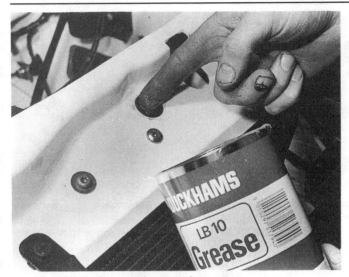
6.1 Lubricating the bonnet release catch

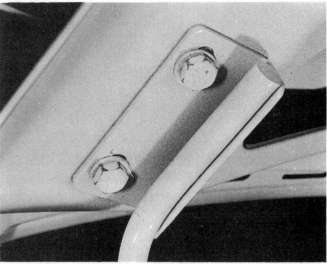
7.3 Bonnet hinge retaining bolts

6 Further adjustment of the bonnet fit is available by loosening the hinges and the locknuts of the bump stops on the front crossmember. The bonnet can now be adjusted to give an even clearance between its outer edges and the surrounding panels. Adjust the front bump stops to align the edges of the bonnet with the front wing panels, then retighten the locknuts and hinge bolts.

8 Bonnet release cable and latch – renewal

Release cable

1 Open and support the bonnet. If the cable is broken, release the latch by reaching up from the underside and release it by hand using a suitably shaped rod (welding rod may suffice).
2 Release the cable nipple from the latch and the outer cable from the inner face of the front panel (photo).
3 Detach the cable from the release handle by sliding it sideways (photo).
4 Withdraw the cable through the engine compartment side complete with the bulkhead grommet.
5 Refit in the reverse order of removal and check for satisfactory operation on completion.

Bonnet latch

6 Detach the cable from the latch, then undo the three retaining screws. Remove the latch from the front panel (Fig.11.1).
7 Refit in the reverse order of removal.

9 Bonnet safety catch – removal and refitting

1 Open the bonnet and support it with its stay rod.
2 Make an alignment mark around the fitted position of the safety catch to bonnet with a pencil or marker pen. Undo the catch retaining bolts and the release lever Torx screw, then remove the two items complete with the connecting rod (photo).
3 Refit in the reverse order of removal but note the following points:

  (a) Align the catch with the positional marking mode when removing the catch
  (b) When tightening the release lever screw, tighten it slowly to the specified torque wrench setting. When tightened check that the lever movement is satisfactory and does not bind.
  (c) Take care not to overtighten the catch or release lever securing bolts/screw or the safety catch may stick in the open or closed position

10 Door trim panels – removal and refitting

Front doors

1 Prise free the trim from the window regulator handle, undo the retaining screw, note the fitted position of the handle on its shaft (with the window fully raised), then withdraw the handle and the escutcheon plate (photos).
2 Undo the retaining screw and remove the remote release handle surround (photo).
3 Prise back the two trim covers from the door pull handle using a

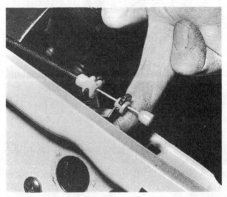
8.2 Bonnet catch release cable attachment

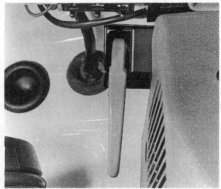
8.3 Bonnet release handle

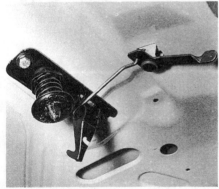
9.2 Bonnet safety catch

Chapter 11 Bodywork

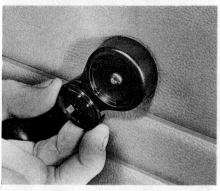
10.1A Window regulator handle and retaining screw with trim removed for access

10.1B Removing the window regulator handle ...

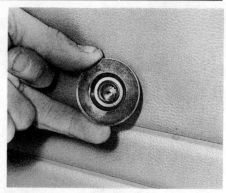
10.1C ... and escutcheon plate

10.2 Removing the remote release handle surround screw

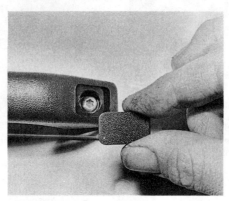
10.3A Remove the trim cover ...

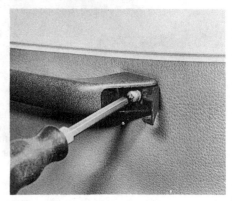
10.3B ... and door pull handle retaining screws

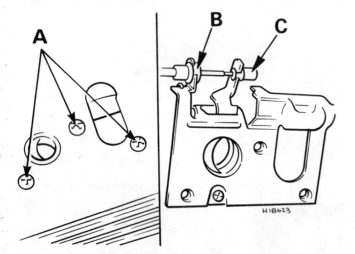

Fig. 11.1 Bonnet latch and cable connection (Sec 8)

   A   Latch retaining screws   C   Inner cable nipple
   B   Cable bush

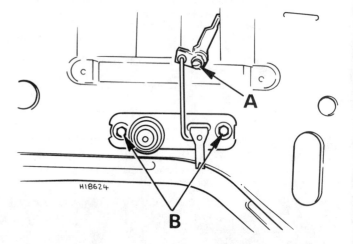

Fig. 11.2 Bonnet safety catch showing lever screw (A) and retaining bolts (B) (Sec 9)

suitable screwdriver or a similar tool. Undo the two retaining screws and remove the handle (photos).
4   Carefully prise free the door trim panel, prising with a suitable tool between the panel around its outer and lower edges at the fixing points. Remove the panel.
5   Refitting is a reversal of the removal procedure.

Rear doors
6   The rear door trim panels are secured by plastic retaining clips, the removal of which requires the use of a forked tool similar to that shown in the photo. These clips are easily broken so take care when prising them free. Remove the trim panel.
7   Refit in the reverse order of removal. Align the panel and press the clips into position.

Chapter 11 Bodywork

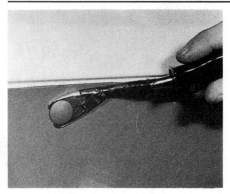

10.6 Prising free a rear trim panel clip using a forked lever tool

10.8 Sliding door bump stop

10.9 Sliding door inner lock

11.5 Door lock barrel and securing clip

11.8 Door inner remote release unit

11.9 Door lock unit retaining screws

Sliding side door

8   Undo the retaining screws and remove the bump stop (photo).
9   Undo the retaining screw and remove the inner lock escutcheon (photo).
10 Carefully prise free the plastic retaining clips using a tool of similar type to that shown in photo 10.6, then remove the panel.
11 Refit in the reverse order of removal.

9   Undo the three retaining screws and remove the door lock complete with connecting rods and the inner remote release unit (photo). Detach the connecting rods from their retaining clips as the assembly is removed from the door.
10 Refit in the reverse order of removal. Do not fully tighten the door lock screws until after the inner remote release unit and the connecting rods are secured.

11 Front door fittings – removal and refitting

Exterior handle
1   Remove the door inner trim panel as described in the previous Section. Peel back the insulation sheet for access to the inner door components.
2   Working through the access apertures in the inner panel, undo the two Torx screws and remove the handle (Fig. 11.3).
3   Refit in the reverse order of removal.

Door lock barrel
4   Remove the door inner trim panel as described in the previous Section. Peel back the insulation sheet for access to the inner door components.
5   Prise free the lock barrel securing clip using a suitable screwdriver as a lever, withdraw the lock unit and detach the operating rod (photo).
6   Refit in the reverse order of removal.

Door lock unit
7   Remove the door inner trim panel as described in Section 10. Peel back the insulation sheet for access to the inner door components.
8   Detach the remote release handle by unscrewing the two Torx screws (photo).

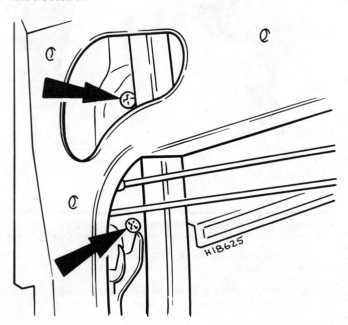

Fig. 11.3 Exterior door handle screws (arrowed) (Sec 11)

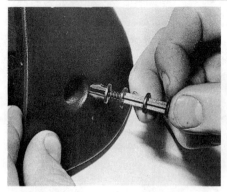
12.2A Door hinge outer trim retaining clip removal

12.2B Door hinge trim retaining clip removal

12.2C Door hinge retaining bolts

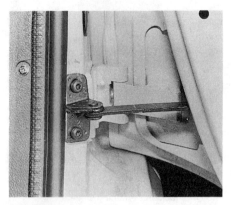
12.3 Door check strap

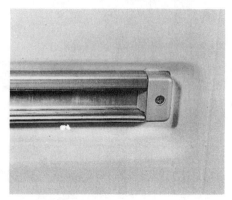
13.1 Sliding door track end stop

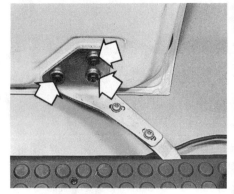
13.2 Sliding door lower guide support attachment to door (arrowed)

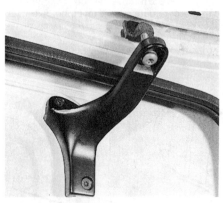
13.3 Sliding door upper guide support

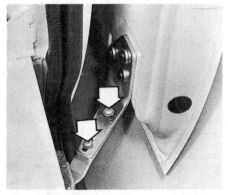
13.6 Sliding door lower guide support flush fitting adjuster screws (arrowed)

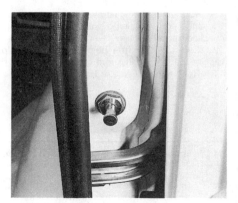
13.7 Sliding door striker plate

12 Front door – removal and refitting

1 Disconnect the battery earth lead. Open the door and position a suitable jack or support blocks underneath it, but do not raise the door, just take its weight.
2 Prise free the plastic clip covers from the door upper hinge cover, remove the screws and the cover. Undo the three retaining bolts (photos).
3 Undo the two Torx screws securing the door check strap (photo).
4 Undo the four retaining screws and remove the front footwell trim for access to the door lower hinge bolts. Get an assistant to support the door, then unscrew the two retaining bolts and single nut from the lower hinge and remove the door (Fig. 11.4).
5 To refit the door, first align the stud of the lower hinge to the body and simultaneously insert a retaining bolt and hand tighten it to secure the top hinge and door.
6 Fit and hand tighten the remaining upper and lower hinge bolts.
7 Reconnect the door check strap, shut the door and align it in its aperture so that it has an even clearance all round, then tighten the upper hinge bolts from within the cab.
8 Tighten the lower hinge bolts and retaining nut, then open the door and shut it to ensure that it does not bind with the body aperture at any point. Adjust the door striker plate if necessary then refit the footwell and hinge trims to complete.

Chapter 11 Bodywork

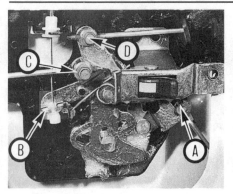

14.5 Sliding door remote release unit showing connections

A Lock rod and clip
B Handle cable
C Latch rod
D Handle rod

14.17A Sliding door unit leading edge screw

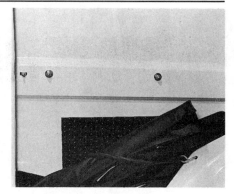

14.17B Sliding door rail retaining nuts (on inside)

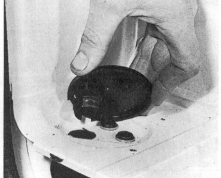

14.19A Remove the circular cover ...

14.19B ... for access to the retaining screws

14.20 Sliding door lower guide attachment to body (arrowed)

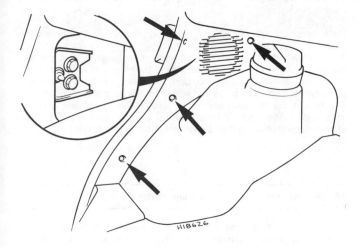

Fig. 11.4 Front footwell side trim panel clips (arrowed) and lower hinge retaining screw locations (Sec 12)

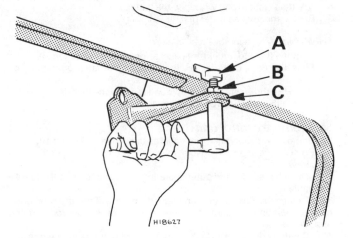

Fig. 11.5 Sliding door height adjustment at the upper guide support (Sec 13)

A Roller unit
B Lock nut
C Upper support

13 Sliding side door – removal and refitting

1 Undo the retaining screw and remove the end stop from the central track (photo).
2 Slide the door open and unscrew the Torx screws securing the door lower guide support (photo).
3 Enlist the aid of an assistant to support the weight of the door on the centre rail, then remove the Torx screws securing the upper guide support (photo).
4 Support the door at each end, slide it to the rear and remove it from the vehicle.

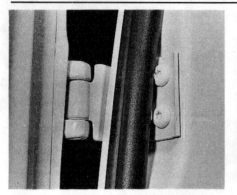  

15.3A Rear door hinge – attachment screws to door

15.3B Rear door hinge – attachment nuts to body

15.4 Rear door check strap

5 Refitting is a reversal of the remove procedure. Align the door and engage it onto the centre track, then reconnect the fittings.
6 When fitted, check the door for satisfactory flush fitting adjustment. Adjust if necessary by loosening off the lower support adjuster screws to reposition the door as required, then tighten them and recheck the fitting (photo). To adjust the height, loosen off the upper support locknut and turn the adjuster bolt as required then retighten the locknut to secure (Fig. 11.5).
7 When fitted, and in the closed position, the door should be aligned flush to the surrounding body and should close securely. If required, adjust the striker plate position to suit (photo).

14 Sliding side door fittings – removal and refitting

Handle
1 Remove the door inner trim as described in Section 10.
2 Detach the outer handle operating rod, then undo the retaining nuts, remove the washers and withdraw the handle.
3 Refit in the reverse order of removal.

Lock unit
4 Remove the door inner trim panel as described in Section 10.
5 Slide the door partly open, then detach the latch operating rod (photo).
6 Undo the three retaining screws and remove the lock unit.
7 Refit in the reverse order of removal.

Lock remote control unit
8 Remove the door inner trim panel as described in Section 10.
9 Remove the caps and unscrew the door latch release handle retaining screws.
10 Detach the inner and outer operating cable and remove the handle and cable.
11 Detach the lock barrel operating rod. Loosen off the latch and outer handle operating rod screws, release the shorter handle rod clip and remove the rod.
12 Carefully drill out the lock remote control unit retaining rivets and remove the unit from the door.
13 Refit in the reverse order of removal. When offering the unit to the door, check that the latch rod fits into the lower retaining boss and engage the outer handle rod. Align the unit with the rivet holes and fit new rivets.
14 Adjust the latch and lock rods to suit, then fit the inner handle cable and the lock barrel rod. Refit the trim.

Guide rails
Upper rail
15 Open the door and then undo the retaining screws and detach the upper guide support. Drill out the pop rivets and remove the rail. Refit in the reverse order of removal. Adjust the upper guide rail to suit, then tighten the screws.

Centre rail
16 Remove the side door as described in Section 13.
17 Undo the retaining screw at the leading edge and the stud nuts on the inside then remove the rail (photos). Note that the rear nut is an expansion nut which also secures the end (stopper)cap.
18 Refit in the reverse order of removal.
Lower rail
19 Open the side door, remove the circular cover from the step and unscrew the two retaining screws (photos).
20 Undo the two screws and detach the lower guide support (photo).
21 Undo the retaining screws and remove the lower rail complete with the lower latch unit.
22 Refit in the reverse order of removal. Check that the door operates smoothly and closes securely.

15 Rear doors – removal and refitting

1 Open the rear doors. If removing the left-hand rear door, remove its inner trim (Section 10).
2 Detach the check strap from the door. Disconnect the wiring from the appropriate fitting(s) in the door (as applicable), and withdraw the loom from the door.
3 Mark around the periphery of each door hinge with a suitable marker pen to show the fitting position of the hinges when refitting the door. Get an assistant to support the door, undo the retaining screws/nuts from each hinge and withdraw the door (photos).
4 Refit in the reverse order of removal. Align the hinges with the previously made marks, then tighten the bolts. Ensure that the check strap is central with the door when reconnected (photo).

16 Rear door fittings – removal and refitting

1 In most instances, the door trim panel will need to be removed to provide access to the item concerned. Refer to Section 10 for details.

Lock barrel
2 Detach the lock lever rod, remove the barrel retaining plate and then withdraw the lock barrel.
3 Refitting is a reversal of the removal procedure.

Latch unit
4 Undo the two retaining bolts, detach the connecting rod and remove the latch. Refit in the reverse order (photo).

Lock and rod unit
5 Hold the upper connecting rod so that the clips can be detached from their rods and the rods disconnected from the lock unit.
6 Undo the three retaining screws and remove the lock unit.

Chapter 11 Bodywork   221

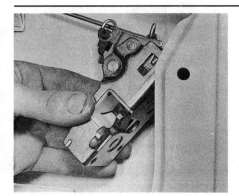
16.4 Rear door latch unit removal

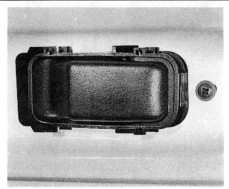
16.8 Rear door inner lock release and retaining screw

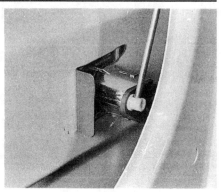
16.10A Rear door lock barrel and retaining clip

16.10B Rear door lock barrel removal from door

16.10C Rear door lock barrel unit and retaining clip

16.10D Rear door release handle, latch and lock unit showing connecting rods and securing screws

16.12 Rear door release handle

16.13A Rear door striker rod (upper)

16.13B Rear door striker rod (lower)

7 Refit in the reverse order of removal.

Lock release and lock unit
8 Undo the retaining screw and then partially withdrawing the release handle, twist it to detach the connecting rod (photo).
9 Disconnect the operating rods from the outer handle and lock barrel, and the upper latch rod from the latch. Undo the three retaining screws and remove the latch unit.
10 If required, the lock barrel can be removed by prising free the retaining clip and withdrawing the lock barrel (photos).
11 Refit in the reverse order of removal. Check the operation of the lock and latch before refitting the inner trim panel.

Inner release handle and striker assemblies
12 To remove the handle, undo the Torx screws and detach the striker rods (upper and lower) (photo).
13 The striker rod guide plates at the upper and lower corners can be removed by unscrewing their retaining screws, as can the striker plates (photos).
14 Refit in the reverse order of removal, but adjust the strikers to suit.

17 Tailgate fittings – removal and refitting

Outer handle
1 Disconnect the latch cover plate or remove the trim panel.
2 Remove the handle cover nuts indicated in Fig. 11.6. Detach the number plate lamp wiring and remove the cover.
3 Disconnect the operating rod from the handle, undo the two retaining screws and remove the outer handle.
4 Refitting is a reversal of the removal procedure.

16.13C Rear door striker guide plate (lower)

16.13D rear door upper striker wedge plate

16.13E Rear door lower striker wedge plate

19.1 Front door window regulator shown with trim panel removed

19.2 Door lock rod clips (arrowed) on vertical stay bar

19.4 Extension channel screw

19.5 Window regulator unit showing securing rivets (arrowed)

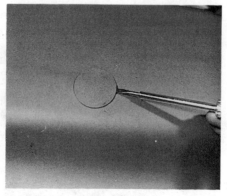
25.3A Prise free the plastic covers ...

25.3B ... for access to bumper screws

25.5 Rear bumper retaining nuts

27.2 Centre spindle nut

27.3 Plastic fastener removal

Chapter 11 Bodywork

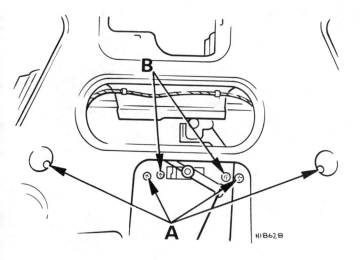

Fig. 11.6 Showing tailgate cover nuts (A) and handle screws (B) (Sec 17)

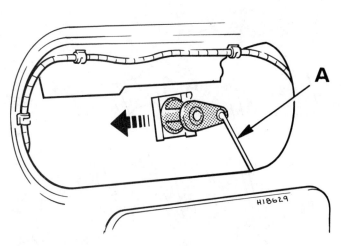

Fig. 11.7 Tailgate lock barrel showing latch connecting rod (A) and securing clip removal direction (arrowed) (Sec 17)

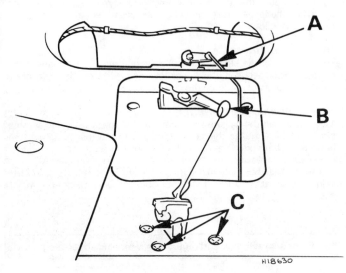

Fig. 11.8 Tailgate lock barrel rod (A) handle rod (B) and latch securing screws (C) (Sec 17)

Fig. 11.9 Supporting the tailgate during removal/refitting (Sec 18)

Lock barrel

5   Release the retainer and detach the lock operating rod. Prise free the U-shaped retaining clip and withdraw the lock barrel (Fig. 11.7).
6   Refit in the reverse order of removal, but ensure that the lock is fitted with the barrel drain hole facing down.

Latch unit

7   Remove the latch cover (three screws) or the trim panel from the tailgate.
8   Detach the latch operating rods, undo the three latch retaining screws and lock the latch rotor. Remove the latch and rods from the tailgate. If a new latch unit is being fitted, detach the operating rods from the old unit and fit them to the replacement latch. Note that although the retaining clip tags may break off during removal, the clips can still be used (Fig.11.8).
9   Refit in the reverse order of removal and check the operation of the latch before refitting the latch cover (or trim panel as applicable).

18 Tailgate – removal and refitting

1   The aid of two assistants will be required to support the tailgate as it is removed. First open the tailgate and detach the wiring to the number plate lamp at the multi-plug connector in the body. Pull the wiring loom through the body and leave it attached to the tailgate.
2   Loosen off the tailgate hinge bolts and get the two assistants to support the weight of the tailgate (Fig. 11.9).
3   Prise up the retaining clips securing the tailgate strut balljoints and detach the balljoint from the stud each side. Take care not to lift the clips by more than 4 mm (0.16 in)
4   Unscrew the hinge bolts and remove the tailgate (Fig. 11.10).
5   If required the weatherstrip can be pulled free from the tailgate.
6   To refit the weatherstrip, first ensure that the joint faces are clean

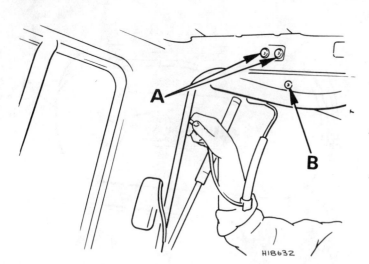

Fig. 11.10 Showing tailgate hinge to body bolts (A) and hinge to tailgate bolts (B). Pull the disconnected wiring loom through as shown (Sec 18)

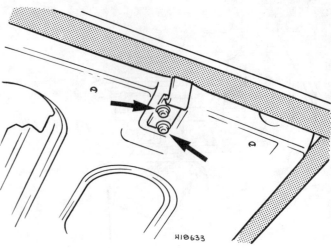

Fig. 11.11 Tailgate side bump adjustment and retaining screws (Sec 18)

balljoints onto their studs using only hand pressure. Note that the struts are gas filled type and therefore cannot be repaired. If renewing them, be sure to obtain the correct replacements (car types will not do).
9   When the tailgate is refitted check its adjustment and if necessary re-adjust as follows.

Height adjustment
10   Loosen off the hinge retaining bolts and reset the tailgate at the required height to suit the latch/striker engagement and the body aperture, then fully retighten the bolts.

Side clearance adjustment
11   Loosen off the tailgate side bump guides, the striker plate and the hinge bolts. Centralise the tailgate in its aperture then retighten the hinge bolts. If required, re-adjust the position of the striker plate so that the tailgate closes securely. Now adjust the position of the side bump guides so that they only just contact the D pillar bumpers when the tailgate is set at the safety catch position, and only make full contact when the tailgate is closed (Fig. 11.11).

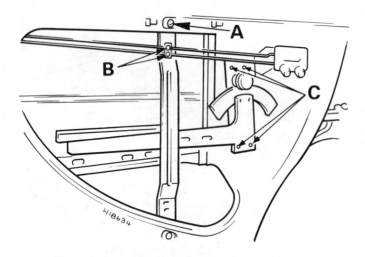

Fig. 11.12 General view of door window regulator components (Sec 19)

  A   Stay bar retaining bolt
  B   Lock rod clips
  C   Regulator unit retaining rivets

---

19   Front door window and regulator – removal and refitting

1   Raise the window then remove the door trim panel as described in Section 10. Peel back the insulation sheet (photo).
2   Detach the door lock rod clips from the vertical stay bar(photo).
3   Remove the upper stay bar bolt and pivot the bar towards the front of the door (Fig. 11.12).
4   Remove the screw securing the extension channel from the latch end of the door, then pull the extension channel from the fixed channel (photo).
5   Prevent the window from dropping by wedging it up with a suitable block of rubber (or get an assistant to support it), then drill out the four rivets securing the regular unit (4.5 mm drill). Withdraw the regulator unit from the door (photo).
6   Lower the door glass. To remove it, tilt it out at the top (with the lock rods behind it) then withdraw it from the door(Fig. 11.13).
7   If a new window is being fitted, locate the regulator channel so that it is 90 mm (3.5 in) from the front edge of the glass. Ease assembly of the channel by lubricating it with liquid soap or French chalk.
8   If required the door weather strip can be removed by gripping it at its top corner and pulling it free from the door. When refitting the channel first locate it in the front lower corner, then fit it progressively to the sides and top of the frame.

then feed the weatherstrip onto the tailgate. First locating it over each corner, then pressing it home at each centre point between the corners, and pushing it by thumb pressure towards the corners.
7   When the weatherstrip is fully located on the tailgate, engage the small sealing lip over the flange of the tailgate (if required). When the tailgate is refitted, it may be necessary to re-adjust the position of the striker plate to enable the tailgate to close correctly.
8   Refit the tailgate in the reverse order of removal. Press the strut

Chapter 11 Bodywork

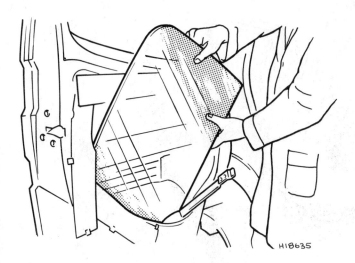

Fig. 11.13 Withdrawing the glass from the door (note stay bar position) (Sec 19)

Fig. 11.14 Withdrawing the door quarter window (Sec 20)

the weatherstrip over the window aperture flange. Apply a progressive and continuous pressure until the window and weatherstrip are fully engaged in the aperture at which point the cord will pull free (Fig. 11.15).

21 Front door sliding window glass and frame – removal and refitting

1   Slide the window open. Working from the inside, press the weatherstrip from the top corner aperture flange, then continue along the top of the aperture and down each side to free the strip (Fig. 11.16).
2   Get an assistant to support the glass from the outside, whilst you press it out from the inside.
3   A strong length of cord and two assistants will be required to refit the window. With the weatherstrip in position round the window, locate the cord into the groove of the weatherstrip so that the cord ends cross at one of the lower corners. Offer the assembly to the door from the outside, and pass the cord ends through the frame aperture. Get the two assistants to press firmly on the window from the outside whilst you pull on the cord at an angle of 90° to the glass so that the weatherstrip progressively unfolds over the aperture flange. As the cord moves around the aperture, the assistants should apply the pressure at the point adjacent to the cord as it unfurls the seal of the weatherstrip (Fig. 11.17).
4   When the glass is in position, tap the weatherstrip with the flat of the hand all round to seat the seal against the door panel.

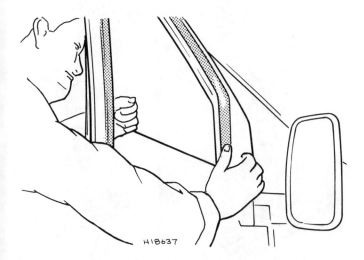

Fig. 11.15 Inserting the door quarter window using cord to locate the weatherstrip (Sec 20)

9   Refit the window and regulator in the reverse order of removal. Before refitting the door trim panel, raise and lower the window to ensure that it operates in a satisfactory manner.

20 Front door quarter glass – removal and refitting

1   Remove the door mirror trim as described in Section 28.
2   Press the glass outwards using firm hand pressure whilst simultaneously pulling free the rubber weatherstrip from the top corner (Fig. 11.14). As the glass is extracted from the door, push it firmly, in a progressive manner, clear along its edges from the inside out until finally it can be removed.
3   To refit the quarter window, first loop a length of strong cord into the weatherstrip groove so that the cord ends are at the lower corners. Passing the cord through the aperture of the window, locate the lower edges of the weatherstrip over the flange of the aperture, press the glass inwards and simultaneously pull the cord to progressively locate

22 Front door sliding window glass and frame – dismantling and reassembly

1   With the glass and frame removed from the vehicle (Section 21), pull free the weatherstrip and lay the unit on a cloth covered work area.
2   Undo the two retaining screws from one side of the joint frame, then prise the frame apart using a suitable screwdriver inserted between the joints (Fig. 11.18).
3   Withdraw the fixed glass from the frame by pulling its top edge. If necessary, cut the silicone seal using a suitable knife to release the glass from the frame.
4   Depress the window catch, undo the pawl retaining screw and remove the pawl, catch, button and spring (Fig. 11.19).
5   Undo the two screws and twist free the catch cover. Remove the threaded plate, rubber gasket and O-rings and keep them somewhere safe. Carefully drive the two plastic guides from the runner using a suitable screwdriver, then pivot the top of the sliding window outwards (window runner disengaged) and withdraw the glass.
6   Withdraw the silent channel from the frame by pulling it free.
7   Undo the retaining screw and remove the end seal and block.

Fig. 11.16 Releasing the window weatherstrip from the flange (Sec 21)

Fig. 11.17 Installing the door sliding windows assembly (Sec 21)

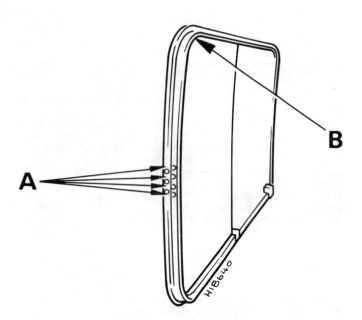

Fig. 11.18 Door sliding window unit showing joint screws (A) and seal block screw (B) (Sec 22)

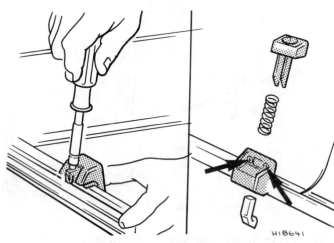

Fig. 11.19 Remove screw to dismantle window catch pawl and cover assembly (Sec 22)

8 Clean all of the old sealant from the fixed glass, renew any parts as necessary and have some clear silicone sealant at hand during the reassembly.

9 Reassembly is a reversal of the removal procedure but note the following points:

  (a) Clean the frame and glass of grease and oil sealant with methylated spirit prior to fitting. Apply a thin bead of clear sealant around the edges of the fixed glass before inserting it into position
  (b) When fitting the window catch, ensure that the O-rings and the rubber gasket are correctly fitted
  (c) When refitting the weatherstrip, align the drain holes in the strip with the corresponding holes in the frame

23 Opening rear quarter window – removal and refitting

1 Unscrew and remove the two catch-to-D pillar retaining screws, partly open the window and then pull on the glass to detach the hinge leaves and withdraw the glass (Fig. 11.21).

2 The weatherstrip can be pulled free from the frame flange if required. The catch and studs can be removed from the glass by detaching the five screw caps then undoing the special retaining nuts.

3 Refitting is a reversal of the removal procedure but note the following special points:

  (a) Ensure that the weatherstrip and frame flange joint surfaces are clean. When fitting the weatherstrip, ensure that the drain holes align
  (b) Use new rubber seal washers when reassembling the catch and hinge studs

Chapter 11 Bodywork

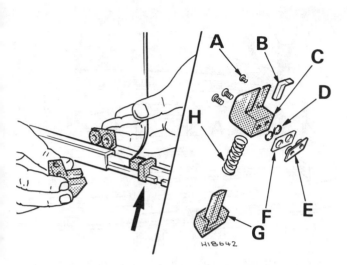

Fig. 11.20 Sliding window catch and runner guide components (Sec 22)

A Pawl screw
B Pawl
C Catch
D O-ring
E Plate
F Seal
G Button
H Spring

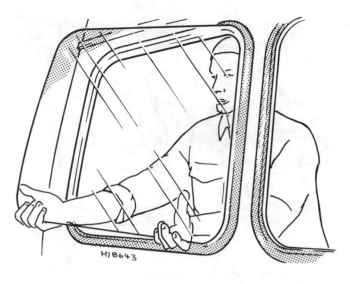

Fig. 11.21 Removing the opening rear quarter window (Sec 23)

24 Windscreen and fixed windows – removal and refitting

The windscreen, tailgate and fixed side windows are direct glazed to the body using special adhesive. Purpose made tools are required to remove the old glass and fit the new glass, therefore this work is best entrusted to a specialist.

25 Bumpers – removal and refitting

Front bumper
1 Unscrew and remove the single Torx bolt retaining the bumper end under each wheel arch (Fig. 11.22).
2 Detach and remove the front number plate from the bumper.
3 Prise free the plastic retaining screw covers, then unscrew the bumper retaining screws (photos). Withdraw the front bumper.
4 Refit in the reverse order of removal. Align the bumper correctly before fully tightening the retaining screws.

Rear bumper
5 Working from the underside of the vehicle at the rear, unscrew and remove the two bumper retaining nuts and bolts each side (photo).
6 Give the plastic end caps a sharp pull to disengage them from the retainers. Withdraw the bumper rearwards from the vehicle.

Note: *If a tow bar is fitted, its wiring harness will need to be detached before removing the bumper.*

7 If required, the plastic end caps can be removed from the bumper by detaching the retaining clips, then sliding the caps free.
8 Refit in the reverse order of removal. Align the bumper correctly before fully tightening the retaining bolts and nuts.

26 Front grille – removal and refitting

1 Open the bonnet and support it in the raised position.
2 Unscrew and remove the retaining screw at each end of the grille.

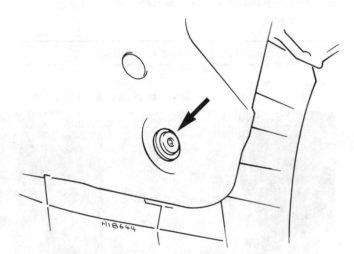

Fig. 11.22 Front bumper retaining bolt under the wheel arch (arrowed) (Sec 25)

3 Detach the headlamp washer hose (if fitted) from the T-piece connector and plug it to prevent fluid leakage, or tie it up above the reservoir fluid level.
4 Prise free the caps from the five fasteners, then turn the fasteners a quarter of a turn to release them and withdraw the grille panel (Fig. 11.23).
5 Refit in the reverse order of removal. Insert the fasteners by simply pushing them into position, then fit their caps.

27 Windscreen grille – removal and refitting

1 Remove the windscreen wiper arms and blades as described in Chapter 12.
2 Unscrew and remove the 32 mm nut from the centre spindle, then open and support the bonnet (photo).
3 Extract the plastic fastener and retaining screw from the cowl end pieces and remove them (photo).
4 Detach the remaining fasteners, withdraw the grille panel and

Chapter 11 Bodywork

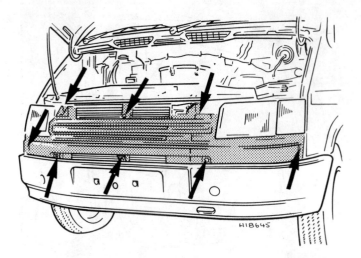

Fig. 11.23 Front grille panel retaining screw locations (arrowed) (Sec 26)

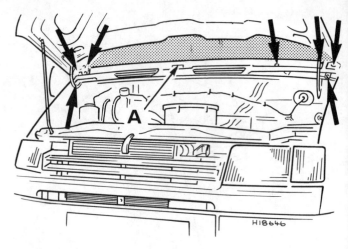

Fig. 11.24 Windscreen grille (cowl) retainers (arrowed) and 32 mm nut (A) (Sec 27)

disconnect the washer hoses (Fig. 11.24).
5 Refit in the reverse order of removal. Check the windscreen washers for satisfactory operation on completion.

28 Door mirror and glass – removal and refitting

Glass renewal
1 If the mirror glass is to be renewed, carefully prise free the outer trim

retainer using a suitable screwdriver as a lever (photo). Remove the retainer and glass (photo).
2 To refit the glass, insert it into the retainer then carefully press the rim of the retainer into position around the rim of the mirror.

Mirror assembly
3 Prise free the plastic caps from the mirror trim fasteners, then unscrew and remove the four fasteners (photos). Lift and remove the trim.
4 Undo the two Torx retaining screws and remove the mirror (photo).
5 Refit in the reverse order of removal.

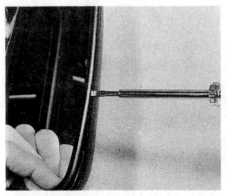

28.1A Prise free and ...

28.1B ... remove the outer trim from the mirror

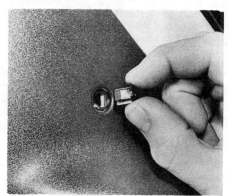

28.3A Extract the plastic caps ...

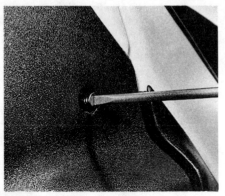

28.3B ... for access to the screw fasteners

28.4 Rear view mirror retaining screws

Chapter 11 Bodywork 229

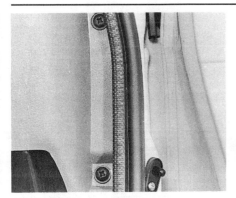
29.8 Facia side mounting screws

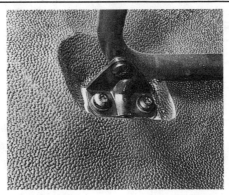
31.5A Bench seat front mounting bracket to floor

31.5B Bench seat rear mounting bracket to floor

31.6 Bench seat restrainer attachment to floor

32.3 Remove trim cover for access to upper anchor bolt

32.6 Seat belt to floor fixings

29 Facia – removal and refitting

1 Disconnect the battery earth lead.
2 Remove the steering wheel as described in Chapter 10.
3 Remove the steering column switches as described in Chapter 12.
4 Undo the four retaining screws and remove the instrument panel surround trim. Prise free and remove the trim switches.
5 Remove the instrument panel as described in Chapter 12.
6 Undo the four screws retaining the heater facia, detach the switch wiring plug and remove the panel.
7 Undo the two retaining nuts and withdraw the heater control panel. Detach the heater warm air ducts.
8 Release the two quarter-turn fasteners in the glove box (Fig.11.25). Unscrew the four facia retaining screws (two each side)(photo) and then release the three quarter-turn fasteners from the facia top edge. Withdraw the facia unit from the vehicle(Fig. 11.26).
9 Refit in the reverse order of removal. Refer to the respective Chapters concerned for the relevant refitting details. Ensure that all wiring connections are securely made. Check the operation of the various switches, instruments and controls on completion.

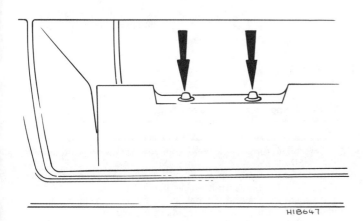

Fig. 11.25 Fastener locations (arrowed) in the glovebox (Sec 29)

Fig. 11.26 Facia fastener location points (arrowed) (Sec 29)

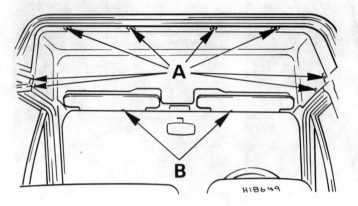

Fig. 11.27 Front heading Securing clips (A) and sun visors (B) (Sec 30)

Fig. 11.28 Front bucket seat retaining nuts (Sec 31)

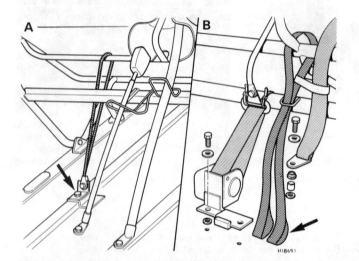

Fig. 11.29 Seat restrainer strap types used (Sec 31)

A Double chassis cab strap fixing
B Single chassis cab strap fixing

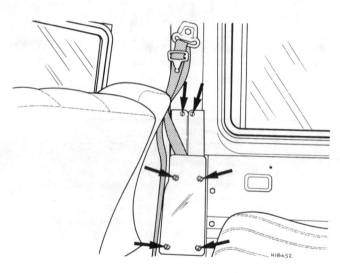

Fig. 11.30 B pillar trim fixing points (Sec 32)

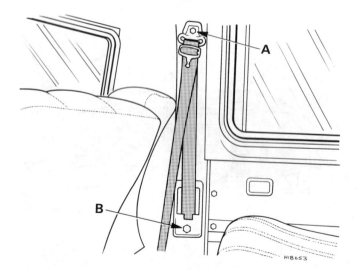

Fig. 11.31 Seat belt upper anchor bolt (A) and lower retractor bolt (B) (Sec 32)

30 Headlining – removal and refitting

1 Unscrew the retaining screws and pull free the A pillar trims.
2 Detach and remove the sun visors. Also remove the interior lamp unit (Chapter 12).
3 Get an assistant to support the headlining, then undo the retaining screw clips and remove the headlining from the vehicle.
4 If required, the centre and rear headlining sections can be detached and removed in the same manner (where applicable).
5 To refit the headlining, locate it in position and secure it at the rear with screw clips (push them into position), feed the interior lamp wires through the headlining at the front and then secure it in position along with the front and sides.
6 Refit the interior lamp and the A pillar trims to complete.

Chapter 11 Bodywork

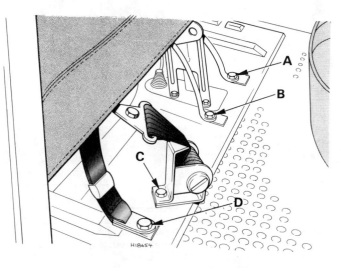

Fig. 11.32 Seat floor fixings (Sec 32)

A Lap belt buckle
B Belt buckle (right-hand)
C Belt buckle (left-hand)
D Lap belt

31 Seats – removal and refitting

Front bucket seat

1 Move the seat to full forward position and then unscrew and remove the two Torx bolts from the seat track.
2 Move the seat to the rearmost setting then unscrew and remove the four retaining bolts (Torx type).
3 Undo the four retaining nuts indicated in Fig. 11.28, and lift out the seat.
4 Refit in the reverse order of removal. Tighten the bolts to the specified torque wrench setting.

Front bench seat

5 Unscrew and remove the four, six or seven (depending on model) bolts which secure the seat to the floor (photos).
6 Where applicable, cut free the plastic outer seat belt buckle to frame ties, and detach the seat restrainer strap (Fig. 11.29)(photo).
7 Remove the seat, passing the seat belt webbing and stalk through as it is withdrawn.
8 Refit in the reverse order of removal. Tighten the retaining bolts to the specified torque wrench setting.

32 Seat belt and stalk – removal and refitting

1 Undo the six retaining screws and remove the B pillar trim, passing the webbing through the trim slot as it is withdrawn (Fig. 11.30).
2 Unscrew the lower anchor bolt but take care not to allow the webbing to retract into the reel.
3 Prise free the cover from the upper anchor bolt, undo the bolt (photo).
4 Unscrew and remove the lower retaining bolt (Fig. 11.31).
5 Pivot the retractor unit to the side and remove it from the B pillar.
6 Remove the stalk to floor bolts to free the stalks, then where applicable, cut free the two tie-straps securing the outer belt buckle. Undo the retaining bolts and remove the seat belts (photo).
7 Renew the belts if they are worn or malfunction.
8 Refit in the reverse order of removal. When refitting the inertia reel, ensure that the tag engages correctly in the B pillar.
9 Note that the centre lap buckle tongue differs, and is connected to the floor stalk near the driver's seat base, the driver's belt stalk being on the inner side of the floor under the passenger seat.
10 Tighten the retaining bolts to the specified torque wrench settings.

Chapter 12 Electrical system

Contents

Alternator – fault finding and testing	7
Alternator – maintenance and special precautions	5
Alternator – removal and refitting	6
Alternator brushes – removal, inspection and refitting	9
Alternator overhaul – general	8
Battery – charging	3
Battery – maintenance	2
Battery – removal and refitting	4
Brake stop-light switch – removal and refitting	20
Bulbs and lamp units – removal and refitting	29
Cigarette lighter – removal and refitting	26
Courtesy light switch – removal and refitting	17
Direction indicator switch – removal and refitting	14
Fault diagnosis – electrical system	42
Fuses and relays – general	13
General description	1
Headlamps – alignment	27
Headlamps and headlamp bulbs – removal and refitting	28
Heater motor switch – removal and refitting	21
Horn – removal and refitting	30
Ignition switch and lock barrel – removal and refitting	16
Instrument panel – removal and refitting	22
Instrument panel components – removal and refitting	23
Instrument panel surround facia and components (tachograph models) – removal and refitting	24
Instrument panel surround switches – removal and refitting	18
Lighting switch – removal and refitting	15
Radio aerials and speakers (standard) – removal and refitting	40
Radio (digital) – removal and refitting	39
Radio (standard) – removal and refitting	38
Reversing light switch – removal and refitting	19
Speedometer cable – removal and refitting	25
Starter motor – overhaul	12
Starter motor – removal and refitting	10
Starter motor – testing in the vehicle	11
Tailgate wiper components – removal and refitting	36
Washer nozzles and hoses – removal and refitting	37
Windscreen/headlamp washer reservoir – removal and refitting	35
Windscreen wiper linkages – removal and refitting	34
Windscreen wiper motor – removal and refitting	33
Wiper arms – removal and refitting	32
Wiper blades – renewal	31
Wiring diagrams – notes for guidance	41

Specifications

General

System type	12 volt, negative earth
Battery type	Lead acid, maintenance-free
Battery capacity	360A/60RC (standard) or 500A/75RC (heavy duty)

Alternator

Make and type	Bosch K1-55A or Lucas A127/55
Nominal output	55 amp
Minimum allowable length of slip ring end brushes	5 mm (0.197 in)
Regulating voltage at 4000rpm (engine speed) and 3 to 7 amp load	13.7 to 14.6 volts
Drivebelt tension	See Chapter 2

Starter motor

Make	Lucas or Bosch
Type:	
Lucas	5M90, 2M100 or M127
Bosch	Long frame 0.80 kW or Long frame 1.1 kW or JF 2.7 kW
Number of brushes (all types)	4
Rotation	Clockwise
Minimum allowable brush length:	
Lucas 5M90 and 2M100	8.0 mm (0.32 in)
Lucas M127	12.0 mm (0.47 in)
Bosch (all models)	10.0 mm (0.39 in)
Minimum thickness of commutator:	
Lucas 5M90 and 2M100	2.05 mm (0.08 in) (segment thickness)
Lucas M127	38 mm (1.5 in) (diameter)
Bosch (all models)	32.8 mm (1.29 in) (diameter)
Armature end float:	
Lucas 5M90 and 2M100	0.25 mm (0.01 in)
Lucas M127	0.6 mm (0.02 in)
Bosch (all models)	0.3 mm (0.01 in)

Chapter 12 Electrical system

233

Bulbs

	Wattage
Headlamp (semi-sealed beam) (halogen)	55/60
Sidelamps	5
Direction indicators	21
Brake lamps (load platform chassis)	21
Tail lamps (load platform chassis)	5
Brake/tail lamps (van, bus)	21/5
Reverse lamp (load platform chassis)	21
Reverse lamp (van, bus)	21
Rear foglamp	21
Number plate lamp (load platform chassis)	10
Number plate lamp (van, bus)	5
Interior lamp	10

Fuses

Fuse number	Rating/colour	Circuits protected
1	25 amp/neutral	Headlamp flasher, heater blower motor
2	10 amp/red	Direction indicators
3	15 amp/blue	Windscreen wiper motor and washer pump, (non-headlamp washer models)
4	10 amp/red	Reverse lamps, stop-lamps and heated rear window
5	10 amp/red	Fuel and temperature gauges, ignition warning lamp, oil pressure warning lamp and low brake fluid level warning lamp
6	15 amp/blue	Tailgate wiper motor and washer pump
7	15 amp/blue	Dim-dip lighting
8 and 9		Spares
10	10 amp/red	Right-hand side and tail lamps, instrument panel lamps
11	10 amp/red	Left-hand side and tail lamps, number plate lamp
12	10 amp/red	Right-hand dip beam
13	10 amp/red	Left-hand dip beam
14	10 amp/red	Rear foglamps, headlamp power washer timer
15		Spare
16	10 amp/red	Right-hand main beam and main beam warning lamp
17	10 amp/red	Left-hand main beam
18	10 amp/red	Front courtesy lamp, rear interior lamps, clock/tachograph, radio, windscreen washer pump (models with headlamp wash)
19	20 amp/yellow	Hazard warning lamps
20	25 amp/neutral	Headlamp power washer, cigarette lighter
21	20 amp/yellow	Heated rear window
22		Spare
23		Positions for five spare fuses

Wiper blades	Champion C55-01

Relays in main fuse box

Identification	Function
A	Automatic transmission inhibitor
B	Not used
C	Dim-dip lighting
D	Dim-dip lighting
E	Not used
F	Heated rear window
G	Direction indicator flasher (Heavy duty type if trailer towed)
H	Headlamp power washer
I	Intermittent windscreen wiper
J	Intermittent windscreen wiper
K	Not used

Torque wrench settings

	Nm	lbf ft
Number plate lamp to lift type rear door	4.5 to 6.7	3.3 to 4.9
Wiper motor driving lever	9 to 11	6.6 to 8.1
Delco wiper pivot housing	8.5 to 12	6.2 to 8.8
Wiper housing to body nut (metal)	2.5 to 3.5	1.8 to 2.5
Wiper housing to plastic cover nut (plastic)	15 to 20	11 to 15

1 General description

The electrical system is of 12 volt negative earth type. The battery is charged by a belt-driven alternator which incorporates a voltage regulator. The starter motor is of pre-engaged type where a solenoid moves the drive pinion into engagement with the ring gear before the starter motor is energised.

An alternator of either Lucas or Bosch manufacture is fitted. The output is controlled by an integral regulator, the maximum output being given in the Specifications.

The alternator is mounted towards the front of the engine and is driven by the V-belt which also serves to drive the water pump, the drive being taken from the crankshaft pulley.

234   **Chapter 12   Electrical system**

Although repair procedures are given for some components in this Chapter, it may well be more economic to renew worn or defective components as a complete unit.

2   Battery – maintenance

1   At the intervals given in *Routine maintenance* at the beginning of this manual disconnect the leads (negative first) from the battery and clean the terminals and lead ends. After refitting the leads (negative last) smear the exposed metal with petroleum jelly.

2   The battery fitted as standard equipment is of the maintenance-free type in which the electrolyte level does not have to be checked. The battery condition can be checked using a DC voltmeter with a 6 to 20 volt range. Before connecting the voltmeter to the battery, stabilize its voltage by turning on the headlamps for a period of 30 seconds, switch them off and then wait a further 30 seconds before attaching the voltmeter. The voltage reading of a battery in a satisfactory condition is 12.65 volts. If required the battery can be charged up in the normal manner used for recharging a conventional battery.

3   If a non-standard conventional battery has been fitted, the following checks should be made at periodic intervals.

4   Check that the plate separators inside the battery are covered with electrolyte. To do this remove the battery covers and inspect through the top of the battery. On batteries with a translucent case it may be possible to carry out the check without removing the covers. If necessary top up the cells with distilled or de-ionized water.

5   At the same time wipe clean the top of the battery with a dry cloth to prevent the accumulation of dust and dampness which may cause the battery to become partially discharged over a period.

6   Also check the battery clamp and platform for corrosion. If evident remove the battery and clean the deposits away. Then treat the affected metal with a proprietary anti-rust liquid and paint with the original colour.

7   Whenever the battery is removed it is worthwhile checking it for cracks and leakage. Cracks can be caused by topping up the cells with distilled water in winter *after* instead of *before* a run. This gives the water no chance to mix with the electrolyte, so the former freezes and splits the battery case. If the case is fractured, it may be possible to repair it with a proprietary compound, but this depends on the material used for the case.

8   If topping up the battery becomes excessive and the case is not fractured, the battery is being over-charged and the voltage regulator may be faulty.

9   If the vehicle covers a small annual mileage it is worthwhile checking the specific gravity of the electrolyte every three months to determine the state of charge of the battery. Use a hydrometer to make the check and compare the results with the following table:

	Ambient temperature above 25°C (77°F)	Ambient temperature below 25°C (77°F)
Fully charged.........	1.210 to 1.230	1.270 to 1.290
70% charged.........	1.170 to 1.190	1.230 to 1.250
Fully discharged ...	1.050 to 1.070	1.110 to 1.139

Note that the specific gravity readings assume an electrolyte temperature of 15°C (60°F); for every 10°C (18°F) below 15°C (60°F) subtract 0.007. For every 10°C (18°F) above 15°C (60°F) add 0.007.

10   If the battery condition is suspect first check the specific gravity of electrolyte in each cell. A variation of 0.040 or more between any cells indicates loss of electrolyte or deterioration of the internal plates.

11   A further test can be made by checking the battery voltage as described in paragraph 2.

3   Battery – charging

1   In winter time when heavy demand is placed upon the battery, such as when starting from cold and much electrical equipment is continually in use, it is a good idea occasionally to have the battery fully charged

from an external source at the rate of 3.5 to 4 amps.

2   The battery can be charged in position in the vehicle, but first disconnect the negative lead then the positive lead from the battery.

3   Connect and operate the charger in accordance with the makers' instructions as to charge duration.

4   Alternatively, a trickle charger charging at the rate of 1.5 amps can be safely used overnight.

5   Specially rapid 'boost' charges which are claimed to restore the power of the battery in 1 to 2 hours are not recommended as they can cause serious damage to the battery plates through overheating.

6   While charging the battery note that the temperature of the electrolyte should never exceed 100°F (37.8°C).

7   Always switch the charger off before disconnecting the leads.

8   When reconnecting the vehicle battery leads, reconnect the positive lead first, then the negative lead.

4   Battery – removal and refitting

1   The battery is located in the engine compartment on the left-hand side.

2   Note the location of the leads then unscrew the nut and disconnect the negative lead.

3   Lift the plastic cover (if fitted) then unscrew the nut and disconnect the positive lead.

4   Unscrew the clamp bolt and lift the battery from the platform, taking care not to spill any electrolyte on the bodywork (photo).

5   Refitting is a reversal of removal. Refit the leads in the order positive first and negative last.

5   Alternator – maintenance and special precautions

1   Periodically wipe away any dirt which has accumulated on the outside of the unit, and also check that the plug is pushed firmly on the terminals. At the same time check the tension of the drivebelt and adjust it if necessary as described in Chapter 2.

2   Take extreme care when making electrical circuit connections, otherwise damage may occur to the alternator or other electrical components employing semi-conductors. Always make sure that the battery leads are connected to the correct terminals. Before using electric-arc welding equipment to repair any part of the vehicle, disconnect the battery leads and the alternator multi-plug. Disconnect the battery leads before using a mains charger. Never run the alternator with the multi-plug or a battery lead disconnected.

6   Alternator – removal and refitting

1   Disconnect the battery negative lead.

2   Loosen the alternator mounting and adjustment nuts and bolts, then swivel the alternator in towards the cylinder block (Fig. 12.1).

3   Slip the drivebelt(s) from the alternator pulley(s).

4   Pull the multi-plug from the rear of the alternator (photo).

5   Remove the mounting and adjustment nuts and bolts, and withdraw the alternator from the engine.

6   Refitting is a reversal of removal, but tension the drivebelt(s) as described in Chapter 2. Where applicable the front mounting bolt should be tightened before the rear mounting bolt, as the rear mounting incorporates a sliding bush.

Chapter 12 Electrical system

235

4.4 Battery location showing lead connections and the clamp bolt

6.4 Rear face of the Bosch alternator showing multi-plug connector point and securing clip (arrowed)

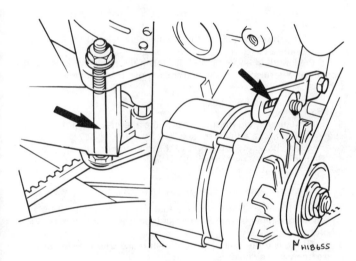

Fig. 12.1 Alternator mounting and adjuster bolts (arrowed) (Sec 6)

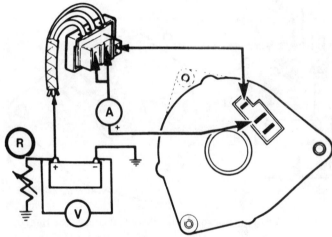

Fig. 12.2 Alternator output test circuit (Sec 7)

7 Alternator – fault finding and testing

Note: *To carry out the complete test procedure use only the following test equipment – a 0 to 20 volt moving coil voltmeter, a 0 to 100 amp moving coil ammeter, and a rheostat rated at 30 amp.*

1 Check that the battery is at least 70% charged by using a hydrometer as described in Section 2.
2 Check the drivebelt tension with reference to Chapter 2.
3 Check the security of the battery leads, alternator multi-plug and interconnecting wire.
4 *To check the cable continuity* pull the multi-plug from the alternator and switch on the ignition being careful not to crank the engine. Connect the voltmeter between a good earth and each of the terminals in the multi-plug in turn. If battery voltage is not indicated, there is an open circuit in the wiring which may be due to a blown ignition warning light bulb if on the small terminal.
5 *To check the alternator output* connect the voltmeter, ammeter and rheostat as shown in Fig. 12.2. Run the engine at 3000 rpm and switch on the headlamps, heater blower and, where fitted, the heated rear window. Vary the resistance to increase the current and check that the alternator rated output is reached without the voltage dropping below 13 volts.
6 *To check the positive side of the charging circuit* connect the

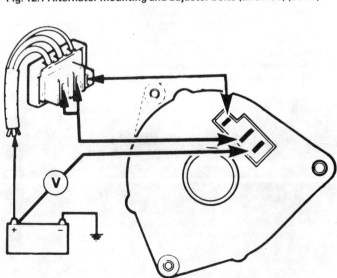

Fig. 12.3 Alternator positive check circuit (Sec 7)

Chapter 12 Electrical system

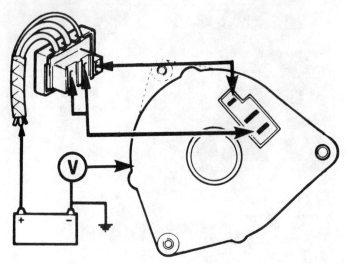

Fig. 12.4 Alternator negative check circuit (Sec 7)

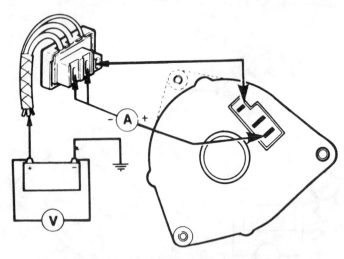

Fig. 12.5 Alternator voltage regulator test circuit (Sec 7)

voltmeter as shown in Fig. 12.3. Start the engine and switch on the headlamps. Run the engine at 3000 rpm and check that the indicated voltage drop does not exceed 0.5 volt. A higher reading indicates a high resistance such as a dirty connection on the positive side of the charging circuit.

7 *To check the negative side of the charging circuit* connect the voltmeter as shown in Fig. 12.4. Start the engine and switch on the headlamps. Run the engine at 3000 rpm and check that the indicated voltage drop does not exceed 0.25 volt. A higher reading indicates a high resistance such as a dirty connection on the negative side of the charging circuit.

8 *To check the alternator voltage regulator* connect the voltmeter and ammeter as shown in Fig. 12.5. Run the engine at 3000 rpm and when the ammeter records a current of 3 to 5 amps check that the voltmeter records 13.7 to 14.5 volts. If the result is outside the limits the regulator is faulty.

8 Alternator overhaul – general

1 Such is the inherent reliability of modern alternators that apart from possible renewal of the brushes, any other form of repair or overhaul is seldom required.
2 Renewal of the brushes for both the Bosch and Lucas units is described in the following Section. Should any other repair be necessary it is recommended that the advice of an auto electrician be sought. Alternatively, an exchange reconditioned unit can be purchased which is likely to be the most economical solution in the long run.

9 Alternator brushes – removal, inspection and refitting

1 Disconnect the battery negative lead.

Bosch type
2 Remove the screws and withdraw the regulator and brush box from the rear of the alternator.
3 If the length of either brush is less than the minimum given in the Specifications, unsolder the wiring and remove the brushes and springs (photos).

Lucas type
4 Pull the multi-plug from the rear of the alternator then remove the screws and withdraw the rear cover.
5 Remove the screws and remove the regulator/brush box unit.
6 Renew the brushes if either is less than the minimum length given in the Specifications.

Both types
7 Wipe clean the slip rings with a fuel-moistened cloth – if they are very dirty use fine glasspaper to clean them then wipe with the cloth (photo).
8 Refitting is a reversal of removal, but make sure that the brushes move freely in their holders.

10 Starter motor – removal and refitting

1 Jack up the front of the vehicle and support on axle stands. Apply the handbrake.
2 Disconnect the battery negative lead.
3 Disconnect the starter motor wiring from the motor unit (photo).

9.3A Alternator regulator and brush box (Bosch)

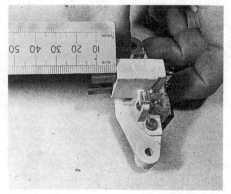

9.3B Checking the length of the alternator brushes (Bosch)

9.7 Alternator slip rings (arrowed) (Bosch)

237

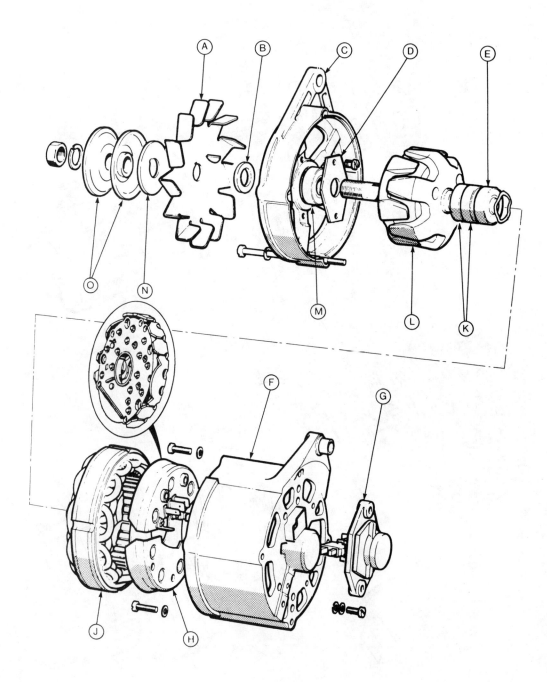

Fig. 12.6 Exploded view of the Bosch alternator (Sec 9)

A Fan
B Spacer
C Drive end housing
D Drive end bearing retaining plate
E Slip ring end bearing
F Slip ring end housing
G Brush box and regulator
H Rectifier (diode) pack (Inset shows N1-70A diode pack)
J Stator
K Slip rings
L Rotor
M Drive end bearing
N Spacer
O Pulley

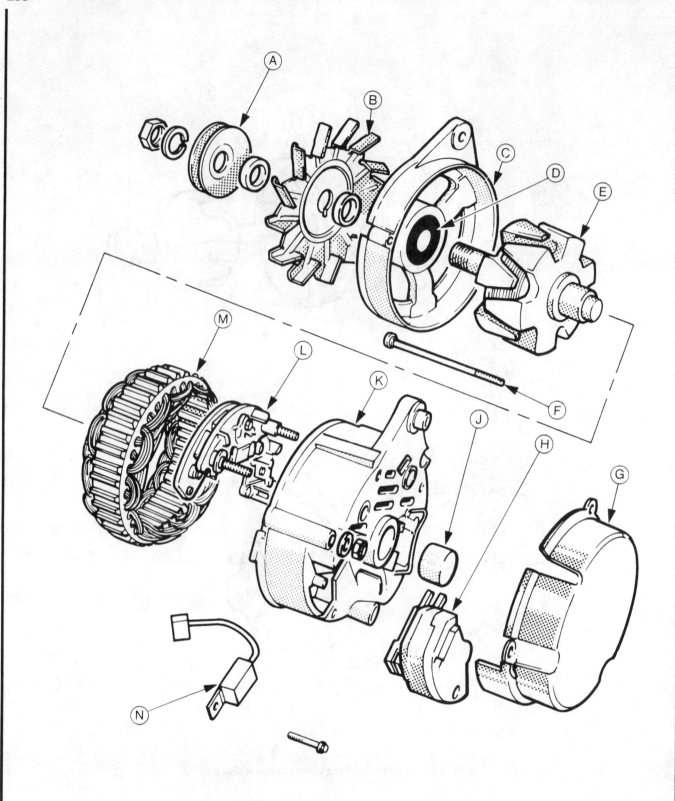

Fig. 12.7 Exploded view of the Lucas alternator (Sec 9)

A Pulley
B Fan
C Drive end housing
D Drive end bearing
E Rotor
F Through bolt
G End cover
H Regulator/brushbox
J Slip ring end bearing
K Slip ring end housing
L Rectifier unit
M Stator
N Suppressor

Chapter 12 Electrical system

10.3 Starter motor and wiring connections

10.4A Starter motor viewed from underneath showing mounting bolts

10.4B Starter motor unit removed from vehicle

4 Unscrew and remove the starter motor retaining bolts from the transmission housing and where applicable, the mounting bracket to the engine. Withdraw the starter motor unit (photos).
5 Refitting is a reversal of removal, but ensure that the wiring connections are securely made.

11 Starter motor – testing in the vehicle

1 If the starter motor fails to operate first check the condition of the battery as described in Section 2.
2 Check the security and condition of all relevant cables.

Solenoid check
3 Disconnect the battery negative lead and all leads from the solenoid.
4 Connect a battery and 3 watt testlamp between the starter terminal on the solenoid body (Fig. 12.8). The testlamp should light. If not, there is an open circuit in the solenoid windings.
5 Now connect an 18 watt testlamp between both solenoid terminals (Fig. 12.9), then energise the solenoid with a further lead to the spade terminal. The solenoid should be heard to operate and the testlamp should light. Reconnect the solenoid wires.

On load voltage check
6 Connect a voltmeter across the battery terminals then disconnect the low tension lead from the coil positive terminal and operate the starter by turning the ignition switch. Note the reading on the voltmeter which should not be less than 10.5 volts.
7 Now connect the voltmeter between the starter motor terminal on the solenoid and the starter motor body. With the coil low tension lead still disconnected operate the starter and check that the recorded voltage is not more than 1 volt lower than that noted in paragraph 6. If the voltage drop is more than 1 volt a fault exists in the wiring from the battery to the starter.
8 Connect the voltmeter between the battery positive terminal and the terminal on the starter motor. With the coil low tension lead disconnected operate the starter for two or three seconds. Battery voltage should be indicated initially, then dropping to less than 1 volt. If the reading is more than 1 volt there is a high resistance in the wiring from the battery to the starter and the check in paragraph 9 should be made. If the reading is less than 1 volt proceed to paragraph 10.
9 Connect the voltmeter between the two main solenoid terminals and operate the starter for two or three seconds. Battery voltage should be indicated initially then dropping to less than 0.5 volt. If the reading is more than 0.5 volt the ignition switch and connections may be faulty.
10 Connect the voltmeter between the battery negative terminal and the starter motor body, and operate the starter for two or three seconds. A reading of less than 0.5 volt should be recorded; however, if the reading is more, the earth circuit is faulty and the earth connections to the battery and body should be checked.

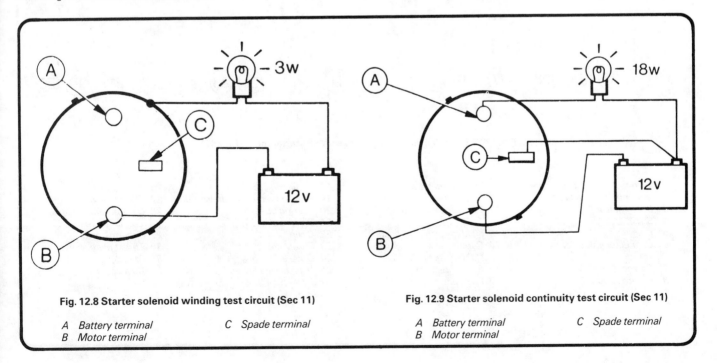

Fig. 12.8 Starter solenoid winding test circuit (Sec 11)

A Battery terminal
B Motor terminal
C Spade terminal

Fig. 12.9 Starter solenoid continuity test circuit (Sec 11)

A Battery terminal
B Motor terminal
C Spade terminal

240

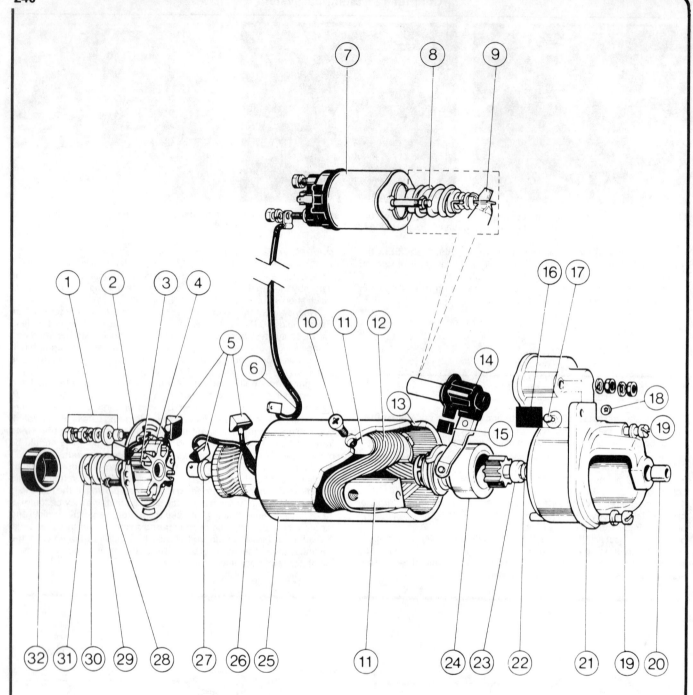

Fig. 12.10 Exploded view of the Lucas 5M90 starter motor (Sec 12)

1 Terminal nuts and washers	9 Engagement lever	18 Retaining clip	26 Armature
2 Commutator end plate	10 Pole screw	19 Housing retaining screws (2)	27 Thrustwasher
3 Brush housing	11 Pole shoe	20 Bearing bush	28 Commutator end plate retaining screws (2)
4 Brush springs	12 Field coils	21 Drive end housing	
5 Brushes	13 Field to earth connection	22 C clip	29 Bearing bush
6 Connector link, solenoid to starter	14 Rubber seal	23 Thrust collar	30 Thrust plate
	15 Rubber dust pad	24 Drive assembly	31 Star clip
7 Solenoid unit	16 Rubber dust cover	25 Main casing (yoke)	32 Dust cover
8 Return spring	17 Pivot pin		

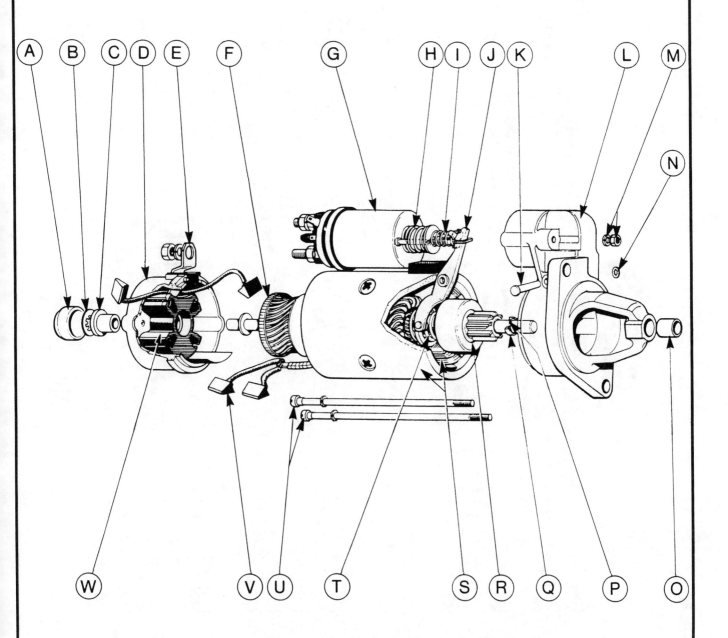

Fig. 12.11 Exploded view of the Lucas 2M100 starter motor (Sec 12)

A	Rubber cover	G	Solenoid
B	Retainer	H	Spring
C	Commutator end housing bush	I	Lost motion spring
D	Commutator end housing	J	Actuator lever
E	Terminal link	K	Pivot pin
F	Armature	L	Drive end bracket

M	Solenoid nut/washer	S	Pole shoe/yoke unit
N	Clip	T	Spacer spring
O	End bracket bush	U	Through bolts
P	C clips	V	Brush
Q	Thrust collar	W	Brush box unit
R	Pinion gear unit		

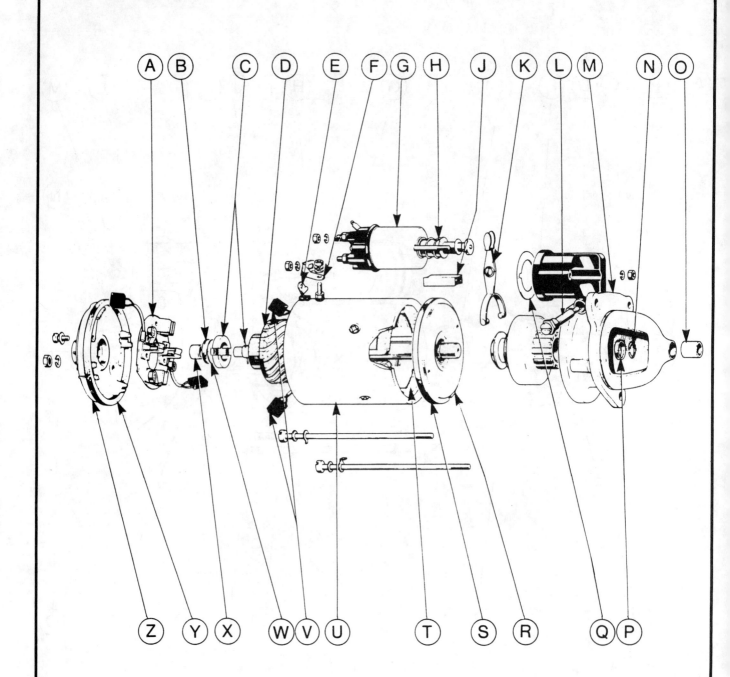

Fig. 12.12 Exploded view of the Lucas M127 starter motor (Sec 12)

- A   Brush gear unit
- B   Fibre washer
- C   Brake shoes and cross peg
- D   Armature
- E   Flexible link
- F   Copper link
- G   Solenoid
- H   Return spring
- J   Seal grommet
- K   Engagement lever
- L   Eccentric pivot pin
- M   Drive end fixing bracket
- N   C clip
- O   Bearing bush
- P   Thrust collar
- Q   Gasket
- R   Intermediate bracket
- S   Seal ring
- T   Field coils
- U   Yoke
- V   Insulated brushes (field coils)
- W   Steel thrustwasher
- X   Bearing bush
- Y   Seal ring
- Z   Commutator end bracket

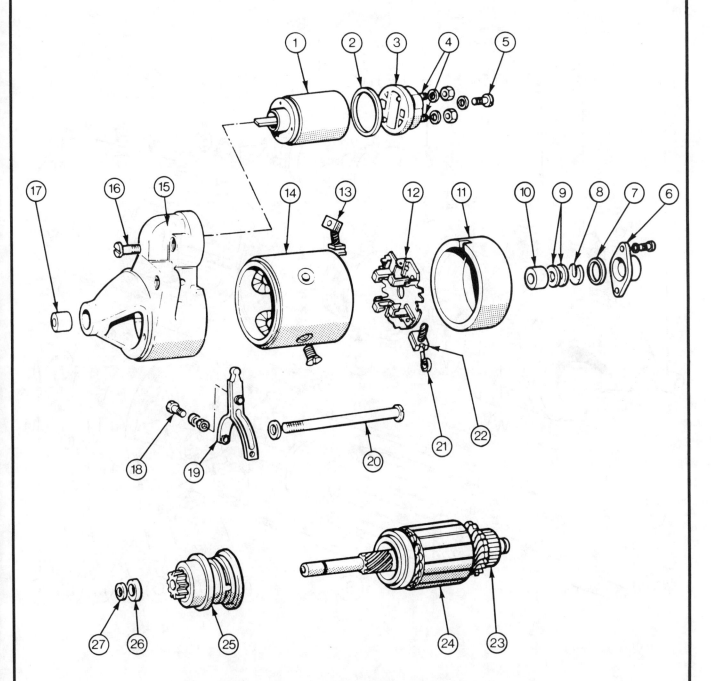

Fig. 12.13 Exploded view of the Bosch long frame starter motor (Sec 12)

1 Solenoid body
2 Gasket
3 Switch contacts and cover
4 Terminals (main)
5 Retaining screw
6 End cover
7 Seal
8 C clip
9 Shim washer
10 Bearing bush
11 Commutator end housing
12 Brushbox assembly
13 Connector link
14 Main casing (yoke)
15 Drive end housing
16 Solenoid retaining screw
17 Bearing bush
18 Pivot screw
19 Actuating lever
20 Through bolt
21 Brush spring
22 Brush
23 Commutator
24 Armature
25 Drive pinion and roller clutch assembly
26 Thrust collar
27 C-clip

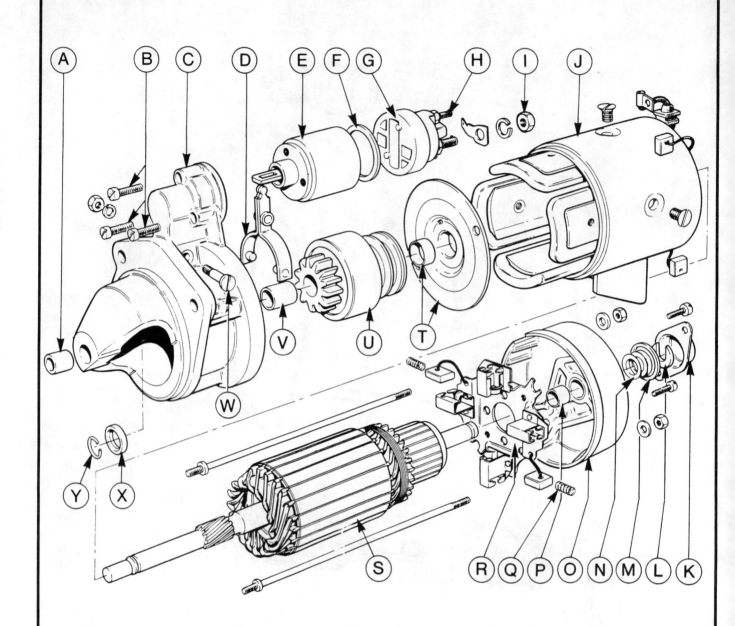

Fig. 12.14 Exploded view of the Bosch type JF (2.7kW) starter motor (Sec 12)

A Bearing bush
B Solenoid screws
C Drive end housing
D Actuating fork
E Solenoid body
F Gasket
G Switch contacts and cover
H Main terminals
I Nut
J Yoke main casing
K End cover
L C clip
M Seal
N Shims
O Commutator end housing
P Bearing bush
Q Brush spring
R Brush box unit
S Armature
T Centre bearing end plate
U Drive pinion/roller clutch unit
V Bearing bush
W Pivot screw
X Thrust collar
Y C clip

Chapter 12 Electrical system 245

12.2A Unscrew the nuts ...

12.2B ... and withdraw the solenoid

12.3A Remove the solenoid spring ...

12.3B ... and unhook the armature

12.4A Remove the plastic cap ...

12.4B ... and star clip from the commutator end plate

12.5A Removing the starter drive end bracket screws

12.5B Starter yoke location cut-out (arrowed)

12.6 Remove the starter motor engagement lever pivot

12 Starter motor – overhaul

Note: *Overhaul of the Lucas 5M90 type starter motor is described in this Section. However, the overhaul of the Lucas 2M100, M127 and Bosch starters is similar. Refer to the appropriate figures for the detail differences, and paragraphs 19 on as applicable for overhaul procedures which differ to those given*

1   Unscrew the nut and remove the connecting wire from the solenoid.
2   Unscrew the nuts and remove the solenoid from the end bracket (photos).
3   Unhook and remove the solenoid armature and spring from the engagement lever (photos).
4   Prise the plastic cap from the commutator endplate and ease off the star clip with a screwdriver (photos).
5   Unscrew the two screws from the drive end bracket and withdraw the yoke. Note the location of the washers on the armature, and the yoke location cut-out (photos).
6   Using a suitable metal drift drive the engagement lever pivot from the end bracket. The retaining clip will distort as the pin is removed and must be renewed (photo).
7   Withdraw the armature together with the engagement lever from the end bracket then separate the two components. Remove the rubber block and seal (photos).
8   Remove the screws and withdraw the end plate from the yoke sufficient to slide the two field brushes from their holders. Note the endplate cut-out and take care not to damage the gasket (photos).
9   Mount the armature in a vice then using a suitable metal tube drive the collar from the C-clip. Extract the C-clip and withdraw the collar followed by the drive pinion assembly.
10 Unscrew the nut from the terminal on the endplate, remove the spring washer, plain washer and insulator, then push the stud and second insulator through the endplate and unhook the brushes.
11 If necessary drill out the rivets and remove the brush box and gasket from the endplate.
12 Clean all the components in paraffin and wipe dry.
13 Check the length of the brushes. If either is worn to less than the minimum length given in the Specifications renew them. To do this cut the brush lead and then solder on the new lead.

Chapter 12 Electrical system

12.7A Remove the rubber block (arrowed) ...

12.7B ... and withdraw the armature from the end bracket

12.8A Remove the screws ...

12.8B ... and withdraw the yoke end plate

12.8C Yoke end plate cut-out (arrowed)

12.18 Using a socket to fit new star clip to the starter armature shaft

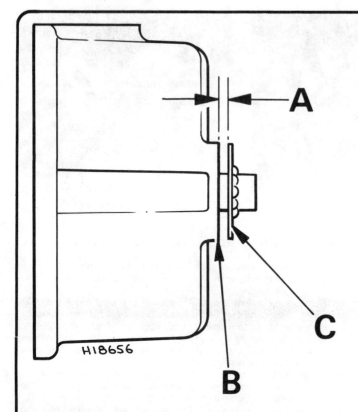
Fig. 12.15 Lucas 2M100 starter motor, showing clearance (A) required between thrust washer (C) and bearing face (B) (Sec 12)

A = 0.25 mm (0.010 in)

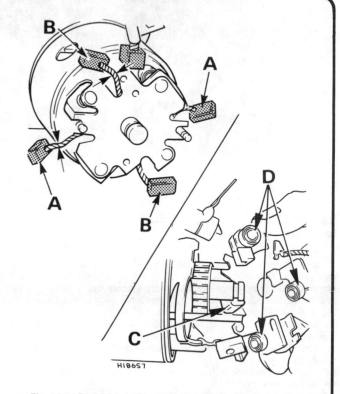
Fig. 12.16 Brush identification – Bosch starter motor (Sec 12)

A Field brushes
B Terminal brushes
C Brush plate
D Brush holder springs

Chapter 12 Electrical system

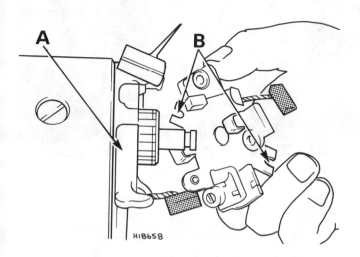

Fig. 12.17 Bosch starter motor brush plate location (Sec 12)

A Field winding loops B Locating cut-outs

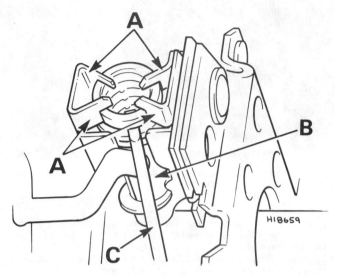

Fig. 12.18 Bosch (2.7 kW) starter motor brush removal method (Sec 12)

A Spring retainer tabs B Brush
C Screwdriver

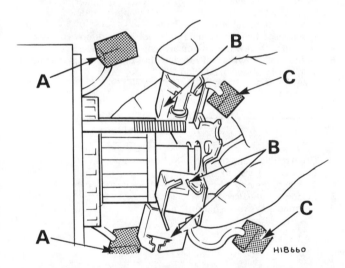

Fig. 12.19 Bosch (2.7 kW) starter motor brush plate removal showing field coil brushes (A), brush retainer tabs (B) and brush plate brushes (C) (Sec 12)

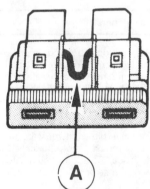

Fig. 12.20 When inspecting a fuse, look for a break in the wire at (A) (Sec 13)

14 Clean the commutator with fine glasspaper then wipe clean with a fuel moistened cloth. If it is excessively worn it may be skimmed in a lathe and then polished provided that it has not reduced to under the minimum diameter given in the Specifications. Clean any burrs from the insulation slots, but do not increase the width of the slots.
15 Check the armature windings for good insulation by connecting a testlamp and leads between each commutator segment in turn and the armature shaft; if the bulb glows the insulation is faulty.
16 Check the field windings in the yoke for security and for the condition of soldered joints. Check the continuity and insulation of the windings using a testlamp and leads.
17 Check the bearing bushes for wear and if necessary renew them using a soft metal drift. The bushes should be immersed in clean SAE 30/40 grade oil for a minimum of 20 minutes before being fitted. The bushes are made of self-lubricating porous bronze and must not be reamed otherwise the self-lubricating quality will be impaired.
18 Reassembly is a reversal of dismantling, but fit a new brush box gasket if necessary. Use a two legged puller to pull the collar onto the

C-clip on the armature shaft. Fit a new star clip to the engagement lever pivot pin and to the end of the armature shaft making sure that all endfloat is eliminated from the shaft (photo).

Lucas 2M100
19 The actuating (engagement) lever on some models is secured by a peened pin rather than a clip. Relieve the peened section of the pin when removing it, then during reassembly, peen over the end of the pin to secure it.
20 During final reassembly, press the armature against the commutator end plate thrustwasher and check that the retaining clip ring to bush face clearance does not exceed 0.25 mm (0.010 in), see Fig. 12.15. Drive the clip further onto the armature shaft to suit if necessary.

Bosch long frame 0.80 kW and 1.1 kW
21 Refer to Fig. 12.16 for the brush identification on the holder and Fig. 12.17 for positional location details.
22 Smear a small amount of lithium based grease onto the end of the armature shaft before refitting the end cap.

Bosch 2.7kW
23 Brush/brushplate removal method: Bend open the brush retainer tabs and remove the brush and spring, but take care not to let the spring

13.2 Fuse and relay box showing identification diagram on inside face of the lid

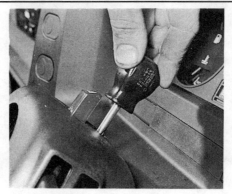
14.2A Removing the upper column shroud retaining screw

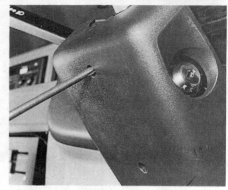
14.2B Removing a screw securing the lower shroud

fly out as it is released. Repeat this procedure to remove the other brushes. When the brushes are removed, the plate can be withdrawn (Fig. 12.18 and 12.19).
24 When refitting the brush plate, align the cut-outs with the studs to ensure correct brush plate position. When the brushes and springs are compressed into their holders, bend the retaining tabs over to secure them in position.
25 Smear a small amount of lithium based grease onto the end of the armature shaft before refitting the end cap.

13 Fuses and relays – general

1 The fuses and relays are located in a common box which is attached to the underside of the facia on the drivers' side of the vehicle.
2 The respective fuses and relays are identified on the diagram on the inside surface of the fuse box lid (photo). Each fuse is also marked with its rating. The circuits that the fuses and relays protect are given in the Specifications at the start of this Chapter.
3 To inspect/renew a fuse or relay, simply unclip the fuse box lid and hinge it open. If a fuse has blown, it must be renewed with one of identical rating and never renew it more than once without finding the source of the trouble (usually a short circuit). Always switch off the ignition before renewing a fuse or relay. All fuses and relays are a push fit.
4 A fusible link is incorporated in the positive lead of the battery, its function being to protect the main wiring loom in the event of a short circuit. When this fuse blows, all of the wiring circuits are disconnected, and will remain so until the cause of the malfunction is repaired and the link renewed.

14 Direction indicator switch – removal and refitting

1 Disconnect the battery negative lead.
2 Remove the screws and withdraw the upper and lower shrouds from the steering column (photos).
3 Remove the two crosshead screws and withdraw the switch from the steering column, also disconnect the wiring multi-plug.
4 Refitting is a reversal of removal.

15 Lighting switch – removal and refitting

The procedure is identical to that for the direction indicator switch in Section 14.

16 Ignition switch and lock barrel – removal and refitting

1 Disconnect the battery negative lead.
2 Remove the screws and withdraw the upper and lower shrouds from the steering column.
3 Insert the ignition key and turn to position 'I' then depress the lock spring using a suitable rod or tool inserted through the access hole in the side of the cylinder, and withdraw the steering lock barrel. Slight movement of the key will be necessary in order to align the cam.
4 With the key fully inserted extract the spring clip taking care not to damage its location, then withdraw the key approximately 5 mm (0.2 in) and remove the barrel from the cylinder.
5 Disconnect the wiring multi-plug (photo) then remove the two grub screws and withdraw the ignition switch.
6 Refitting is a reversal of removal, but check the operation of the steering lock in all switch positions.

17 Courtesy light switch – removal and refitting

1 Open the door and unscrew the cross head screw.
2 Remove the switch from the door pillar and pull the wire out sufficiently to prevent it from springing back into the pillar (photo).
3 Disconnect the wire and remove the switch.
4 Refitting is a reversal of removal.

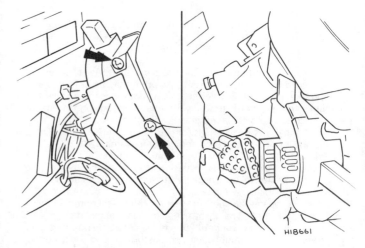
Fig. 12.21 Direction indicator switch removal. Arrows indicate the retaining screws (Sec 14)

Chapter 12 Electrical system

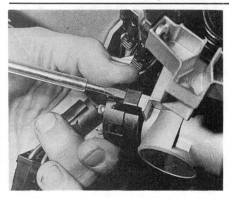

16.5 Disconnecting the ignition switch wiring multi-plug

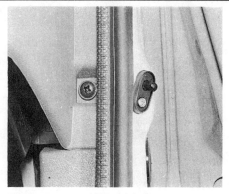

17.2 Door courtesy light switch

18.2A Undo the retaining screws ...

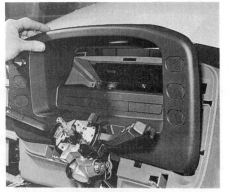

18.2B ... and remove the instrument panel surround (steering wheel removed for photographic purposes)

18.3 Switch removal from the instrument panel surround

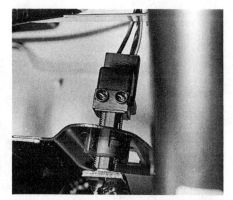

20.1 Brake stop light switch

3 Disconnect the wiring harness plug from the rear of the appropriate switch, then press it from the panel to remove it.
4 Refit in the reverse order of removal. Tighten the column upper mounting nuts to the specified torque setting (see Specifications in Chapter 10). Check the switches for satisfactory operation on completion.

19 Reversing light switch – removal and refitting

1 Jack up the front of the car and support on axle stands. Apply the handbrake.
2 Working beneath the car disconnect the wiring and unscrew the switch from the gearbox.
3 Refitting is a reversal of removal, but make sure that the wiring is secured clear of the exhaust system.

20 Brake stop-light switch – removal and refitting

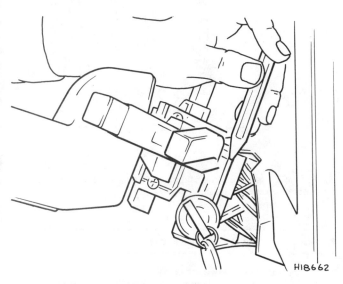

Fig. 12.22 Removing the lock barrel (Sec 16)

1 Disconnect the wiring connector from the switch unit, then twist the switch anti-clockwise and remove it from its retaining bracket (photo).
2 Refit in the reverse order of removal.

21 Heater motor switch – removal and refitting

18 Instrument panel surround switches – removal and refitting

1 Disconnect the battery earth lead, then undo the retaining screws and remove the upper and lower steering column shrouds.
2 Loosen off the upper steering column mounting nuts to slightly lower the column, then unscrew the retaining screws within the surround, tilt out the surround from its base and withdraw it (photos).

1 Disconnect the battery earth lead.
2 Undo the four screws retaining the heater control panel to the facia and partially withdraw the panel to allow access to the rear of the switch (photo).
3 Detach the switch wiring connector, compress the retainer tabs and withdraw the switch from the panel (photo).

250 Chapter 12 Electrical system

21.2 Removing the heater panel retaining screws

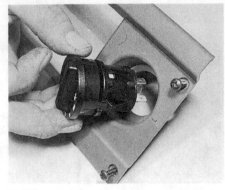
21.3 Heater motor control switch removal

22.4 Instrument panel retaining screws (arrowed)

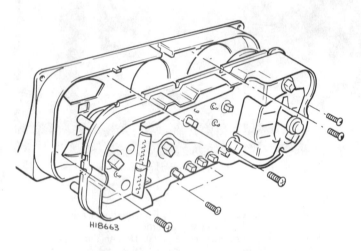
Fig. 12.23 Separate the front surround unit from the main unit (Sec 23)

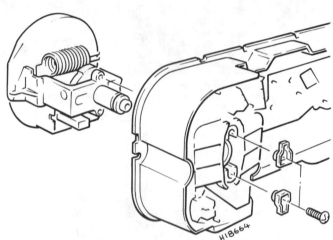
Fig. 12.24 Speedometer removal (Sec 23)

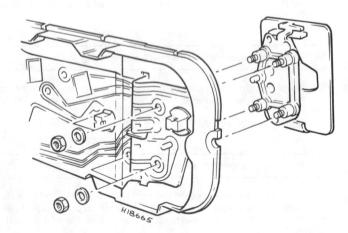
Fig. 12.25 Fuel gauge removal (Sec 23)

4 Refit in the reverse order of removal. Check the switch for satisfactory operation on completion.

22 Instrument panel – removal and refitting

1 Disconnect the battery earth lead.
2 Undo the retaining screws and remove the steering column shrouds.
3 Loosen off the upper column mounting nuts and allow the column to drop a fraction. Undo the instrument panel surround retaining screws, tilt the panel out at the bottom and withdraw the surround. Disconnect the switch wiring connectors to remove the surround.
4 Unscrew and remove the four instrument panel retaining screws (photo), withdraw the panel sufficiently to enable the speedometer cable and the wiring block connectors to be detached from the rear of the panel. Remove the instrument panel.
5 Refit in the reverse order of removal. Check for satisfactory operation of the various instruments and associate components on completion.

Chapter 12 Electrical system

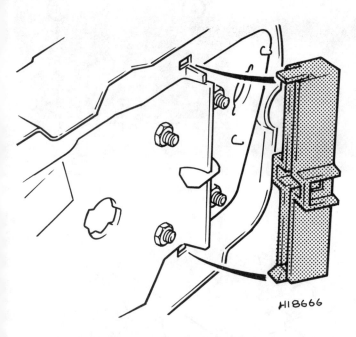

Fig. 12.26 Remove the multi-plug retainer (Sec 23)

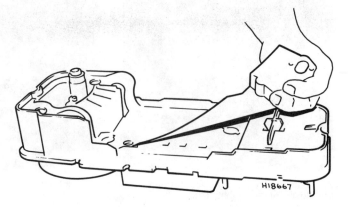

Fig. 12.27 Remove the printed circuit from the panel (Sec 23)

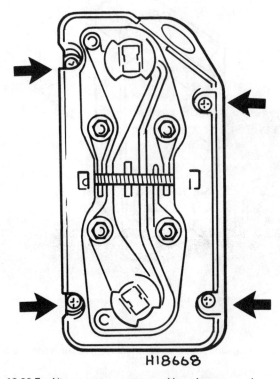

Fig. 12.28 Fuel/temperature gauge and housing screws (arrowed) on tachograph models (Sec 24)

23 Instrument panel components – removal and refitting

1 Remove the instrument panel as described in the previous Section. When handling the instrument panel and removing or refitting its components, take care not to damage the printed circuit and avoid knocking or dropping the unit as it can easily be damaged (photo).

Warning and illumination bulbs
2 Untwist the bulbholder and withdraw it from the rear face of the panel. Remove the bulb from its holder (photo).
3 Refit in the reverse order of removal.

Panel glass
4 Undo the two retaining screws at the top and detach the two retaining clips at the bottom, then remove the glass (photo).
5 Refit in the reverse order of removal.

Speedometer head
6 Undo the six retaining screws and remove the front surround unit from the main unit.
7 Undo the two retaining screws and withdraw the speedometer head from the panel.
8 Refit in the reverse order of removal.

Fuel gauge (non-tachograph models)
9 Undo the six retaining screws and remove the front surround unit from the main unit.
10 Unscrew the four retaining nuts and remove the washers. Withdraw the fuel/temperature gauge unit from the front face side of the unit (Fig. 12.25).
11 Refit in the reverse order of removal.
Note: *To remove the fuel gauge sender unit from the fuel tank, refer to the appropriate Section in Chapter 3.*

Clock
12 Undo the six retaining screws and remove the front surround unit from the main unit.
13 Unscrew the three retaining nuts and withdraw the clock from the main unit.
14 Refit in the reverse order of removal.

Printed circuit
15 Remove the bulbholders, the clock and the fuel/temperature gauge.
16 Detach the multi-plug retainer (Fig. 12.26) then carefully remove the printed circuit.
17 Refit in the reverse order of removal.

24 Instrument panel surround facia and components (tachograph models) – removal and refitting

1 Disconnect the battery earth lead.
2 Undo the retaining screws and remove the upper and lower shrouds from the steering column.
3 Pull free the heater control switch knob, and where fitted, the intermittent wiper control knob.
4 Undo the four retaining screws, and remove the facia surround from the instrument panel sufficiently enough to detach the switch wires.

Chapter 12 Electrical system

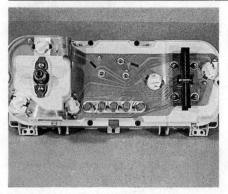

23.1 Instrument panel rear face showing printed circuit and bulbholders

23.2 Bulbholder removal

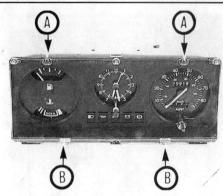

23.4 Panel glass retaining screws (A) and clips (B)

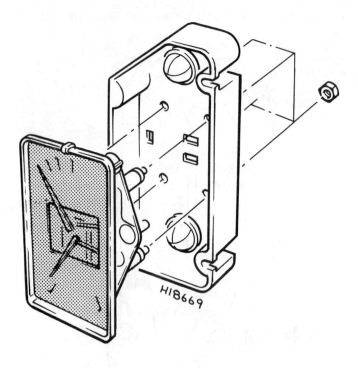

Fig. 12.29 Fuel gauge separation from housing (tachograph models) (Sec 24)

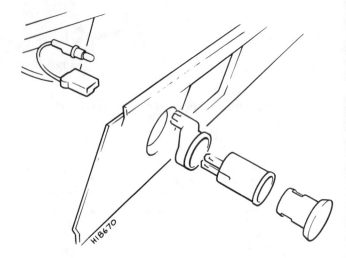

Fig. 12.30 Cigarette lighter components (Sec 26)

5 Disconnect the switch wire connectors. Untwist the illumination bulbs to remove them. Remove the panel.
6 To remove the combined fuel/temperature gauge unit, press them from the rear and detach the multi-connector plug. Undo the four retaining screws and remove the gauge and housing. Undo the four nuts and remove the gauge from the housing.
7 Refit in the reverse order of removal.

25 Speedometer cable – removal and refitting

1 Remove the instrument panel as described in Section 22.
2 Raise the vehicle at the front end and support it securely on axle stands.
3 Working underneath the vehicle, disconnect the speedometer cable from the transmission, according to type, as described in Chapter 6.
4 Withdraw the cable through the bulkhead and remove it from the engine compartment.
5 Refit in the reverse order of removal. When inserting the cable through the bulkhead, pass it through the grommet and align the colour band on the cable with the grommet. Re-route the cable so that it is clear

of any moving parts on which it could chafe and do not bend it too much.

26 Cigarette lighter – removal and refitting

1 Disconnect the battery earth lead.
2 Undo the four retaining screws and remove the heater control panel from the facia a sufficient amount to allow access to the rear of the unit.
3 Detach the wiring connector and illumination bulbholder from the rear of the cigarette lighter, release and remove the lighter unit from the panel. To remove the lighter unit from the illumination ring, compress the retaining lugs and withdraw the unit (photo).
4 Refit in the reverse order of removal. Check for satisfactory operation of the lighter on completion.

27 Headlamps – alignment

1 It is recommended that the headlamp alignment is carried out by a Ford garage using modern beam setting equipment. However in an emergency the following procedure will provide an acceptable light pattern.
2 Position the vehicle on a level surface with tyres correctly inflated

Chapter 12 Electrical system

26.3 Cigarette lighter and illumination bulbholder

28.2 Prise free the rubber cap from the headlamp

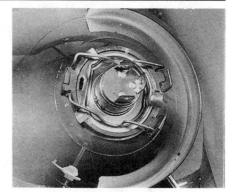

28.3A Headlamp bulb and retaining clip

28.3B Headlamp bulb removal

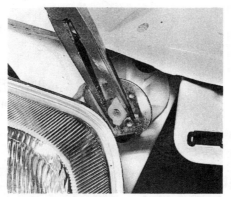

28.8A Rotate the plastic retainers ...

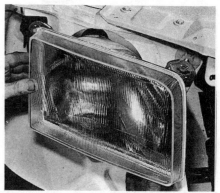

28.8B ... and withdraw the headlamp unit

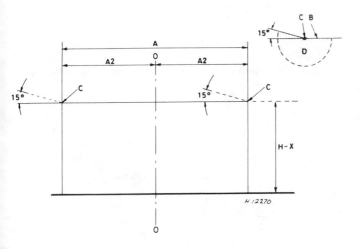

Fig. 12.31 Headlamp alignment chart (Sec 27)

- A Distance between headlamp centres
- B Light-dark boundary
- C Beam centre dipped
- D Dipped beam pattern
- H Height from ground to centre of headlamps
- X All variants — 16.0 cm (6.3 in or 1.0°)

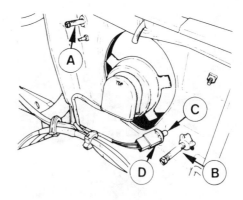

Fig. 12.32 Rear view of headlamp showing horizontal (A) and vertical (B) beam adjuster screws, sidelamp bulb (C) and holder (D) (Sec 27)

approximately 10 metres (33 feet) in front of, and at right-angles to, a wall or garage door.
3   Draw a vertical line on the wall corresponding to the centre line of the vehicle. The position of the line can be ascertained by marking the centre of the front and rear screens (doors) with crayon then viewing the wall from the rear of the vehicle.

4   Complete the lines shown in Fig. 12.31.
5   Switch the headlamps on dipped beam and adjust them as necessary using the knobs located behind the headlamps (Fig. 12.32). Cover the headlamp not being checked with cloth.
6   Holts Amber Lamp is useful for temporarily changing the headlight colour to conform with the normal usage on Continental Europe.

28 Headlamps and headlamp bulbs – removal and refitting

Headlamp bulb
1   To remove a headlamp bulb, open and support the bonnet.
2   Pull free the wiring connector from the rear of the headlamp unit and

Chapter 12 Electrical system

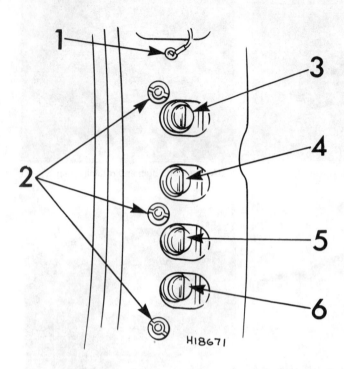

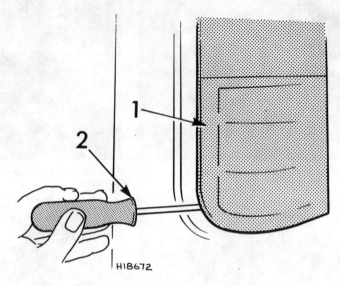

Fig. 12.34 Rear lamp cluster (1) removal using screwdriver (2) (Sec 29)

Fig. 12.33 Rear lamp cluster bulb identification (Sec 29)

1 Earth lead
2 Wing nuts (plastic)
3 Tail/stop lamp
4 Rear indicator
5 Reversing lamp
6 Foglamp

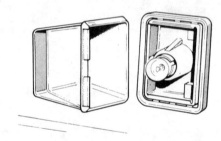

Fig. 12.36 Rear number plate lamp – Chassis cab models (Sec 29)

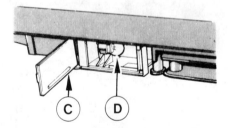

Fig. 12.35 Rear number plate lens (C) and bulb (D) – Tailgate models (Sec 29)

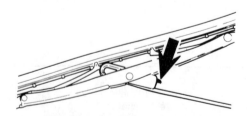

Fig. 12.37 Depress clip (arrowed) to release wiper blade from the arm (Sec 31)

remove the rubber cap (photo).
3 Release the bulb holder retaining clip and withdraw the bulb. Do not touch the bulb glass with the fingers; however, if it is touched, wipe it clean with a tissue soaked in methylated spirit (photos).
4 Refit in the reverse order of removal and check for satisfactory operation on completion.

Headlamp unit
5 Detach the wiring connector from the rear of the headlamp unit.
6 Twist the indicator bulbholder and wire to detach the indicator bulb and holder. Similarly remove the sidelamp bulbholder.
7 Press the lock tongue of the plastic retaining clip at the top of the indicator unit to the side and then use a suitable screwdriver to detach it. Withdraw the indicator unit from the front.
8 Using suitable pliers, turn the two plastic headlamp retainers a quarter of a turn, withdraw the headlamp unit from the top mountings

and disengage it from the lower adjuster (photos).
9 Refit in the reverse order of removal. Refer to Section 27 for details on headlamp beam adjustment. Check the headlamps and indicators for satisfactory operation on completion.

29 Bulbs and lamp units – removal and refitting

Sidelamp bulbs
1 Open the bonnet, untwist the sidelamp holder from the rear of the

Chapter 12 Electrical system

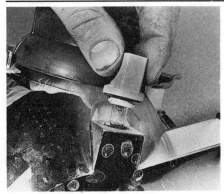
29.1 Sidelamp bulb and holder removal

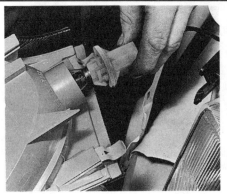
29.2 Front indicator bulb and holder removal

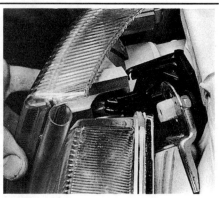
29.3 Front indicator unit removal

29.4 Side repeater lamp

29.6A Rear lamp cluster bulbs and holders are accessible through the respective apertures

29.6B Removing a rear lamp cluster bulb and holder

headlamp and withdraw the bulb from its holder (photo).

Front direction indicator bulbs
2   Open the bonnet, reach down and untwist the bulbholder from the rear of the indicator unit. Remove the bulb from the holder (photo).
3   If required, the indicator unit can be removed by compressing the plastic retainer clip with a screwdriver and releasing the two inner clips (photo).

Side repeater lamp bulbs
4   Pull the bulbholder unit from the rear of the lamp, then remove the bulb from its holder (photo).
5   To remove the side repeater lamp unit, simply compress the retaining lugs and simultaneously withdraw the unit.

Rear lamp cluster bulbs – Van and Bus models
6   Remove the cover flap from the rear body corner to gain access to the lamp cluster (if fitted). Untwist and withdraw the appropriate bulb and holder from the lamp unit. Press and untwist the bulb to remove it from its holder (photos).
7   To remove the lamp cluster, undo the retaining nuts on the inside, withdraw the unit and detach the wiring connector (photo).

Rear lamp cluster bulbs – Chassis cab models
8   Release the lamp lens retaining clip and move the lens out of the way. Remove the appropriate bulb from the holder in the lamp unit.
9   To remove the lamp unit, detach the wiring connectors, unscrew the retaining nuts and remove the lamp cluster.

Rear number plate lamp bulb – double rear door model
10   Prise free the lamp cap using a suitable screwdriver, then remove the bulb from its holder by pulling it free (photos).

Rear number plate lamp bulb – Tailgate model
11   Slide the lamp lens towards the left and pivot it out of the way.

Push and untwist the bulb to remove it from its holder (Fig. 12.35).

Rear number plate lamp bulb – chassis cab models
12   Pull free the lamp cap, then press and untwist the bulb to remove it from its holder (Fig. 12.36).

Interior lamp bulb
13   Insert a small electrical screwdriver blade into the indent in the lamp unit and carefully prise it free (photo). Press and untwist the bulb from its holder. The rear interior lamp has a festoon type bulb and this type is simply prised free from its holder.

Hazard warning switch bulb
14   Pull free the hazard warning switch lens/button then remove the bulb (photo).

Bulbs – refitting
15   Refit all bulbs in the reverse order of removal. Check for satisfactory operation on completion.

30 Horn – removal and refitting

1   The horn is located at the front end of the vehicle, behind the grille panel and underneath the left-hand headlamp. Access is improved by removal of the grille (Chapter 11) (photo).
2   Detach the two lead connectors from the horn, unscrew the horn mounting bracket bolt and withdraw the horn.
3   If required, the horn can be separated from the mounting bracket by undoing the retaining nut. The horn cannot be adjusted or repaired and therefore if defective, it must be renewed.
4   Refit in the reverse order of removal. Check for satisfactory operation on completion.

Chapter 12 Electrical system

29.7 Detaching the wiring connector from the rear lamp cluster unit

29.10A Rear number plate lamp removal

29.10B Separate the lens/cover from the holder for access to bulb

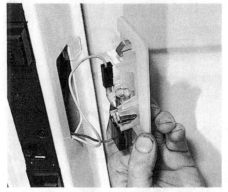
29.13 Interior lamp unit removal

29.14 Hazard warning switch button/lens removal gives access to bulb

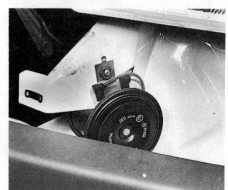
30.1 Horn location and wiring connectors

32 Wiper arms – removal and refitting

1   Remove the wiper blades as described in Section 31.
2   Lift the hinged covers and remove the nuts and washers securing the arms to the spindles (photo).
3   Mark the arms and spindles in relation to each other then prise off the arms using a screwdriver. Take care not to damage the paintwork.
4   Refitting is a reversal of removal.

33 Windscreen wiper motor – removal and refitting

1   Disconnect the battery earth lead.
2   Remove the wiper arms and blades as described in the previous Sections.
3   Unscrew and remove the plastic retaining nut from the centre pivot (photo).
4   Remove the windscreen grille panel as described in Chapter 11.
5   Undo the steel pivot nut and remove it together with its shim.
6   Compress the locktab and detach the wiring connector from the wiper motor (photos).
7   Use a screwdriver as a lever and prise free the linkage arms from the balljoint on the wiper link arm (photo). Collect the felt spacer washers.
8   Unscrew the retaining screws and remove the wiper motor unit and its support bracket.
9   If the wiper motor link arm is to be removed, first mark the relative positions of the arm and shaft, then unscrew the retaining nut and remove the arm. Unscrew the three bolts to remove the wiper motor from its bracket.
10 Refit in the reverse order of removal, but note the following special points:

(a) Ensure that the motor link arm is correctly realigned as it is refitted to the drive shaft
(b) Reconnect the link arms onto their balljoints by pressing them into position by hand. Locate the felt washers as shown (photo)
(c) One of two pivot housing types will be fitted, being of Delco or SWF manufacture. The Delco housing is plastic, the SWF steel.

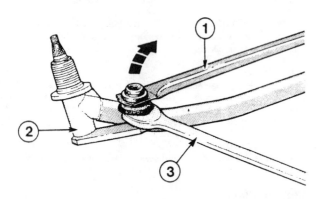
Fig. 12.38 Levering the link arm (1) from the pivot (2) using an open jaw spanner (3) as the lever (Sec 34)

31 Wiper blades – renewal

1   The wiper blades should be renewed when they no longer clean the windscreen or tailgate window effectively.
2   Lift the wiper arm away from the windscreen or tailgate window.
3   With the blade at 90° to the arm depress the spring clip and slide the blade clear of the hook then slide it up off the arm (photo).
4   If necessary extract the two metal inserts and unhook the wiper rubber.
5   Fit the new rubber and blade in reverse order making sure where necessary that the cut-outs in the metal inserts face each other.

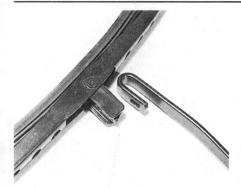
31.3 Wiper blade removed from arm

32.2 Wiper arm to spindle retaining nut

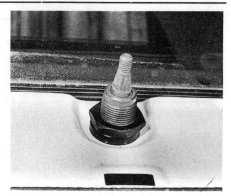
33.3 Wiper arm central pivot

33.6A General view of the windscreen wiper motor (grille removed)

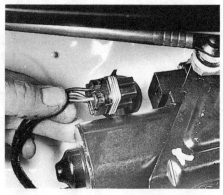
33.6B Detach the wiring connector from the wiper motor

33.7 Disconnect the link arms from the balljoint

33.10 Link arms and felt washers on balljoint connection

35.1 Windscreen washer reservoir and pump connections

37.2 Windscreen washer hose connections to nozzle and T-piece connector

Depending on the type used, it is most important that the pivot nuts are correctly fitted and tightened to the specified torque wrench setting according to type. On both types it is important that the steel nut is fitted first and secures the pivot housing to cowl. The plastic nut (which is very similar to the steel type in appearance) must secure the central pivot plastic cover. The Delco type pivot housing steel nut must not be tightened in excess of 12 Nm (8.8 lbf ft)

(d) *On completion, check the wipers for satisfactory operation*

34 Windscreen wiper linkages – removal and refitting

1 Remove the wiper arms and the windscreen grille panel as described in the previous Section.
2 Lever the wiper link arms from the wiper motor link balljoint, and note the position of the felt washers.
3 Unscrew and remove the steel pivot nuts. Lever the link arms from the pivots and remove them.
4 To renew the pivot housing on the Delco type housing, drill a 6.5 mm (0.25 in) hole to a depth of 10 mm (0.4 in) at the pre-marked points each side of the housing and then separate the housing from the tube.
5 To renew the pivot housing on the SWF type, carefully grind off the two rivet heads at the points shown, and then drive the rivets out using a suitable pin punch (Fig. 12.40). Separate the housing from the tube.
6 To reassemble the housing and tube on both types, slide the two together and slot in the end of the tube. Where necessary, tap the housing home using a plastic hammer and align the retaining pin or rivet holes (as applicable).
7 Insert the rivets or retaining pins to secure. When fitted correctly, the pins should project slightly at each end.
8 Refitting is a reversal of the removal procedure. Refer to the previous Section for details concerning the wiper motor and linkage refitting.

35 Windscreen/headlamp washer reservoir – removal and refitting

1 If the washer pump is faulty, it can be removed from the reservoir by draining or syphoning off the fluid from the reservoir, disconnecting the feed hose and the wiring connector, then prising the pump from the reservoir. Renew the seal bush if it is defective (photo).

2 To remove the reservoir, drain and detach the items mentioned in paragraph 1, but leave the pump(s) in position unless they are to be removed. Undo the retaining bolts and remove the reservoir.

3 Refit in the reverse order of removal. Lubricate the seal bush with

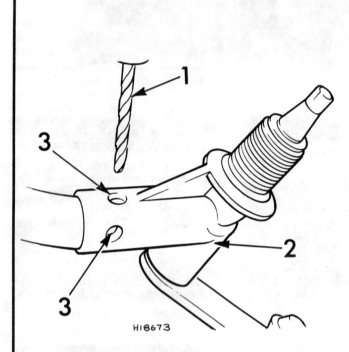

Fig. 12.39 Delco pivot housing renewal (Sec 34)

1 Drill
2 Plastic housing
3 10 mm deep holes

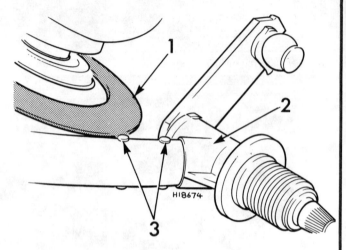

Fig. 12.40 SWF pivot housing renewal (Sec 34)

1 Grinding disc
2 Steel housing
3 Rivet heads

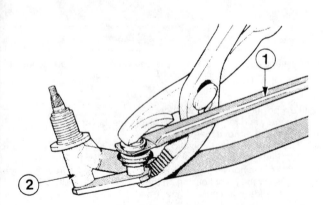

Fig. 12.41 Reconnecting the wiper link (1) to the pivot (2) using suitable grips (Sec 34)

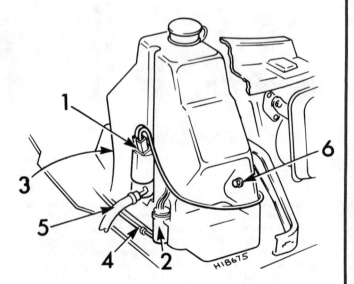

Fig. 12.42 Windscreen/headlamp washer reservoir showing: (Sec 35)

1 Pump unit – headlamp
2 Pump unit – windscreen
3 Securing bolt
4 Hose (windscreen washer)
5 Hose and clip (headlamp washer)
6 Securing bolt

Chapter 12 Electrical system

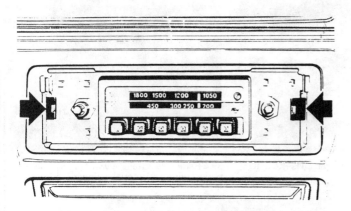

Fig. 12.43 Pull retaining tangs (arrowed) inwards (Sec 38)

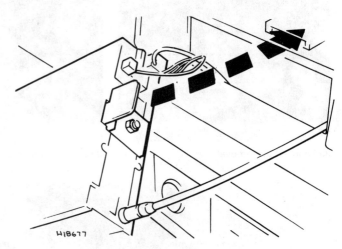

Fig. 12.44 Where applicable, realign the plastic support bracket as shown when installing the radio (Sec 38)

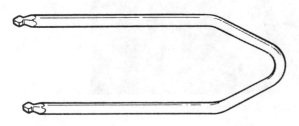

Fig. 12.45 Radio/cassette extractor tool (Sec 39)

glycerine or a soapy solution to ease fitting. Engage the reservoir locating pin in the hole in the body side member as it is fitted.
4   On completion top up the reservoir with the required water/washer solution mix (do not use antifreeze) and check for leaks and satisfactory operation.

36 Tailgate wiper components – removal and refitting

Wiper arm and blades
1   Proceed as described for the windscreen wiper arms and blades removal and refitting (Section 31). Note the position of the wiper arm in the parked position before removing it to be sure of correctly repositioning it during refitting.

Wiper motor
2   Detach the battery earth lead, then remove the tailgate trim panel for access.
3   Remove the wiper arm and blade.
4   Detach the wiring connector from the wiper motor.
5   Undo the retaining screws and remove the wiper motor unit.
6   Refit in the reverse order of removal, then check for satisfactory operation on completion.

Washer reservoir
7   The procedure is similar to that for the removal and refitting of the windscreen/headlamp washer reservoir. Remove the right-hand rear trim panel for access.

37 Washer nozzles and hoses – removal and refitting

1   The washer hoses and nozzles can be separated at the appropriate joint by pulling them free from the compression joint. Allow for a small amount of leakage as they are separated.
2   To remove the nozzles and hoses it may be necessary to first remove the mounting panel for access. Removal of the windscreen grille panel gives access to the windscreen washer hose-to-nozzle connections. Removal of the front grille gives access to the headlamp washer nozzles and hoses (photo).
3   The nozzles are retained to the mounting panel by expanding lugs which have to be compressed with suitable pliers to allow removal, or by retaining screws (headlamp washers).
4   Refit in the reverse order of removal. Ensure that the joints are securely made and check them for any signs of leakage on completion.
5   If required, the nozzle jets can be adjusted to suit by inserting a pin into the jet orifice and swivelling it to adjust as required.

38 Radio (standard) – removal and refitting

1   Disconnect the battery earth lead.
2   Pull free the radio control knobs, the plastic tone control lever and the spacer from the tuning control.
3   Unscrew the two retaining nuts and remove the facia plate.
4   Referring to Fig. 12.43, pull the retaining tangs inwards using a suitable hooked tool (welding rod bent to suit will suffice), then withdraw the radio. Detach the wiring and aerial from the rear of the unit and then remove the radio.
5   Refit in the reverse order of removal. When it is fully installed, trim (tune) it for the best reception in an interference free zone. Using a small electrical screwdriver, insert it into the trim screw on the front face of the unit and turn the screw progressively to obtain the best possible reception.

39 Radio (digital) – removal and refitting

1   Disconnect the battery earth lead.
2   To withdraw the radio unit from its aperture, you will need two extractors similar to the type shown in Fig. 12.45. These are readily

Chapter 12 Electrical system

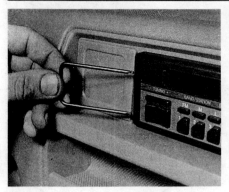
39.3A Insert the removal tools ...

39.3B ... to withdraw the radio/cassette unit

39.4 Radio/cassette unit withdrawn to show typical wiring and aerial connections on rear face

40.2 Remove cap for access to the aerial retaining nut (roof mounted type)

40.9 Standard speaker unit location

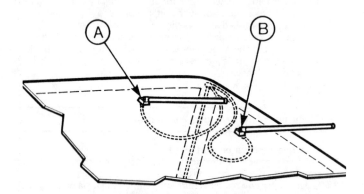

Fig. 12.46 Roof mounted aerial installation (Sec 40)

A High roof fitting
B Low roof fitting

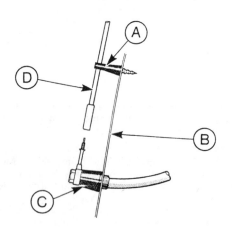

Fig. 12.47 Side mounted aerial installation (Sec 40)

A Aerial support   C Aerial base
B Body   D Aerial

available from radio accessory outlets. Alternatively, they can be fabricated from wire rod of suitable gauge.

3 Insert the removal tools into the slots in the front face of the unit each side as shown (photo). When they are felt to engage, push each one outwards simultaneously, then pull them evenly to withdraw the radio unit from its aperture (photo).

4 With the radio clear of the facia, detach the aerial and wiring connectors from the rear of the unit. Remove the radio (photo).

5 Undo the retaining nut and remove the support bracket from the rear of the unit.

6 Refit the radio in the reverse order of removal. When fitted, it will need to be key-coded to operate. Unless the code is known, this must be entrusted to a Ford dealer.

Chapter 12 Electrical system

2 Prise free the retaining nut cover from the aerial, unscrew the nut and remove the washer (photo).
3 Remove the radio and detach the aerial from its rear face (Section 38 or 39 as applicable).
4 The aerial and lead can now be withdrawn from the vehicle. As the lead is routed up the windscreen A pillar, it is advisable to remove its trim and/or attach a length of strong cord to the aerial lead and pull the lead through until clear, then detach the cord. Leave the cord in position so that it can be re-attached to pull and guide the lead back through the original route.
5 Refit in the reverse order of removal. Arrange the lead in the A pillar as shown in Fig. 12.46.

Side mounted aerial

6 This type of aerial is usually fitted to Luton body variants to improve reception. Start by removing the radio from its aperture and disconnect the aerial lead from its rear face.
7 Unscrew and remove the aerial base mounting bolt, detach the aerial from the mounting and pull the lead through. Withdraw the aerial downwards from the top support, then remove the aerial and lead.
8 Refit in the reverse order of removal. When feeding the lead through the bulkhead, ensure that the grommet is securely fitted and keep the lead away from any moving parts (Figs. 12.48 and 12.49).

Speakers

9 Undo the retaining screws and remove the side cowl trim (photo).
10 Undo the retaining screws and withdraw the speaker unit. Detach the wiring.
11 Refit in the reverse order of removal.

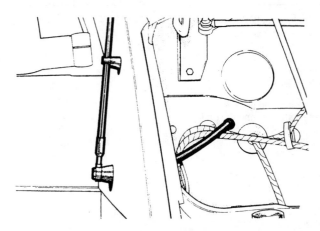

Fig. 12.48 Route the side mounted aerial cable as shown (Sec 40)

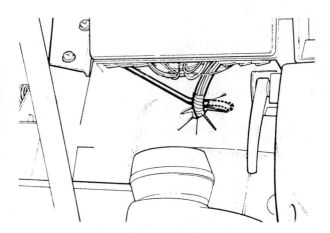

Fig. 12.49 Side mounted aerial cable route in cab (Sec 40)

41 Wiring diagrams – notes for guidance

1 Wiring diagrams are illustrated by schematic representations of the vehicle looms. Engine looms, main looms and rear looms are shown on separate pages to enable diagrams to be selected to suit a particular vehicle.
2 For ease of use, components are identifiable by numbering in a circle. Reference should be made to the appropriate key to the figure for a description of the component. Separate keys are provided for engine, main and rear looms.
3 Internal wiring joints are also numbered and shown in their approximate locations. These are readily identifiable by the hexagonal shape around the number. For clarity, internal joint wiring details are shown with the keys to the figures.
4 Connections between the looms are highlighted by arrowheads containing the same letter (A, B or C).
5 For further information, refer to the page about wiring diagram layout.
6 Low level instrumentation is easily identifiable on vehicles by the absence of a cigarette lighter.

40 Radio aerial and speakers (standard) – removal and refitting

Aerial – roof mounted

1 Remove the cab headlining to gain access to the underside of the aerial (Chapter 11).

42 Fault diagnosis – electrical system

Symptom	Reason(s)
Starter fails to turn engine	Battery discharged or defective Battery terminal and/or earth leads loose Starter motor connections loose Starter solenoid faulty Starter brushes worn or sticking Starter commutator dirty or worn
Starter turns engine very slowly	Battery discharged Starter motor connections loose Starter brushes worn or sticking Battery terminal and/or earth leads loose
Starter noisy	Pinion or flywheel ring gear teeth badly worn Mounting bolts loose

262　Chapter 12　Electrical system

Symptom	Reason(s)
Battery will not hold charge for more than a few days	Battery defective internally Electrolyte level too low Battery terminals loose Alternator drivebelt(s) slipping Alternator or regulator faulty Short circuit
Ignition light stays on	Alternator faulty Alternator drivebelt(s) broken
Ignition light fails to come on	Warning bulb blown or open circuit Alternator faulty
Fuel or temperature gauge gives no reading	Wiring open circuit Sender unit faulty Gauge faulty
Fuel or temperature gauge gives maximum reading all the time	Wiring short circuit Gauge faulty
Lights inoperative	Bulb blown Fuse blown Battery discharged Switch faulty Wiring open circuit Bad connection due to corrosion
Failure of component motor	Commutator dirty or burnt Armature faulty Brushes sticking or worn Armature bearings seized Fuse blown Wiring loose or broken Field coils faulty

Chapter 12 Electrical system

263

Wiring diagram layout

Wire codes
The wire from each terminal contains a code which details the wire destination, colour code(s), wire size and function. A typical code and its meaning is shown:

Typical code 100-502 G/BL-0.75 (31)
100-502   *Wire destination code*
G/   *Wire colour code – Primary colour*
/BL   *Wire colour code – Secondary colour(s)*
0.75   *Wire cross-section in mm²*
(31)   *Function code*

Wire destination code
This indicates the destination points of the wire using the component or internal joint numbers from the key. For example wire 100-502 runs between items 100 and 502. A destination code XX indicates that a short length of wire has been fitted to effect a weatherproof seal at a multi-plug. To aid identification, component numbering has been allocated as follows:

100 to 138 – engine looms   *500 to 519 – Internal joints – engine looms*
200 to 300 – main looms   *520 to 599 – Internal joints – main looms*
400 to 435 – rear looms   *600 to 699 – Internal joints – rear looms*

Wire Colour Code

BR	*Brown*	*R*	*Red*	*G*	*Green*	*Y*	*Yellow*
P	*Purple*	*BL*	*Blue*	*W*	*White*	*LG*	*Light Green*
BK	*Black*	*S*	*Slate*	**PK*	*Pink*	*O*	*Orange*

*If a pink wire is found and not detailed as shown on the wiring diagram, this indicates a product modification. The wire should be traced at least 150 mm (6 ins) beyond the splice joint to identify the wire colour within the loom.

Function code
This code, which appears in brackets as part of the wire code, defines the wire function. The meaning of the codes used are detailed below:

Function Code Definition Table

1	Ignition system, low tension wires, distributor	54	Positive (+) fused – general (key in drive position)
15	Battery positive (+) via ignition switch (key in drive position)	56	Headlamp circuits
		56A	Headlamp circuits – main beam
19	Diesel pre-heat circuits	56B	Headlamp circuits – dip beam
30	Battery positive (+) direct from battery	58	Side lamps, interior lamps and lamps general
31	Battery negative (-) direct to battery	61	Charging system warning circuits
31B	Negative switched circuits	71	Horn circuits
49	Direction indicator circuits (positive feed to relay)	75	Radio circuits
49A	Direction indicator circuits (feed to switch)	76	Loudspeaker circuits
50	Starting system circuits	C	Trailer direction indicator circuits
51	Charging system circuits	*R	Right-hand direction indicator circuits
53	Wiper system circuits	*L	Left-hand direction indicator circuits
53C	Washer system circuits		

*Because the same rear loom may be fitted to either LHD or RHD vehicles, the meaning of the codes R and L may be reversed. They cannot therefore be used to differentiate between LH and RH rear direction indicator circuits.

Key to Figs. 12.50 to 12.53 inclusive

100.	Main loom multi-plug		112.	Alternator
101.	Main loom multi-plug		113.	Reverse lamp switch connector
102.	Chassis earth		114.	Fuel tank sender unit connector
103.	Fuel shut-off solenoid (UK minibus)		119.	Battery
104.	Ignition module		121.	Automatic transmission inhibitor switch
105.	Starter motor sub loom connector		122.	Selector quadrant illumination sub loom connector
106.	Ignition coil		132.	Chassis earth
107.	Distributor		133.	Fuel tank sender unit
108.	Temperature sender unit		134.	Reverse lamp switch
109.	Oil pressure switch		135.	Selector quadrant illumination
110.	Anti-dieseling solenoid		137.	Engine earth
111.	Starter motor		138.	Engine block

Internal joint detail – Engine loom

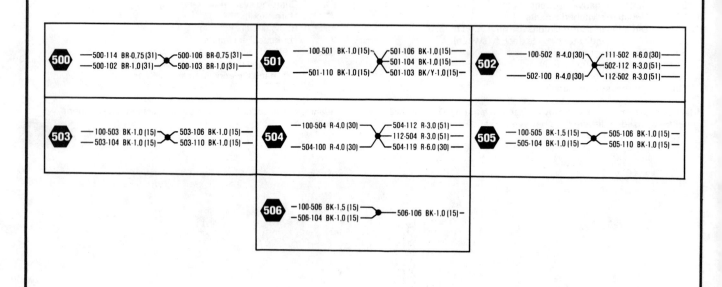

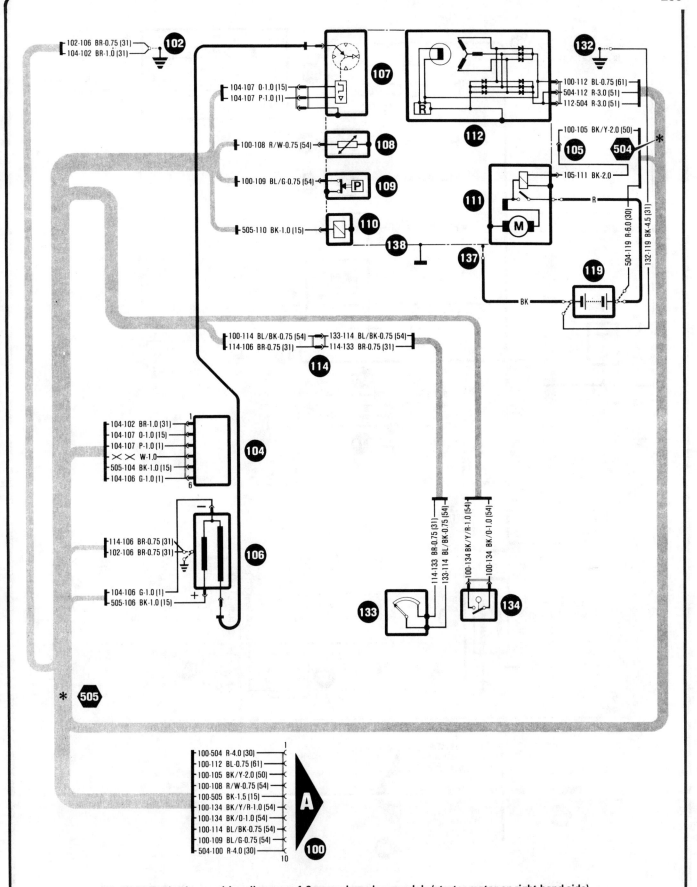

Fig. 12.50 Engine loom wiring diagram – 1.6 manual gearbox models (starter motor on right-hand side)

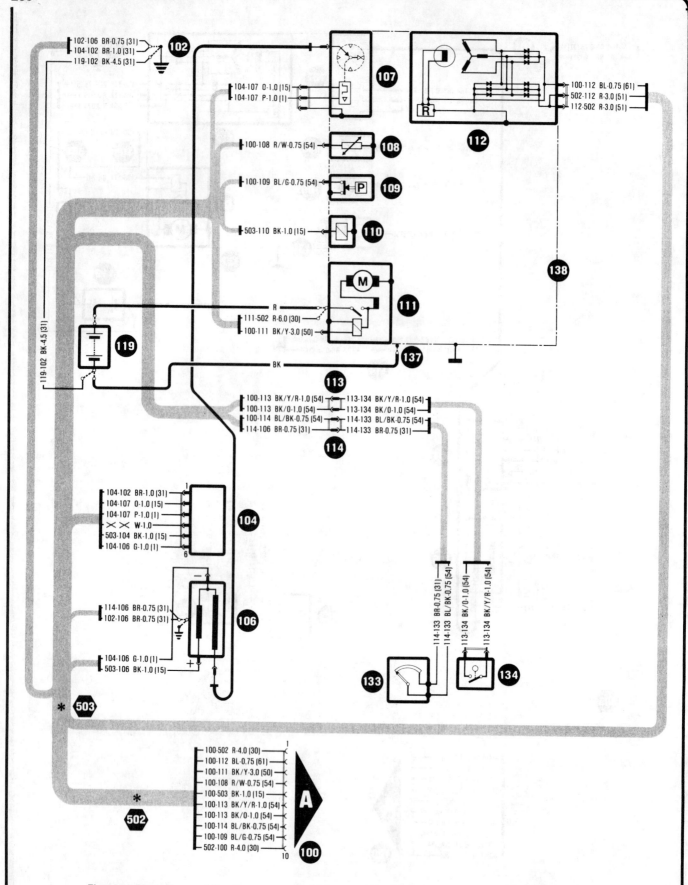

Fig. 12.51 Engine loom wiring diagram – 1.6 and 2.0 manual gearbox models (starter motor on left-hand side)

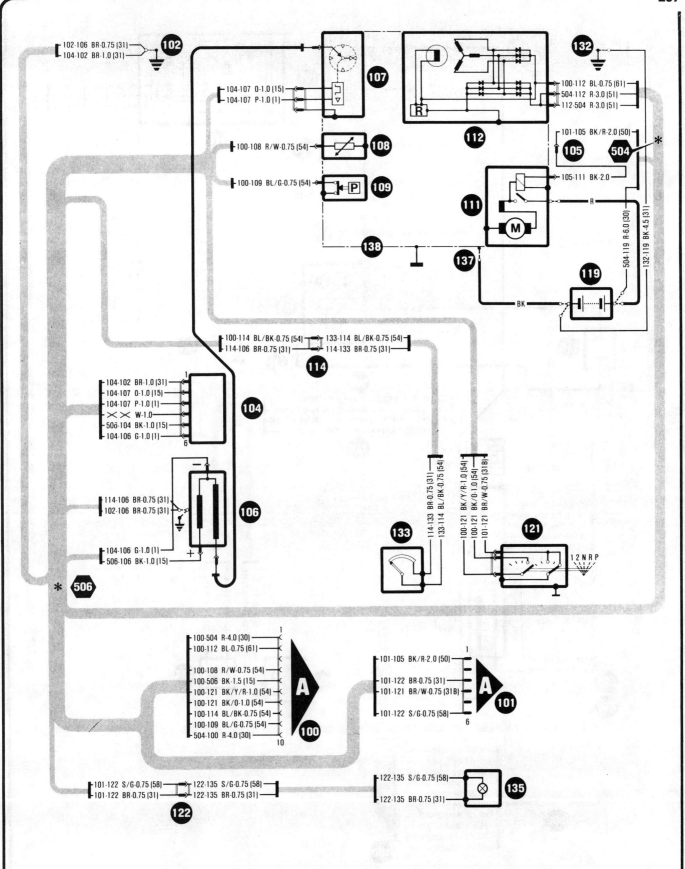

Fig. 12.52 Engine loom wiring diagram – 2.0 automatic transmission models

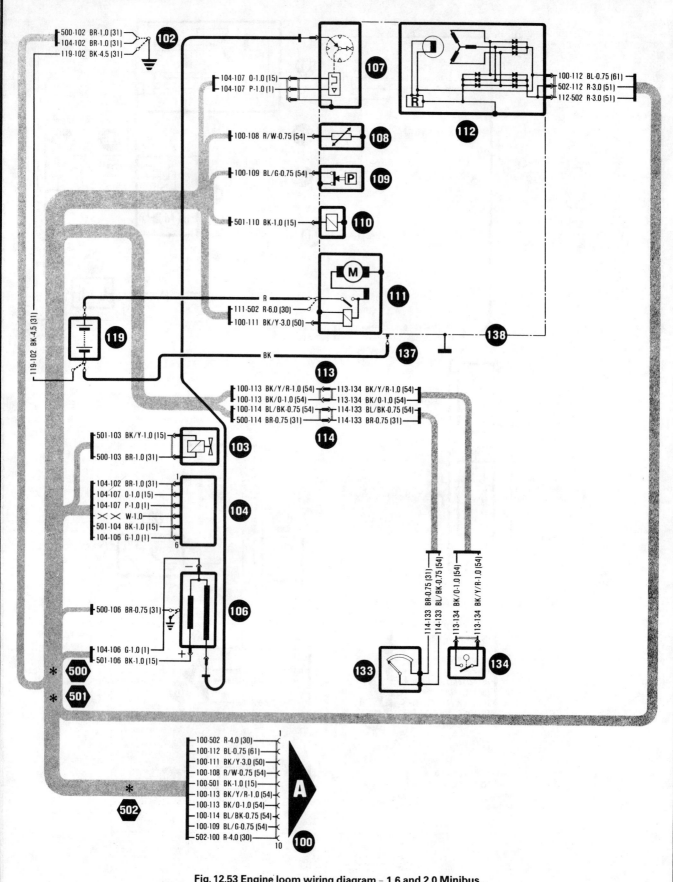

Fig. 12.53 Engine loom wiring diagram – 1.6 and 2.0 Minibus

Key to Figs. 12.54 and 12.55

200.	Engine loom multi-plug	237.	Warning lamp connector (for tachograph installation)
201.	Engine loom multi-plug	238.	Instrument cluster (lamps not annotated are illumination lamps)
202.	Rear loom multi-plug	239.	Low brake fluid warning lamp
203.	Rear accessory loom multi-plug	240.	Ignition warning lamp
204.	LH direction indicator	241.	Main beam warning lamp
205.	LH sidelamp	242.	Low oil pressure warning lamp
206.	LH headlamp	243.	Direction indicator warning lamp
207.	Heater motor resistor	244.	Voltage stabiliser
208.	Horn	245.	Fuel gauge
209.	RH headlamp	246.	Temperature gauge
210.	RH sidelamp	248.	Trailer direction indicator warning lamp
211.	RH direction indicator	249.	Stop lamp switch
212.	LH direction indicator side repeater	250.	Exterior light and windscreen wiper switch
214.	Headlamp washer pump	251.	Heated rear screen switch (black connector)
215.	Windscreen washer pump	252.	Tailgate wiper switch (green connector)
216.	RH direction indicator side repeater	253.	Tailgate wiper switch (yellow connector)
218.	Low brake fluid level switch	254.	This connector is not used
219.	Windscreen wiper motor	255.	Radio
220.	LH courtesy lamp switch	256.	Radio sub loom connector
221.	Radio speaker	257.	Heater connector
222.	RH courtesy lamp switch	258.	Heater motor switch
223.	Fuse module	259.	Heater control illumination
224.	Automatic transmission inhibitor relay	260.	Heater motor
227.	Heated rear screen relay	261.	Cigar lighter
228.	Direction indicator unit	262.	Main loom earth
229.	Headlamp wash relay	266.	Side stepwell lamp sub loom connector (UK minibus)
230.	Intermittent wipe relay – front	267.	Side stepwell lamp (UK minibus)
231.	Intermittent wipe relay – rear	268.	Trailer direction indicator warning lamp connector for service use
232.	Rear foglamp switch (red connector)		
233.	Rear interior lamp switch (blue connector)	273.	Radio earth
234.	Ignition switch	300.	Bulkhead
235.	Hazard, horn, direction indicator and dipswitch		
236.	Instrumentation connector (for tachograph installation)		

Internal joint detail – Main loom

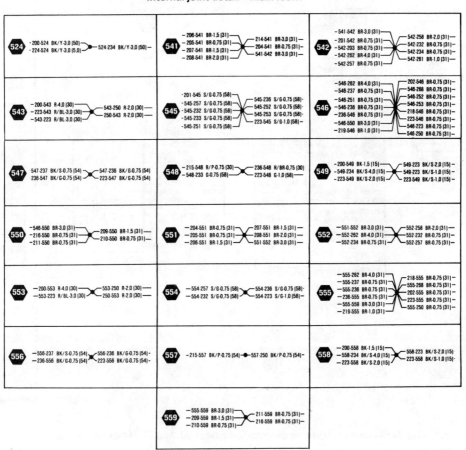

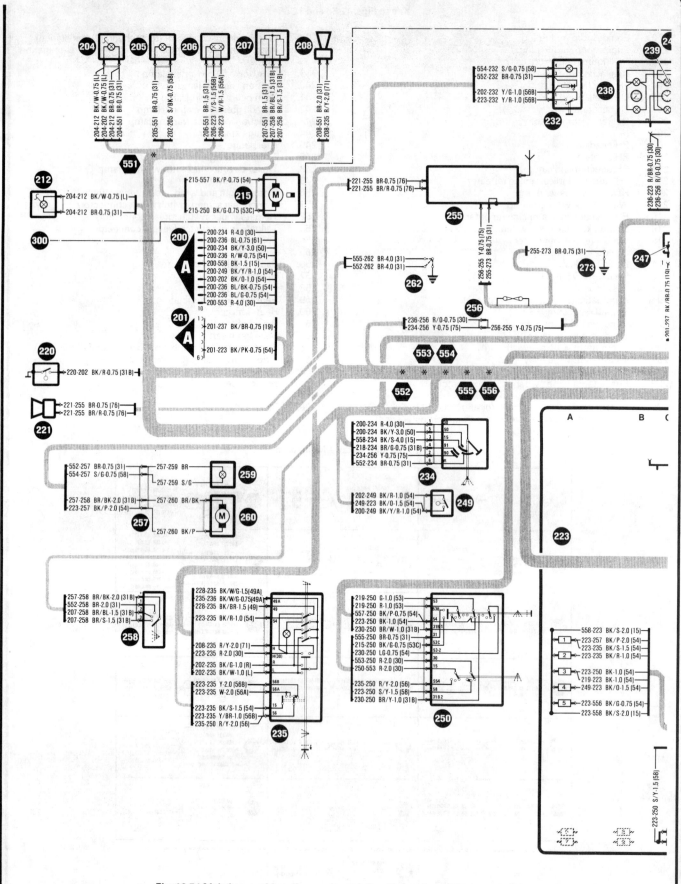

Fig. 12.54 Main loom wiring diagram – models with low level instrumentation

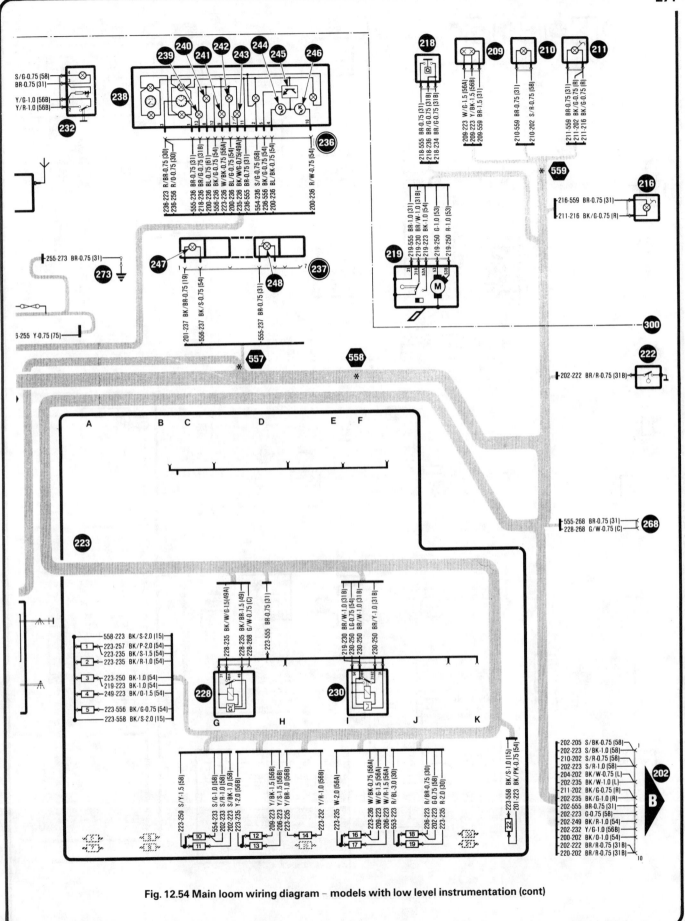

Fig. 12.54 Main loom wiring diagram – models with low level instrumentation (cont)

Fig. 12.55 Main loom wiring diagram – models with high level instrumentation

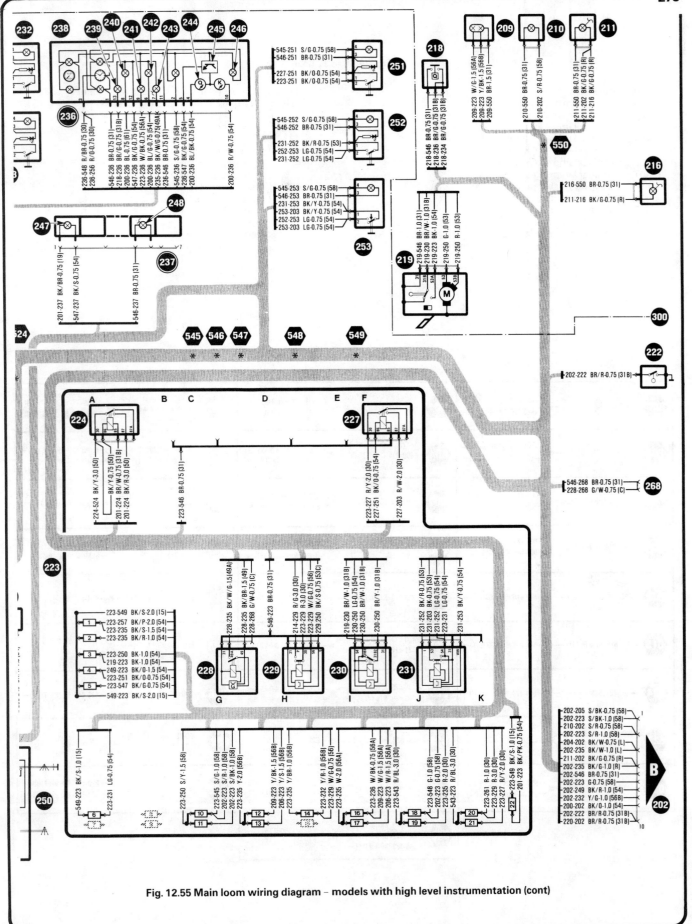

Fig. 12.55 Main loom wiring diagram – models with high level instrumentation (cont)

Key to Figs. 12.56 to 12.60 inclusive

300.	Bulkhead	419.	Number plate lamp sub loom connector
400.	Main loom multi-plug (rear loom)	420.	Passenger's side rear lamp cluster
401.	Main loom multi-plug (rear accessory loom)	421.	Number plate lamp
403.	Interior lamp – Front	422.	Number plate lamp
404.	Interior lamp – LCY only	423.	Tailgate wiper motor
405.	Interior lamp – LCX only	424.	Heated rear screen
406.	Interior lamp – LCY only	425.	Heated rear screen
407.	Interior lamp – Rear	426.	Rear door(s) sub loom connector
413.	Tailgate washer pump	428.	Rear stepwell lamp (UK minibus)
414.	Earth point, door pillar	429.	Passenger rear door sub loom connector
415.	Earth point, door pillar	433.	Interior lamp connection/stepwell lamp sub loom connection
416.	Earth point, crossmember	434.	Earth point, door
417.	Driver's side rear lamp cluster	435.	Earth point, door
418.	Number plate lamp sub loom connector		

Internal joint detail – Rear loom

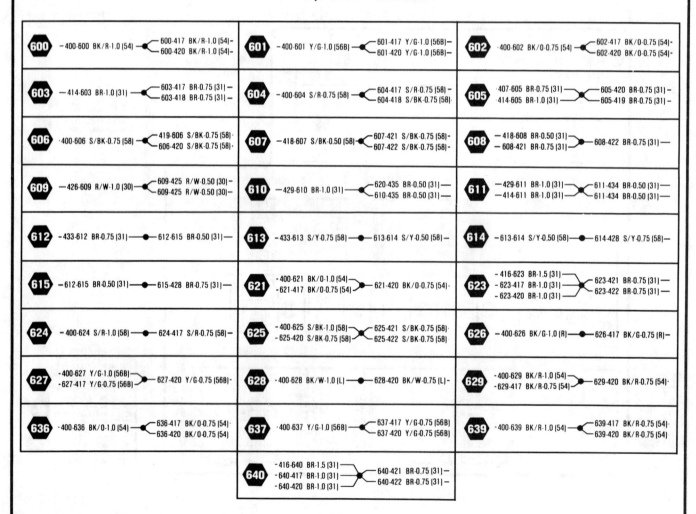

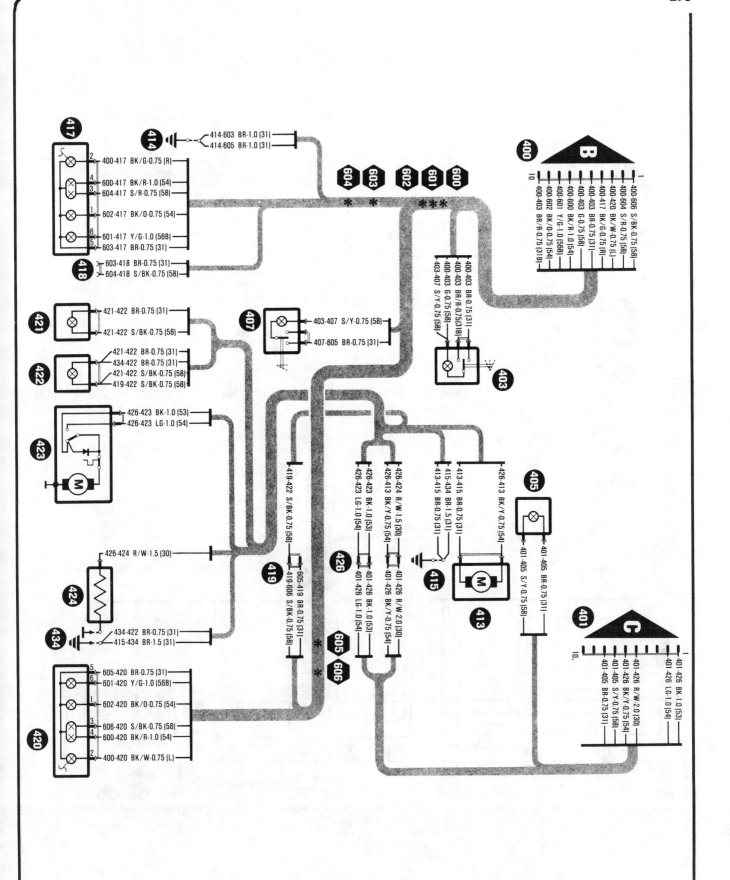

Fig. 12.56 Rear loom wiring diagram – short wheelbase Van/Bus/Kombi models with tailgate

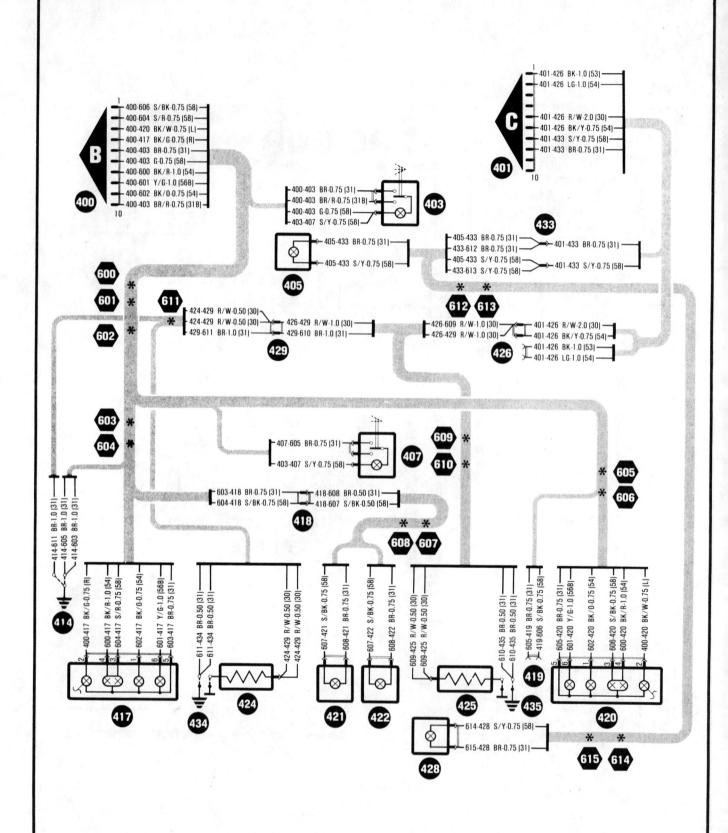

Fig. 12.57 Rear loom wiring diagram – short wheelbase Van/Bus/Kombi models with twin rear doors

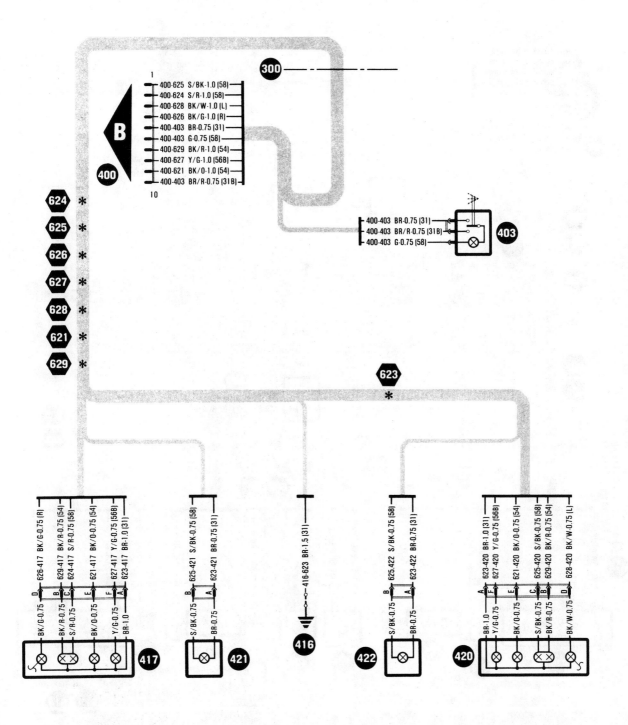

Fig. 12.58 Rear loom wiring diagram – short wheelbase Chassis cab models

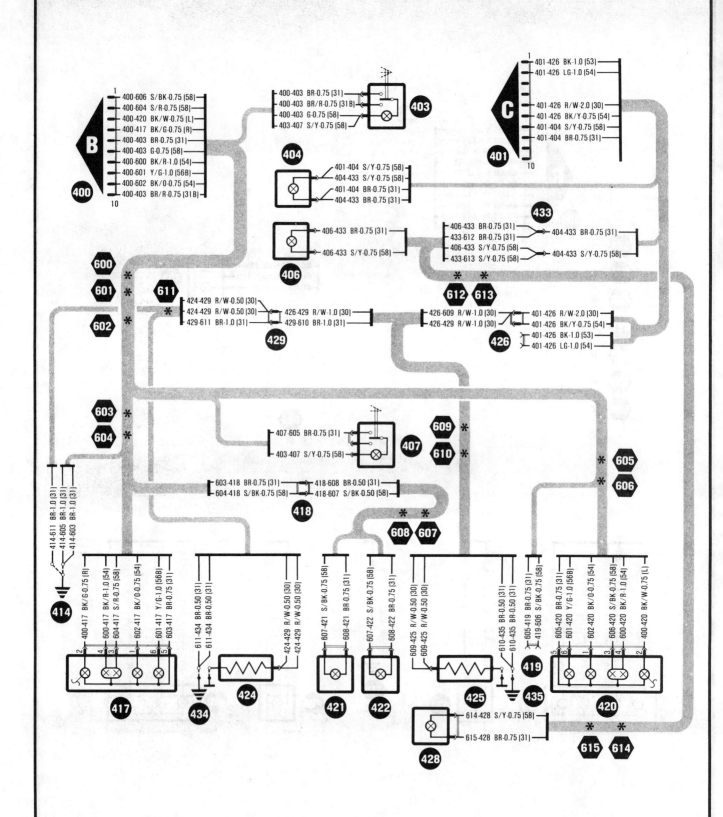

Fig. 12.59 Rear loom wiring diagram – long wheelbase Van/Bus/Kombi models with twin rear doors

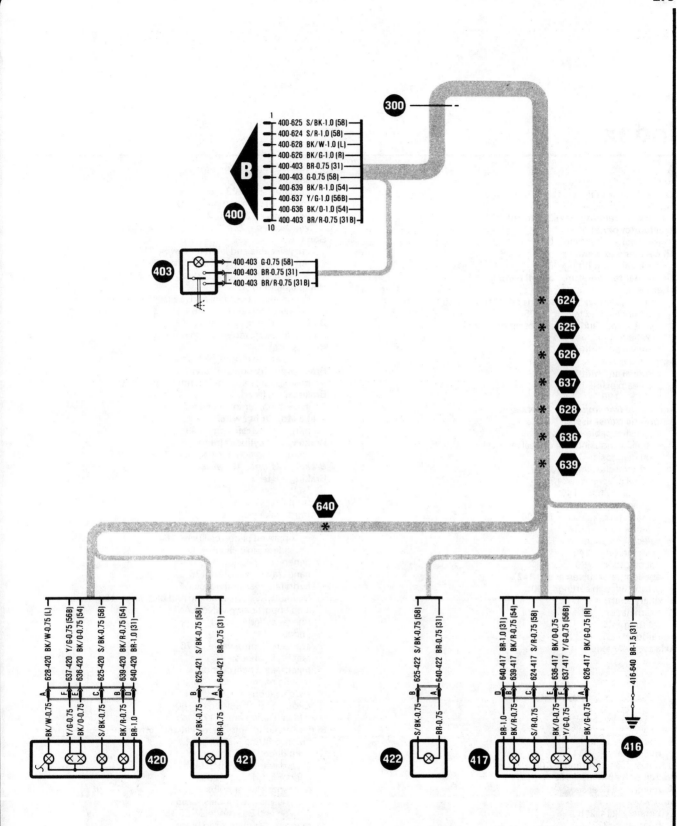

Fig. 12.60 Rear loom wiring diagrams – long wheelbase Chassis cab models

Index

A

About this manual – 5
Accelerator cable
   removal, refitting and adjustment – 66
Accelerator pedal
   removal and refitting – 67
Air cleaner and element
   removal and refitting – 64
Air cleaner temperature control testing – 64
Alternator
   brushes, removal, inspection and refitting – 236
   fault finding and testing – 235
   maintenance and special precautions – 234
   overhaul
   removal and refitting – 234
Anti-roll bar
   removal and refitting
      beam axle – 194
      IFS – 190
Automatic transmission – 88 *et seq*
Automatic transmission
   downshift cable – 139
   extension housing rear oil seal renewal – 142
   fault diagnosis – 143
   fluid level checking – 137
   general description – 136
   inhibitor switch – 140
   removal and refitting – 142
   routine maintenance – 15, 136
   selector cable – 137
   selector mechanism – 139
   selector rods – 140
   specifications – 90
   speedometer driven gear – 142
   torque wrench settings – 90
Auxiliary shaft
   examination – 44
   refitting – 46
   removal – 37
Axle *see* **Rear axle**

B

Battery
   charging – 234
   maintenance – 234
   removal and refitting – 234
Bleeding the brakes – 162
Bodywork – 212 *et seq*
Bodywork
   bonnet – 214, 215
   bumpers – 227
   damage repair – 213, 214
   doors – 215, 217, 218, 219, 220
   facia – 229
   front grille – 227
   general description – 212
   headlining – 230
   maintenance – 15, 20, 212, 213, 214
   mirror – 228

   seat belt and stalk – 231
   seats – 231
   tailgate – 221, 223
   torque wrench settings – 212
   windows – 224, 225, 226
   windscreen – 227
Bonnet
   release cable and latch renewal – 215
   removal, refitting and adjustment – 214
   safety catch removal and refitting – 215
Brake caliper (front)
   dismantling, overhaul and reassembly – 166
   removal and refitting – 166
Brake disc (front)
   examination, removal and refitting – 167
Brake pedal
   removal and refitting – 177
Brake pressure control valve
   description, removal and refitting – 176
Brake shoes (rear)
   inspection, removal and refitting – 168
Brake stop-light switch
   removal and refitting – 249
Brake wheel cylinder (rear)
   dismantling and reassembly – 172
Braking system – 159 *et seq*
Braking system
   fault diagnosis – 178
   front brake calliper – 166
   front brake disc – 167
   front disc pads
      opposed piston calliper – 164
      sliding pin calliper – 164
   general description – 160
   handbrake – 174, 175, 176
   hydraulic lines and hoses – 164
   hydraulic system, draining and bleeding – 162
   load apportioning valve (LAV) – 177
   master cylinder – 173
   pedal – 177
   pressure control valve – 176
   rear brake shoes – 168
   rear brake wheel cylinder – 172
   routine maintenance – 15, 20, 161
   specifications – 159
   torque wrench settings – 160
   vacuum servo unit – 174
Bulbs and lamp units, removal and refitting
   bulbs, refitting – 255
   front direction indicator bulbs – 255
   hazard warning switch bulb – 255
   interior lamp bulb – 255
   rear lamp cluster bulbs
      Chassis cab models – 255
      Van and Bus models – 255
   rear number plate lamp bulbs
      Chassis cab models – 255
      double rear door model – 255
      Tailgate model – 255
   side repeater lamp bulbs – 255
   sidelamp bulbs – 254
Bumpers
   removal and refitting – 227
Buying spare parts – 9

Index

C

Camshaft
  and cam followers, examination – 42
  refitting – 48
  removal – 36
Calliper *see* **Brake calliper**
Capacities – 6
Carburettor
  general description – 67
  overhaul – 70
  removal and refitting – 69
  slow running adjustment – 69
Cigarette lighter
  removal and refitting – 252
Clutch – 83 *et seq*
Clutch
  adjustment checking – 83
  cable – 84
  fault diagnosis – 87
  general description – 83
  inspection – 86
  pedal – 85
  refitting – 87
  release bearing and lever – 86
  removal – 86
  routine maintenance – 83
  specifications – 83
  torque wrench settings – 83
Clutch cable
  removal and refitting – 84
Clutch pedal
  removal and refitting – 85
Clutch release bearing and lever
  removal and refitting – 86
Coil (ignition)
  description and testing – 80
Connecting rods
  examination and renovation – 42
  refitting – 45
  removal – 39
Conversion factors – 13
Coolant mixture, general – 55
Cooling, heating and ventilation systems – 52 *et seq*
Cooling, heating and ventilation systems
  cooling fan – 57
  demister and air vent hoses – 60
  draining – 55
  drivebelt – 58
  fault diagnosis – 62
  filling – 56
  flushing – 56
  general description – 52
  heater unit – 59, 60
  routine maintenance – 15, 20, 53
  specifications – 52
  temperature gauge and sender unit – 59
  thermostat – 57
  torque wrench settings – 52
  water pump – 58
Courtesy light switch
  removal and refitting – 248
Crankcase ventilation system
  description and maintenance – 41
Crankshaft and main bearings
  examination and renovation – 41
  refitting – 45
  removal – 40
  renewal
    front oil seal – 38
    rear oil seal – 39
Cylinder block and bores
  examination and renovation – 42
Cylinder head
  decarbonising, valve grinding and renovation – 44
  dismantling – 36
  reassembly – 47
  refitting – 48
  removal – 34

D

Demister and air vent hoses
  removal and refitting
    demister hose – 61
    face vent nozzle – 62
Differential carrier (H type axle)
  removal and refitting – 158
Differential unit (G and F type axles)
  repair and overhaul – 158
Dimensions – 6
Direction indicator switch
  removal and refitting – 248
Disc brake *see* **Brake disc**
Disc pad (front)
  inspection, removal and refitting
    opposed piston calliper – 165
    sliding pin calliper – 164
Distributor
  removal, examination and refitting – 78
Doors
  front door fittings, removal and refitting – 217
  front, removal and refitting – 218
  rear door fittings, removal and refitting – 220
  rear, removal and refitting – 220
  sliding side, removal and refitting – 219
  sliding side door fittings, removal and refitting – 220
  trim panels removal and refitting – 215
  windows – 224, 225, 226
Downshift cable (automatic transmission)
  removal, refitting and adjustment – 139
Drivebelt
  adjustment and renewal
    power-steering drivebelt – 58
    water pump/alternator drivebelt – 58
Driveplate
  refitting – 46
  removal – 38
Drive pinion oil seal (all axle types)
  renewal – 157

E

Electrical system – 232 *et seq*
Electrical system
  alternator – 234, 235, 236
  battery – 234
  bulbs and lamp units – 254
  cigarette lighter – 252
  direction indicator switch – 248
  fault diagnosis – 261, 262
  fuses and relays – 233, 248
  general description – 233
  headlamps – 252, 253
  horn – 255
  instrument panel – 250, 251
  radio – 259, 260, 261
  routine maintenance – 20
  specifications – 232
  speedometer cable – 252
  starter motor – 236, 239, 245
  switches – 248, 249
  tailgate wiper components – 259
  torque wrench settings – 233
  washer nozzles and hoses – 259
  windscreen/headlamp washer reservoir – 258
  windscreen wiper linkages – 257
  windscreen wiper motor – 256
  wiper arms – 256
  wiper blades – 256
  wiring diagrams – 261, 263 to 279

Index

Electronic amplifier module
removal and refitting – 81
Engine – 26 *et seq*
Engine
ancillary components removal – 34, 49
and transmission
reconnection, all models – 33
separation, automatic transmission models – 33
separation, manual gearbox models – 33
auxiliary shaft – 37, 44, 46
camshaft – 36, 42, 48
crankcase ventilation system – 41
crankshaft – 38, 39, 40, 41, 45
cylinder block and bores – 42
cylinder head – 34, 36, 44, 47, 48
dismantling – 34
examination and renovation – 41
fault diagnosis – 51
flywheel/driveplate – 38, 44, 46
general description – 29
initial start-up after repair – 50
major operations possible with engine in vehicle – 31
major operations requiring engine removal – 31
mountings – 40
oil filter – 39
oil pump – 39, 41, 45
pistons and connecting rods – 39, 42, 45
reassembly – 45
removal and refitting (without transmission) – 31
removal and refitting (with transmission) – 32, 33
removal methods – 31
routine maintenance – 15, 20, 29, 31
specifications – 26 to 28
sump – 38, 46
timing belt and sprockets – 37, 44, 48
torque wrench settings – 28, 29
valve clearances – 50
Exhaust manifold
removal and refitting – 74
Exhaust system
checking, removal and refitting – 74

F

Facia
removal and refitting – 229
Fan (cooling)
removal and refitting – 57
Fault diagnosis
automatic transmission – 143
braking system – 178
clutch – 87
cooling, heating and ventilation systems – 62
electrical system – 261, 262
engine – 51
fuel and exhaust systems – 75
general – 22
ignition system – 81
manual gearbox – 136
propeller shaft – 148
rear axle – 158
suspension and steering – 211
Flywheel
refitting – 46
removal – 38
ring gear examination and renovation – 44
Fuel and exhaust systems – 63 *et seq*
Fuel and exhaust systems
accelerator cable – 66
accelerator pedal – 67
air cleaner and element – 64
carburettor – 67, 69, 70
exhaust manifold – 74
exhaust system – 74
fault diagnosis – 75

fuel gauge sender unit – 66
fuel pump – 65
fuel tank – 65
fuel tank filler pipe – 66
general description – 63
inlet manifold – 74
routine maintenance – 15, 20, 64
specifications – 63
torque wrench settings – 63
unleaded fuel – 75
Fuel gauge sender unit
removal and refitting – 66
Fuel pump
testing, removal, servicing and refitting – 65
Fuel tank
removal, servicing and refitting – 65
Fuel tank filler pipe
removal and refitting – 66
Fuses – 233, 248

G

Gearbox *see* **Manual gearbox**
Grille
front, removal and refitting – 227
windscreen, removal and refitting – 227

H

Handbrake
adjustment – 174
cable removal and refitting – 176
lever and primary rod, removal and refitting – 175
Headlamps
alignment – 252
and headlamp bulbs, removal and refitting – 253
Headlining
removal and refitting – 230
Heater motor switch
removal and refitting – 249
Heater unit
dismantling and reassembly – 60
removal and refitting – 59
Horn
removal and refitting – 255
HT leads, general – 80
Hydraulic brake lines and hoses
inspection, removal and refitting – 164
Hydraulic system
draining and bleeding (brakes) – 162

I

Ignition switch and lock barrel
removal and refitting – 248
Ignition system – 76 *et seq*
Ignition system
coil – 80
distributor – 78
electronic amplifier module – 81
fault diagnosis – 81
general description – 15, 20, 77
routine maintenance – 15, 20, 77
spark control system – 81
spark plugs and HT leads – 80
specifications – 76
timing – 79
torque wrench settings – 76
Ignition timing
adjustment – 79
Inhibitor switch (automatic transmission)
removal, refitting and adjustment – 140

Index

Inlet manifold
   removal and refitting – 74
Instrument panel
   components removal and refitting – 251
   removal and refitting – 250
   surround facia and components (tachograph models)
      removal and refitting – 251
   surround switches
      removal and refitting – 249
Introduction to the Ford Transit – 5

J

Jacking – 7

L

Lighting switch
   removal and refitting
Load apportioning valve (LAV), general – 177
Lubricants and fluids – 21

M

Main bearings
   examination and renovation – 41
   refitting – 45
   removal – 40
Manual gearbox and automatic transmission – 80 *et seq*
Manual gearbox, general
   engine and gearbox, removal and refitting – 32
   extension housing rear oil seal (type F, G and N) – 135
   fault diagnosis – 136
   general description – 90
   inspection (all types) – 98
   overdrive unit – 135
   overhaul requirements – 94
   removal and refitting – 91
   routine maintenance – 15, 91
   specifications – 88
   speedometer driven gear – 135
   torque wrench settings – 89
Manual gearbox (type F)
   dismantling – 94
   countershaft gears – 96
   input shaft – 96
   mainshaft – 96
   reassembly – 99
   countershaft assembly – 100
   gearbox general assembly – 100
   input shaft – 100
   mainshaft assembly – 99
   selector housing
      removal, dismantling, reassembly and refitting – 102
Manual gearbox (type G)
   dismantling – 104
   countershaft gear – 105
   input shaft – 105
   mainshaft – 105
   reassembly – 106
   countershaft assembly – 108
   gearbox general assembly – 108
   input shaft assembly – 107
   mainshaft assembly – 106
   selector housing
      removal, dismantling, reassembly and refitting – 110
Manual gearbox (type N)
   dismantling into major sub-assemblies – 110
   input shaft and mainshaft dismantling and reassembly – 114
   reassembly – 117
Manual gearbox (type MT75)
   dismantling – 121

   gearbox housings overhaul – 123
   geartrains and selectors overhaul
      countershaft – 125
      gear lever unit – 130
      input shaft and guide sleeve – 125
      mainshaft – 126
      reverse idler gear unit – 126
      selector shafts and forks – 130
   reassembly – 131
   special overhaul requirements – 120
Master cylinder
   dismantling, overhaul and reassembly – 173
   removal and refitting – 173
Mirror, door and glass
   removal and refitting – 228
Mountings (engine)
   renewal – 40

O

Oil filter
   renewal – 39
Oil pump
   examination and renovation – 41
   refitting – 45
   removal – 30
Overdrive unit, general – 135

P

Pads *see* **Disc pads**
Pistons
   examination and renovation – 42
   refitting – 45
   removal – 39
Plastic components repair – 214
Power-assisted steering system
   fluid level check and system bleeding – 209
Power-steering gear unit
   removal and refitting – 209
Power-steering hoses
   removal and refitting – 210
Power-steering pump
   removal and refitting – 209
Propeller shaft – 144 *et seq*
Propeller shaft
   centre bearing renewal – 146
   fault diagnosis – 148
   general description – 148
   'Guibo' rubber joint – 148
   removal and refitting – 145
   routine maintenance – 15
   specifications – 144
   torque wrench settings – 144
   universal joints and centre bearing – 144

R

Rack and pinion steering gear unit
   dismantling, overhaul and reassembly – 203
   removal and refitting – 201
Radiator
   removal and refitting – 56
Radio removal and refitting
   standard – 259
   digital – 259
Radio aerial and speakers (standard)
   removal and refitting – 261

284 Index

Rear axle – 149 *et seq*
Rear axle
    differential carrier (H type axle) – 158
    differential unit (G and F type axles) – 158
    drive pinion oil seal (all axle types) – 157
    fault diagnosis – 158
    general description – 150
    rear axleshaft (halfshaft) – 153
    rear hub
        F and H type axles – 155
        G type axle – 156
    removal and refitting – 151
    routine maintenance – 150
    specifications – 149
    torque wrench settings – 149
Rear axle shaft (halfshaft)
    removal and refitting – 153
Rear hub (F and H type axles)
    removal, overhaul and refitting – 155
Rear hub (G type axle)
    removal, overhaul and refitting – 156
Recommended lubricants and fluids – 21
Relays – 233, 248
Repair procedures, general – 10
Reversing light switch
    removal and refitting – 249
Routine maintenance
    automatic transmission – 15, 136
    bodywork – 15, 20, 212, 213, 214
    braking system – 15, 20, 161
    clutch – 83
    cooling, heating and ventilation systems – 15, 20, 53
    electrical system – 20
    engine – 15, 20, 29, 31
    fuel and exhaust systems – 15, 20, 64
    general – 15
    ignition system – 15, 20, 77
    manual transmission – 15, 91
    propeller shaft – 15
    rear axle – 150
    suspension and steering – 15, 20, 182

S

Safety first! – 14
Seat belt and stalk
    removal and refitting – 231
Seats
    removal and refitting – 231
Selector cable (automatic transmission)
    adjustment – 137
Selector mechanism (automatic transmission)
    dismantling and reassembly – 139
    removal and refitting – 139
Selector rods (automatic transmission)
    removal, refitting and adjustment – 140
Shock absorber
    removal and refitting
        front (beam axle) – 194
        front (IFS) – 190
        rear – 196
Spare parts buying – 9
Spark control system
    components, removal and refitting – 81
    description – 81

Spark plugs, general – 80
Speedometer cable
    removal and refitting – 252
Speedometer driven gear
    removal and refitting
        automatic transmission – 142
        manual gearbox – 135

Starter motor
    overhaul – 245
    removal and refitting – 236
    testing in the vehicle – 239
Steering column
    dismantling, overhaul and reassembly – 199
    lower universal coupling (IFS)
        removal and refitting – 199
    removal and refitting – 199
    upper bearing renewal – 199
Steering gear (worm and nut)
    dismantling, overhaul and reassembly – 204
    removal and refitting – 204
    rocker shaft pre-load adjustment – 207
    steering drag link removal and refitting – 208
    steering drop arm removal and refitting – 207
Steering wheel
    removal and refitting – 196
Sump
    refitting – 46
    removal – 38
Suspension and steering – 179 *et seq*
Suspension and steering
    fault diagnosis – 211
    front wheel alignment – 210
    front wheel hub – 185, 186
    general description – 181
    power-assisted steering system
        fluid level check and system bleeding – 209
    power-steering gear unit – 209
    power-steering hoses – 210
    power-steering pump – 209
    rack and pinion steering gear unit – 201, 203
    rear axle leaf spring – 195
    rear shock absorber – 196
    routine maintenance – 15, 20, 182
    specifications – 179
    steering column – 199
    steering gear rubber bellows – 204
    steering wheel – 196
    torque wrench settings – 180
    track rod end – 204
    wheels and tyres – 210
    worm and nut steering gear – 204, 207, 208
Suspension (beam axle type)
    anti-roll bar – 194
    front leaf spring – 193
    front shock absorber – 194
    removal and refitting – 191
    stub axle – 192
    suspension bump stop – 194
Suspension (IFS type)
    anti-roll bar – 189
    coil spring and lower suspension arm – 186
    front shock absorber – 190
    front suspension bracket bush – 187
    lower arm balljoint – 189
    suspension bump stop – 186
Switches
    removal and refitting
        brake stop-light – 249
        courtesy light – 248
        direction indicator – 248
        heater motor – 249
        ignition and lock barrel – 248
        instrument panel surround – 249
        lighting – 248
        reversing light – 249

T

Tailgate
    fittings removal and refitting – 221
    removal and refitting – 223
    wiper components removal and refitting – 259

Index

285

Temperature gauge and sender unit
testing, removal and refitting – 59
Thermostat
removal, testing and refitting – 57
Timing belt and sprockets
examination – 44
refitting – 48
removal – 37
Tools – 11
Towing – 8

U

Universal joints and centre bearing
testing for wear – 144
Unleaded fuel, general – 75

V

Vacuum servo unit
removal and refitting – 174
Valve clearances
adjustment– 50
Vehicle identification numbers – 9

W

Washer nozzles and hoses
removal and refitting – 259
Water pump
removal and refitting – 58
Weights – 6
Wheel alignment (front) – 210
Wheel hub (front)
dismantling and reassembly – 186
removal, refitting and adjustment – 185
Wheels and tyres
general care and maintenance – 210
Windows, removal and refitting
front door (and regulator) – 224
front door quarterglass – 225
front door sliding window glass and frame – 225
dismantling and reassembly – 225
opening rear quarter window – 226
Windscreen and fixed washers
removal and refitting – 227
Windscreen grille
removal and refitting – 227
Windscreen/headlamp washer reservoir
removal and refitting – 258
Windscreen wiper linkages
removal and refitting – 257
Windscreen wiper motor
removal and refitting – 256
Wiper arms
removal and refitting – 256
Wiper blades renewal – 256
Wiring diagrams – 261, 263 to 279
Working facilities – 11